Understanding Evolution

Current books on evolutionary theory all seem to take for granted the fact that students find evolution easy to understand when, actually, from a psychological perspective, it is a rather counter-intuitive idea. Evolutionary theory, like all scientific theories, is a means to understanding the natural world.

Understanding Evolution is intended for undergraduate students in the life sciences, biology teachers, or anyone wanting a basic introduction to evolutionary theory. Covering core concepts and the structure of evolutionary explanations, it clarifies both what evolution is about and why so many people find it difficult to grasp. The book provides an introduction to the major concepts and conceptual obstacles to understanding evolution, including the development of Darwin's theory, and a detailed presentation of the most important evolutionary concepts.

Bridging the gap between the concepts and conceptual obstacles, *Understanding Evolution* presents evolutionary theory with a clarity and vision students will quickly appreciate.

Kostas Kampourakis is a researcher at the University of Geneva, where he is presently working on projects relevant to the teaching and the public understanding of genetics. His main areas of interest are evolution and genetics education, as well as the teaching of science concepts and nature of science in the context of the history and philosophy of science.

Understanding Evolution

KOSTAS KAMPOURAKIS

University of Geneva, Switzerland

CAMBRIDGE
UNIVERSITY PRESS

CAMBRIDGE
UNIVERSITY PRESS

University Printing House, Cambridge CB2 8BS, United Kingdom

Published in the United States of America by Cambridge University Press, New York

Cambridge University Press is part of the University of Cambridge.

It furthers the University's mission by disseminating knowledge in the pursuit of education, learning and research at the highest international levels of excellence.

www.cambridge.org
Information on this title: www.cambridge.org/9781107034914

First published 2014

Printed in the United Kingdom by TJ International Ltd. Padstow Cornwall

A catalogue record for this publication is available from the British Library

Library of Congress Cataloguing in Publication data

Kampourakis, Kostas, author.
Understanding evolution / Kostas Kampourakis, University of Geneva, Switzerland.
    pages   cm
ISBN 978-1-107-03491-4 (Hardback) – ISBN 978-1-107-61020-0 (Paperback)   1. Evolution (Biology)
I. Title.
QH366.2.K326 2014
576.8–dc23   2013034917

ISBN 978-1-107-03491-4 Hardback
ISBN 978-1-107-61020-0 Paperback

Additional resources for this publication at www.cambridge.org/9781107034914

To my wife, Katerina, and our children, Mirka and Giorgos, for turning an inherently purposeless life into a deeply meaningful one.

Contents

Prolegomena *page* ix
Acknowledgments xviii

1 An evolving world 1

How we know what we know about evolution 3
Questions answered by evolutionary biology 13
Domestication 19
Epidemic infectious disease 23
Conclusions 29
Further reading 29

2 Religious resistance to accepting evolution 31

Creation and design in nature 33
Evolution and worldviews: perceived conflicts 42
Evolution and religion: scientists' views 46
Distinguishing between knowing and believing 52
Conclusions 60
Further reading 60

3 Conceptual obstacles to understanding evolution 62

Conceptual change in science 64
Design teleology as a conceptual obstacle to understanding evolution 72
Psychological essentialism as a conceptual obstacle to understanding evolution 80
Conceptual change in evolution 89
Conclusions 96
Further reading 97

4 Charles Darwin and the *Origin of Species:* a historical case study of conceptual change 98

The development of Darwin's theory 100
Darwin's conceptual change 108
The publication of the *Origin of Species* 115

Science and religion in the reviews of the *Origin of Species*   121
Conclusions   125
Further reading   125

5   Common ancestry   127

The evolutionary network of life   129
Homology and common descent   139
Homoplasy and convergence   148
Evolutionary developmental biology   157
Conclusions   167
Further reading   167

6   Evolutionary change   169

Adaptation and natural selection   172
Stochastic events and processes in evolution   184
Speciation, extinction, and macroevolution   191
Evolutionary explanations and the historicity of nature   200
Conclusions   206
Further reading   206

Concluding remarks   208

The virtues of evolutionary theory   208
Questions not answered by evolutionary theory   212

Glossary   218
References   228
Index   251

Prolegomena

Evolutionary theory is the central theory of biology. It explains the unity of life by documenting how extant and extinct species share a common ancestry. It also explains the diversity of life by describing how species have evolved from ancestral ones through natural processes. Charles Darwin laid the foundations for current evolutionary theory in his book *On the Origin of Species by Means of Natural Selection, or the Preservation of Favoured Races in the Struggle for Life* (1859), where he argued for the common ancestry of all life and proposed natural selection as the mechanism by which evolution proceeds. Darwin briefly described the process as "descent with modification." This phrase still accurately describes the core of evolutionary theory. However, evolutionary biology has itself evolved since then, incorporating other disciplines such as genetics, **systematics**, or paleontology during the Modern Synthesis of the 1940s (Huxley, 1942), as well as others like developmental biology and genomics later in the twentieth century (which is described as the **Extended Synthesis**; see Pigliucci and Müller, 2010). Today scientists consider evolution to be a fact of life. An evolutionary perspective is dominant in many of the most active fields of biological research, such as genomics and evolutionary developmental biology, and also provides important insights in medical, agricultural, and conservation studies and applications. All in all, evolutionary theory is a powerful theory that organizes and provides coherence to our understanding of life.

Yet evolutionary theory and the idea of biological evolution more generally have been, and continue to be, enormously debated in the public sphere. Various polls taken around the world have shown that there is a rather low public acceptance of evolution (see, for example, Miller *et al.*, 2006). This low acceptance of evolution is often related to a high acceptance of Creationism in various forms (e.g., Intelligent Design [ID] is often considered as the most recent version of Creationism grabbing public attention – see Numbers, 2006, but also Numbers, 2011), and to the attempt to introduce an alternative, religiously founded "explanation" for the origin of species in biology courses (Branch and Scott, 2009). However, Creationism, in any form, does not exhibit the necessary prerequisites for inclusion in the biology curriculum (Sober, 2007; Audi, 2009; Brigandt, 2013). While Creationism is certainly an issue in the United States (see Berkman and Plutzer, 2010; Coyne, 2012), it is by no means restricted to there alone. It exists in the Muslim world, and it seems to be emerging in Europe as well (Graebsch and Schiermeier, 2006; Hameed, 2008; Curry, 2009; Numbers, 2009a; Blancke *et al.*, 2013). Interestingly enough, even literate citizens in countries like China and Japan seem to doubt that evolutionary theory can explain Earth's biodiversity (Cyranoski *et al.*, 2010).

Research on undergraduate students' (both those pursuing degrees in biology and those pursuing degrees in other fields) understanding and acceptance of evolution suggests that they also face similar problems. Students from various countries and religious backgrounds often perceive a conflict between their worldviews and what evolutionary theory suggests (e.g., Brem *et al.*, 2003; Sinatra *et al.*, 2003; Ingram and Nelson, 2006; Deniz *et al.*, 2008; Hokayem and Boujaoude, 2008; Athanasiou and Papadopoulou, 2011; Winslow *et al.*, 2011). This raises serious concerns, as it is important that future scientists and other scholars acquire a clear understanding of what evolution is. This is especially crucial for students who intend to undertake studies in the life sciences, because evolution is its central unifying theory and, as Theodosius Dobzhansky (1973) famously stated, without evolution biology is a pile of sundry facts that make no meaningful picture as a whole. But it is also important that students in other sciences such as physics and chemistry, or even the social sciences and the humanities, also acquire a clear understanding of evolution. Scientific literacy is a demand of our times, particularly since some research fields of biology, such as genetics, genomics, stem cell biology, biotechnology, or conservation ecology, have enormous implications for our lives. Therefore, it is important that all literate people understand the central unifying theory of biology.

Rationale and aims

Many books on evolution have been published, written by evolutionary biologists, philosophers, or historians of science. Some books present the history of evolutionary thought (Larson, 2004; Bowler, 2009a; Ruse, 2009), analyze Darwin's theory in detail (Kohn, 1985a; Hodge and Radick, 2009a; Ruse and Richards, 2009, Ruse, 2013), explain what evolution is (Gould, 2002; Mayr, 2002; Pigliucci and Kaplan, 2006; Ruse and Travis, 2009), provide evidence for evolution (Prothero, 2007; Coyne, 2009; Dawkins, 2009; Rogers, 2011), or explain why it is important for our life (Dupré, 2003; Mindell, 2007; Stamos, 2008; Ayala, 2010a; Vermeij, 2010). Other books explain how evolution is related to religion (Ruse, 2001, 2010; Wilson, 2002; Ayala, 2007; Kitcher, 2007; Miller, 2007), describe the history of the evolution–creation struggle (Ruse, 2005; Numbers, 2006; Bowler, 2007), or explain why Creationism and ID cannot be considered as alternatives to evolution (Eldredge, 2000; Pennock, 2000; Pigliucci, 2002; Ayala, 2006; Sarkar, 2007; Avise, 2010). These are all valuable books, and they present sound arguments and suggestive evidence that shows not only that evolution is a fact of life, but also that evolutionary theory provides the best scientific explanation (so far) for all biological phenomena. However, in most of these books it seems to be taken for granted that it is simple for their readers to understand what evolution is. Therefore, it seems to be assumed that all people need is books which present arguments and evidence for evolution and/or against Creationism and ID. This is what readers will find in many of the excellent books sampled above. But if these books provide ample arguments for both purposes, why then do the public debates about evolution persist? Why is it the case that many people reject evolution or question its validity, despite the evidence for it and its enormous explanatory power in contemporary biological research?

In my view, there is a gap in the existing literature on this topic. Evolution is a rather counter-intuitive idea (from a psychological point of view), and it should not be taken for granted that it is easy for all, or even most, people to understand it. In general, resistance to scientific theories may be due to intuitions that generate preconceptions about the natural world and often make scientific findings seem unnatural and counter-intuitive. For example, children's intuitions make it as difficult for them to accept that organisms may become adapted through natural, evolutionary processes, as it is to accept that the Earth is a sphere. In many cases, these intuitions persist into adulthood (Bloom and Weisberg, 2007). Moreover, it seems that preconceptions related to biology (e.g., the basic living/non-living distinction) are never completely overwritten, despite even a deep understanding of biological processes or expert scientific knowledge (Goldberg and Thompson-Schill, 2009). Such preconceptions make evolutionary concepts difficult to understand. Furthermore, people may misinterpret the implications of evolutionary theory for their lives, and they may also extend these to questions beyond the realm of science. What is necessary is that people realize that evolutionary theory, like all scientific theories, is a means to understand the natural world, and nothing beyond that. It is also a theory which can be put to the test and not something to which we should dogmatically subscribe.

I have written this book in an attempt to fill this gap in the literature, while also trying to present evolutionary theory in a comprehensible manner. To achieve this, I rely not only on evolutionary biology, but also on conceptual development research and on scholarship from both the history and the philosophy of biology. My main intention is to clearly describe the core concepts of evolutionary theory and the features of evolution-ary explanations. However, before attempting this, I am being explicit about the obstacles that affect understanding of evolution, suggesting that the low percentage of acceptance of evolution among students is in part due to a lack of the required understanding. This book explains both what evolution is and why it is difficult to understand. Understanding evolution is neither simple, nor easy to achieve; it is a rather counter-intuitive idea given human intuitions and how we tend to perceive the world around us. Thus, I argue that whether people understand evolution or not *is* a major issue and one that may have been overlooked in the debates surrounding evolution. To the best of my knowledge only two edited books discussing conceptual issues relevant to evolution in some detail have been published (Taylor and Ferrari, 2011; Rosengren *et al.*, 2012), but they are more technical and quite different from this one.

There is another reason for writing this book. Too much ink has been devoted to writing books against ID/Creationism, which has attracted public attention through court cases in the United States. This seems to be a major (political, not strictly religious) issue which, in my view, has misleadingly attracted most attention and as a result other important issues have been overlooked. An insightful research project by Michael Berkman and Eric Plutzer shows why this is the case (see Berkman and Plutzer,

2010, which is a must-read book for anyone interested in the teaching and the public understanding of evolution; see also Berkman and Plutzer, 2011, 2012 for overviews). They estimate that about 28% of US teachers are advocates of evolution and teach it in an appropriate manner; they also estimate that 13% of US teachers somehow advocate Creationism and ID by spending at least one hour of class time on it. Berkman and Plutzer argue that attention should be paid to the 60% of teachers that they call "the cautious 60%," who do not belong to either group of advocates, who cautiously (and reasonably in my view) want to avoid any kind of controversy and of whom 85% accepts evolution. Berkman and Plutzer rightly argue that this "cautious 60%" may do more in hindering scientific literacy than the 13% of explicit advocates of Creationism or ID. An important finding in their survey is that teachers' content knowledge can have a "dramatic effect" on their views and consequently on their teaching practices, as teachers belonging to the "cautious 60%" do not feel confident to teach evolution, although they do accept it. This is a very important point and this is part of the rationale for this book. Instead of trying to show that ID/Creationism is wrong, I have tried to provide the majority of teachers anywhere in the world with a book that explains the conceptual obstacles and the core concepts of evolutionary theory. This book could be used in an undergraduate or a teacher preparation course on evolution, but it could also be read by any biology teacher on his or her own.

I should note at this point that I do not overlook the cultural, religious, worldview, and other issues implicated in the problem of the public acceptance of evolution. I am aware that there are powerful social factors at work, especially among fundamentalist religious believers, that may have nothing to do with conceptual issues. These people usually associate evolutionary theory with a set of liberal values which they perceive as a threat to their own conservative values. They also usually perceive evolutionary theory as a threat to important social and moral issues – and militant modern atheists like Richard Dawkins are in part responsible for this (see Chapter 2 and my Concluding remarks on this). This notwithstanding, context seems to be important for how science is conducted, what conclusions are made, and what its implications are perceived to be. Thus, whether and why people perceive science in general, and evolutionary theory in particular, as a threat to their religious beliefs depends largely on context; generalizations cannot be made. David Livingstone (2003) has argued about the significance of place for the conduct of science, referring to "geographies of scientific knowledge." How science and religion relate to one another also varies around the world (Brooke and Numbers, 2010). However, many excellent treatments of the interplay between science and cultural, social, religious, and worldview factors have already been published. Thus, I have decided not to write much about these issues but to rather focus on conceptual ones, which in my view have not been given the required attention in the literature.

Let me now make clear where I come from and what the specific aims of this book are. I have worked for 12 years as a secondary biology teacher. I have taught evolution to secondary students (in a social context without any serious objection to evolutionary theory, I must note) and I have also conducted research on pre-school, elementary, and secondary students' preconceptions that are relevant to evolutionary theory

(Kampourakis and Zogza, 2007, 2008, 2009; Kampourakis *et al.*, 2012a, 2012b). As a result I am quite aware of students' difficulties in understanding evolution. My main aim with this book is to explain to undergraduate students in the life sciences, some of whom may become biology teachers, why evolution is difficult to understand, and the minimum level of knowledge they should acquire. To achieve this, in this book I first discuss religious resistance to accepting and conceptual obstacles to understanding evolution; I then present some central evolutionary concepts in the light of these obstacles. Throughout the book I have tried to write in a comprehensible manner and I have included several figures which will hopefully contribute to a better understanding of the topics discussed. Reference is always made to articles in books and professional journals from various fields: science, history of science, philosophy of science, and cognitive psychology. In doing so, I am trying to fulfill a secondary aim of this book, which is to serve as a guide to a further and more detailed reading. Bringing together conclusions and insights from research in evolutionary biology, history and philosophy of biology, biology education, and conceptual development, this book might also serve as a guiding light to those wishing to learn more in some or all of these domains. The interested reader will find his or her way to additional literature of interest while reading the chapters of this book.

Consequently, this book is intended primarily for students in the life sciences, either at the undergraduate or graduate level. It provides an introduction to evolutionary theory by presenting not only the core concepts, but also the major conceptual obstacles to understanding evolution. The primary audience of this book also includes biology teachers and educators, as the presentation of concepts and conceptual obstacles is directly relevant to teaching about evolution. Students and teachers could read this book on their own, but it could also be used as a textbook in an introductory evolution course. The book will also be useful to curriculum developers, textbook authors, policy makers, journalists, and anyone interested in evolution or involved in the teaching of evolution and/or its public presentation. Finally, it will be of interest to historians and philosophers of science, as well as cognitive scientists who might be interested in reading how their disciplines can contribute to a proper understanding of science.

I hope the presentation of concepts that takes into account the respective conceptual obstacles will be effective in promoting an appropriate understanding of evolution. Since research suggests that adult resistance to science in general, and to evolutionary biology in particular, may originate in childhood, the various conceptual obstacles are addressed in this book by taking into account students' intuitions, especially those related to teleology and essentialism, which generate preconceptions that in turn make evolutionary theory seem counter-intuitive. Readers of the book will realize which obstacles make evolution difficult to understand, as well as why they persist. Hopefully, they will even be guided to overcome these obstacles themselves. Having understood evolution, readers may then realize that science studies the natural world only. If a supernatural realm exists, it cannot be studied by the rational tools of science. Science does not deny the supernatural, but only acknowledges that it has nothing to say about it. Most importantly, science in general and evolutionary theory in particular is a useful tool in our quest to explore nature and understand life; we should not expect more than that. Consequently, this book is explicit not only about the content of evolutionary science, but also about the nature of science in the wider sense: what science is about, and what its aims are.

Overview of contents

The book consists of six chapters and is divided into two parts. The first part includes the first four chapters which address wider issues relevant to understanding and accepting evolution, such as the nature of evolutionary biology (Chapter 1), religious worldviews and how they relate to evolutionary theory (Chapter 2), conceptual issues and obstacles to understanding evolutionary theory (Chapter 3), and the development of Darwin's theory as a historical case study of conceptual change (Chapter 4). The second part consists of two chapters that are more technical than the earlier ones and which present the core concepts of evolutionary theory along with contemporary knowledge about the evolution of life on Earth, focusing on common ancestry (Chapter 5) and evolutionary change (Chapter 6). Each chapter can be read independently; however, it will be useful for the reader to be aware of the discussion of the conceptual obstacles and conceptual change before reading about concepts.

As students are the main target audience of this book, it includes suggestions for further reading at the end of each chapter. Most major books on evolution published so far are included and their contents are briefly described. Students will thus have a guide for exploring further the issues raised in this book. The book also includes a glossary. Although all concepts will be defined and/or explained in the main text, detailed definitions are also included in the glossary. This can be a useful reference tool that, although is intended to complement the text of the book, also stands on its own. Readers will thus be able to read definitions of the most important evolutionary concepts, and in the main text of the book they will also find references to articles and books that provide further analyses of these. In what follows, I outline the contents of each chapter.

In Chapter 1 I explain how evolutionary biologists work in order to obtain data and what conclusions they can make from it. I then go on to elucidate which questions are answered by evolutionary biology, and how it provides understanding of the world around us, focusing on domestication and infectious disease as examples. Particular cases are described in detail, such as the diversity of dog breeds and the AIDS epidemic, and I argue that evolutionary theory provides a sound explanation for what we observe. This introductory chapter outlines the main features of evolutionary processes and shows that the same basic propositions and models can be used to explain a variety of phenomena. The cases described in this chapter are just some representative ones, discussed for illustrative purposes. The logic of evolutionary theory applies to a lot more.

In Chapter 2 I focus on the relationship between evolutionary theory and religion, in an attempt to explain why many people reject evolution. First, I show that human intuitions about design may not stem from religious beliefs, but rather from our understanding of artifacts. People may think of God as the Creator of our world not (only) because they are religious, but due to their intuitions that make them think of the world as an artifact that consequently demands an artificer. I suggest that people may consider evolution incompatible with their beliefs and worldviews not only because they mistakenly perceive the world as an artifact, but also because they inappropriately extend the applications of evolutionary theory to domains beyond the realm of science. To illustrate how even scientists may do this, the views of three evolutionary biologists – Richard Dawkins, Stephen Jay Gould, and Simon Conway Morris – are compared. An atheist, an agnostic, and a religious person, respectively, make conclusions about the implications of evolutionary theory which are influenced by their worldviews and beliefs. I conclude that in order to seek answers to "big questions" it is necessary to distinguish between what one *knows* and what one *believes*. I suggest that making this distinction clear and achieving conceptual understanding of evolutionary theory are prerequisites for accepting it.

In Chapter 3 I focus on obstacles related to understanding evolution. Having already argued in the previous chapter that the conflict with religious views is only part of the problem and that the real problem may be that people intuitively think of the world as an artifact, I focus on conceptual change in evolution. After explaining what conceptual change in science consists of, I discuss in detail two major conceptual obstacles relevant to evolution – namely, design teleology and psychological essentialism. I analyze these from philosophical and psychological perspectives in order to explain why people tend to think intuitively about the world in teleological and essentialist terms and why thinking this way can make the idea of evolution seem counter-intuitive. I argue that conceptual change in evolution can only take place if these obstacles are properly addressed. To make my case, organisms are compared to non-living natural objects and artifacts, and I explain how organisms differ from artifacts and why organisms therefore require different kinds of explanations than artifacts. Artifacts are objects intelligently designed for some purpose; consequently they have fixed essential proper-ties (as a result of their being designed) and they may be said to exist for some purpose (because this is what they were intentionally created for). This is not the case for organisms. If organisms have essences, these are not fixed; if organisms seem to have purposes, these are evolved, natural ones. I conclude that thinking in essentialist and teleological terms for organisms as if they were artifacts is a major issue that may impact understanding of evolutionary theory. Understanding the differences is crucial for overcoming the obstacles and consequently for understanding evolution.

In Chapter 4 I describe the development of Darwin's theory and I also provide an overview of what he actually wrote in the *Origin*. The chapter starts with the context in which Darwin's theory was developed, taking into account the theories and debates before the *Origin*. By the time the *Origin* was published in 1859, Darwin himself had undergone a conceptual change from his initial views in the 1830s and had developed his theory as an alternative to the views of his times, providing a new explanation for both the common features and the distinctive adaptations of organisms. The important point here is that it took Darwin himself a significant amount of time to develop his theory and to overcome his own initial views. Then the conceptual foundations of the *Origin* are presented, focusing on the influences on Darwin's central arguments (transmutation, common descent, and natural selection). What I also emphasize is that, religious reaction notwithstanding, there were important scientific criticisms of Darwin's theory which were well grounded, and that Darwin was well aware of them and had even sincerely admitted some of them in the *Origin*. These criticisms came both from Darwin's supporters such as Huxley, as well from less sympathetic critics such as Owen and Wilberforce. Consequently, there was more in the reaction to the *Origin* than just religious sentiment, and this chapter also aims to show that disagreement on scientific grounds is possible despite personal views.

In Chapters 5 and 6 I provide a philosophical analysis of some core concepts of contemporary evolutionary theory. Chapter 5 focuses on common ancestry. First, I provide an overview of the evidence that supports the common ancestry of life on Earth, describing what the evolutionary network of life is. I also describe the important insights that the study of microbial life brings to our understanding of evolution in particular, and of life more generally. However, since the main problem with understanding evolution is how complex, multicellular organisms have evolved, I turn my attention to vertebrates (the group which includes humans) to show how evolutionary theory can account for the similar characters we find in organisms. These similarities are either homologies due to common descent or homoplasies due to convergence. There seems to be a continuum of phenomena from homology to homoplasy, and it seems that the study of how characters develop is crucial. This is why I then turn to evolutionary developmental biology, which provides novel insights to the evolution of life on Earth by showing how apparently large morphological transitions may not be so difficult to achieve due to shared underlying molecular networks and mechanisms. Thus, similarities between different organisms may be deeper than was previously thought.

Having described what we know about the common ancestry of all life on Earth, in Chapter 6 I describe the processes of evolutionary change. Adaptations, features or properties that facilitate the survival and reproduction of their possessors in a given environment, are outcomes of natural selection. I describe the various definitions of adaptation and the various perceptions of the process of natural selection. I also argue that stochastic processes have an important role in evolution. There is an important component of unpredictability in evolution, which makes it inherently purposeless. History matters, and one problem we have is how to understand macroevolutionary phenomena, such as speciation and extinction. Epistemic access to the past is difficult to achieve, and so in large part evolutionary explanations have a historical dimension. I conclude that the crucial element for historical explanations is antecedent conditions; particular conditions may have a causal influence on natural processes and turn evolution to one or the other direction.

In my Concluding remarks I describe the virtues of evolutionary theory, and I argue that it cannot, and should not, be used to answer all kinds of questions. My final suggestion is that one should try to understand evolutionary theory without worrying about its religious, metaphysical, or other implications. Having achieved this, one could then decide what these implications are. I believe that evolutionary theory has such implications, but these depend on one's worldview; and this is why there is a variety of reactions to the theory, from dogmatic acceptance of it as a form of secular religion to outright rejection as a form of atheistic dogma. I believe that evolutionary theory shows that life has no inherent purpose, but at the same time it has nothing to say about whether one can find purpose or meaning in life. In contrast, I take evolutionary theory to suggest that each of us can find his or her own meaning and purpose in life. Actually, that humans are able to do this seems to me to be a triumph of evolution; believing that I have achieved this myself makes it a joy. This is, in my view, what an understanding of evolutionary theory can offer: liberate one from fatalistic notions and let one understand the world around us and then find meaning in life through religion, philosophy, art, or any other means one wants.

Acknowledgments

Words are not enough to express my gratitude to the many scholars who kindly offered useful comments and suggestions while I was writing this book. I am indebted to Sandro Minelli and Alan Love, who read the whole book manuscript diligently and made very useful and detailed suggestions. I am very grateful to John Avise, Francisco Ayala, and Michael Ruse for their comments on the whole manuscript. I am also very grateful to Jim Lennox, who helped me clarify (as much as I could) my account of conceptual change. Finally, I want to acknowledge the significant contribution of many scholars who read individual chapters as soon as they were written and made extremely useful suggestions and comments: John Hedley Brooke, David Depew, Patrick Forber, Robert Nola, Kevin McCain, Greg Radick, Karl Rosengren, Mike Shank, Elliott Sober, Paul Thagard, John Wilkins, and Tobias Uller. Authors usually write that any remaining errors are their own. This will be especially true in my case given the high-quality feedback I have received.

I have been working on conceptual issues relevant to understanding evolution for more than ten years. I am indebted to Vasso Zogza, my PhD dissertation advisor, who guided me as a graduate student to understand that conceptual development research has a lot to contribute to understanding science concepts. I am also indebted to my old friend Giorgos Malamis, who guided me through my first forays into the vast literature of philosophy and history of science when I was an undergraduate student. All my research in evolution education was conducted in Geitonas School in Athens, Greece, where I worked as a biology teacher for 12 years. I am grateful to Eleftherios Geitonas, founder and director of this school, and to all my former colleagues there who supported this research in many ways.

While I was writing this book, I was also editing a volume entitled *The Philosophy of Biology: A Companion for Educators* (Kampourakis, 2013a). That book includes important contributions from professional philosophers of biology, and editing it has been an intellectually rewarding experience. I benefited enormously for writing this book by editing that one. Throughout the present book are references to chapters of that book which contain extremely useful analyses of important philosophical topics, all written in a very accessible manner.

Last, but not least, I am indebted to the Cambridge University Press staff. I am very grateful to Martin Griffiths, who supported this book right from the start and guided it in the right direction. I am also grateful to Ilaria Tassistro, Beata Mako, Gary Smith, and

Lynette Talbot for their support during the preparation of this book. Finally, I thank Simon Tegg for drawing those figures I could not draw myself.

The first three sections of Chapter 4 draw in part on sections 3 and 4 of Kampourakis and McComas (2010). The first section of Chapter 6 draws in part on section 2 of Kampourakis (2013b). I was able to draw on and use part of that research with the kind permission from Springer Science + Business Media B.V.

Many authors say that writing a book is a very lonely experience. I managed to write most of this book in my home, surrounded by my family. Much of the writing took place late at night, when they were all asleep. However, in many instances I was also writing with my wife and our children around me, during weekends and holidays. Although having them around may sound as if they were a potential source of distraction, seeing them was a kind of inspiration for me. Over the years I have extensively discussed many of the issues raised in the book with my wife, Katerina, my best friend and companion in life who has a background in the life sciences. Her thoughts, comments, and fierce criticism have always been valuable. Moreover, as I was writing I was thinking that this book should be appropriate for our children, Mirka and Giorgos, to read when they grow up. Existential questions will come up at some point and I wanted to be able to give them this book in order to read about how scientists study the natural world and what they can, and cannot, conclude about it. Thus, I have written this book with my own children and their intellectual/conceptual development in mind.

For being a source of inspiration and for making me feel sentimentally rich, I dedicate this book to my family: my wife and our children for turning an inherently purposeless life into a deeply meaningful one.

1 An evolving world

What is **evolution**? One might define it in many different ways. The term "evolution" might refer either to the fact that organisms have changed over the course of eons, or to the process by which this has taken place or to the outcome of this process, which includes both the exquisite adaptations of organisms and their outstandingly common features. As I do many times in this book, I rely on Charles Darwin's *The Origin of Species* (1859),[1] the foundational text of current **evolutionary theory**,[2] to define evolution. Darwin proposed a "theory of descent with modification through **natural selection**"[3] (Darwin, 1859, p. 343), as an explanation for "the origin of species – that mystery of mysteries" (p. 1). In particular, he aimed to explain the origin of the adaptations of organisms: "how the innumerable **species** inhabiting this world have been modified, so as to acquire that perfection of structure and coadaptation which most justly excites our admiration" (p. 3). The phrase "descent with modification" includes the two central ideas of evolution: All organisms are related to each other because they have descended from a common ancestor through a process of modification that has produced new life forms from pre-existing ones. Thus, evolution might briefly be defined as the natural process by which new species[4] emerge as the modified descendants of pre-existing ones. Evolutionary theory is the scientific theory that explains how this process has

---

[1] The full title of the book was: *On the Origin of Species by Means of Natural Selection, or the Preservation of Favoured Races in the Struggle for Life*. In the rest of this book I refer to Darwin's book simply as the *Origin*.

[2] It should be noted right from the start that the word theory has an entirely different meaning in science compared to the colloquial use. Thus, in science a theory is not simply a hypothesis, a thought, or a speculation (this is what is usually implied with the everyday use of the word), but rather an area of inquiry with widely accepted principles, methods, and foundations or a body of explanatory hypotheses which are strongly supported empirically (Rosenberg, 2005, p. 69).

[3] One major problem that non-experts face with natural selection is to clearly understand what is selected: genes, individuals, or groups? Different views exist on this and experts describe this as the debate about the **levels of selection** (Okasha, 2006). However, it should be made clear that when experts are talking about natural selection, they are referring to an unconscious process of selection taking place in nature, and not to nature consciously selecting anything. Why non-experts tend to favor the latter sense over the former will be discussed in Chapters 2 and 3. An alternative metaphor to describe this process is environmental filtration (Rosenberg and McShea, 2008, p. 18). However, in this book I will stick with Darwin's metaphor, having clarified that natural selection refers to an unconscious process of selection taking place in nature (which is discussed in detail in Chapter 6).

[4] It is difficult to provide a single definition for this concept (see Wilkins, 2009; Ereshefsky, 2010b; Richards, 2010). I describe these difficulties in some detail in Chapter 6. This concept is used throughout this book, rather loosely, to refer to a group of individuals which are reproductively isolated from other groups and/or

taken, and still takes, place on Earth, with reference to particular, old and current, aspects of life on Earth and to particular episodes of its history. What is most important is that evolutionary theory can account for both the unity and the diversity of life. Life has evolved from one or a few universal common ancestor(s) to many different forms of various shapes, sizes, colors, behaviors, and habits. This notwithstanding, they all share some major **characters**,[5] inherited from the common ancestor(s). Evolutionary theory provides the best explanations (so far) for all these phenomena.

In this chapter I provide a broad overview of how evolutionary biologists work to understand both the common origin and the divergence of various life forms. I focus on how evolutionary biologists study nature and obtain data to construct such explanations and reconstruct past events of the history of life on Earth, based on what is often called the "evidence for evolution," e.g., fossils, biogeography, and DNA evidence. Several books presenting the **evidence** for evolution have been published recently (e.g., Coyne, 2009; Dawkins, 2009; Rogers, 2011). Consequently, in this chapter I only provide some illustrative examples. Then, I turn to particular questions about issues relevant to **domestication** and epidemic infectious disease, which serve as case studies. I argue that evolutionary theory provides rational and legitimate answers to these questions, providing sufficient explanations for what is observed.

Before turning to how scientists study evolution, let me make clear an approach which is central in this book. The study of genes and of gene-related phenomena (changes of gene frequencies; changes of gene sequences, etc.) is central in the study of evolution. However, it is difficult to give a single definition for the *gene* concept (see Burian and Kampourakis, 2013 for an overview). Most problematic is the notion of "*genes for*," i.e., genes that *control/encode* **phenotypes**. Genes do not control anything on their own, but operate within cellular environments which affect their expression. If you and I own the same cookbook (DNA) and cook some food, the outcome (phenotype) could be very different even though we have both followed the same recipe (genes). The expression of the **information** in the cookbook (DNA or genes) depends on the cook (developmental system) that will implement it. Consequently, it is useful to mention **development** alongside **heredity**, particularly for multicellular organisms, as developmental processes may produce outcomes different to those expected by reading the DNA sequences alone. To achieve this, throughout the book I refer to DNA sequences which are implicated in phenomena instead of using the overly genetically deterministic language of *genes for* (see Moore, 2002, 2013; Keller, 2010; Burian and Kampourakis, 2013). In a way, this book serves as an experiment to see whether a scientific text can be accurate without any reference to gene **concepts** or "genes for."

---

genetically distinct. For sexually reproducing organisms, a species is defined as a number of, usually similar, organisms that can interbreed and produce fertile offspring.

[5] To avoid inconsistencies by referring to features, traits, characteristics, etc. interchangeably, I will be using the term "character" throughout this book, defined as any recognizable feature of an organism that can exist in a variety of character states, at several levels from the molecular to the organismal (Arthur, 2004, p. 212).

How we know what we know about evolution

Evolution has been taking place on Earth for billions of years. Consequently, although it is still taking place now, much of the information about it comes from the past. In Chapter 6 I describe the importance of history for evolutionary explanations. For now, let me provide an illustration of how evolutionary biologists work. Imagine that you turn on your TV and start watching an episode of a series you have never watched in the past, although its premiere was 20 years ago. You realize that you know nothing about the characters or their relationships, and the plot is too complicated and you can hardly understand what is going on. However, you find it interesting and decide that it is worth the effort to try finding out more about previous episodes. What you might do is try to find them on DVD, or find some information about them on the producer's official website. You might also look for someone who watched the series for a long time and who might thus give you a narrative of past episodes. Eventually, you might end up with much information that would help you follow the plot and keep watching what has become your favorite TV series.

   Unfortunately, studying evolution and obtaining evidence from the past is much more difficult than this. Scientists only have access to what they currently observe; there is no complete record of what happened in the past and, of course, no one was there to witness it. Imagine that in your quest to uncover the plot of previous episodes of your favorite TV series you were unable to find a complete DVD boxset, a website on which the script was available for download, or a friend who had watched it from the very beginning or at least for some time in the past. Imagine that you were only able to find some old episodes from different seasons, a couple of torn pages with a critique of some of the first episodes, some video clips of different episodes uploaded on YouTube without indicating the respective season, and an old interview with one of the members of the cast. What you would have to do would be to watch or read what you managed to obtain and look for clues to events that had taken place in past episodes. But you could also keep watching the current episodes and note down any references to past events that would help you reconstruct the story up to the point that you started watching the series. This is, in part, what evolutionary biologists do. They do not have a direct view of the past, but they can infer past events from what they currently observe. There are three distinct, complementary lines of evidence. The first is quite similar to the one you might try to obtain in your quest to learn more about your favorite series. The other two are more characteristic of doing science.[6]

   What evolutionary biologists do is look for evidence of the past, analogous to the torn pages or the YouTube clips. They look for remnants of the history of life on Earth;

---

[6] Another, perhaps more commonly used, analogy is between an evolutionary biologist and a criminal investigator (e.g., see Cleland, 2002). However, criminal investigators usually investigate individual events (crimes) and do not aim to reveal general patterns (unless a serial killer is involved). Most importantly, they may not be interested in finding out more about sequences of events which may or may not be related. In contrast, to understand what is happening in a TV series, one should try to learn as much as possible about the whole story and not about single events or ones involving individual characters.

these usually exist in rocks and in DNA molecules. For example, human evolution is currently very well understood thanks to both fossil and DNA evidence. This, of course, does not mean that biologists have resolved everything or that no unanswered questions remain. For example, scientists do not agree yet on how exactly humans should be classified. Some scientists use the term Hominini for both chimpanzees/bonobos and humans, whereas others use the term Hominini to refer to the human **clade** only. But this does not mean that any of them questions the fact that the genera *Gorilla*, *Pan*, and *Homo* are closely related.[7] Quite the contrary! Until recently the human clade was distinguished from that of non-human great apes (chimpanzees, bonobos, gorillas, and orangutans) as the Hominidae and the Pongidae family, respectively. However, some scientists now include both humans and great apes under the family Hominidae (Harrison, 2010; Wood, 2010).

   Despite the differences between the skeletons of humans, chimpanzees, and gorillas, there also exist some marked similarities noticed since Darwin's time. Darwin refrained from discussing human evolution in the *Origin*, but was aware that his theory would have relevant implications:

The whole history of the world, as at present known, although of a length quite incomprehensible by us, will hereafter be recognised as a mere fragment of time, compared with the ages which have elapsed since the first creature, the progenitor of innumerable extinct and living descendants, was created. In the distant future I see open fields for far more important researches. Psychology will be based on a new foundation, that of the necessary acquirement of each mental power and capacity by gradation. Light will be thrown on the origin of man and his history. (Darwin, 1859, p. 488)

Darwin's biographers, Adrian Desmond and James Moore (2009), have made the interesting suggestion that Darwin's hatred for slavery made him want to show that all humans had the same ancestry. However, it was not until 1871 that Darwin made public his views on human evolution by suggesting that "It would be beyond my limits, and quite beyond my knowledge, even to name the innumerable points of structure in which man agrees with the other Primates" (Darwin, 1871, p. 191). He then quoted Huxley who, after studying the available evidence, concluded that:

The structural differences between Man and the Man-like apes certainly justify our regarding him as constituting a family apart from them; though, inasmuch as he differs less from them than they do from other families of the same order, there can be no justification for placing him in a distinct order. And thus the sagacious foresight of the great lawgiver of systematic zoology, Linnaeus, becomes justified, and a century of anatomical research brings us back to his conclusions, that man is a member of the same order (for which the Linnaean term PRIMATES ought to be retained) as the Apes and the Lemurs. (Huxley, 1863, p. 124)

---

[7] In many cases those who oppose evolution, for whatever reason, present such disagreements as evidence that science cannot provide conclusive answers. In this case they might consider the fact that some scientists distinguish the human clade from that of the apes, whereas others do not as a controversy pointing to the insufficiency of science, overlooking the fact that all of these scientists consider humans and apes as closely related in an evolutionary sense.

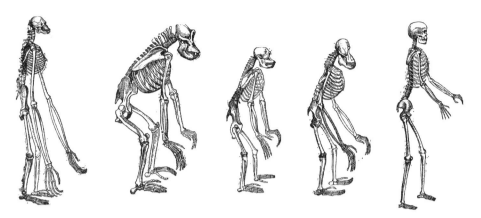

Figure 1.1 The skeletons of gibbons, gorillas, chimpanzees, orangutans, and humans. A picture like this was included in Huxley's book, serving as evidence for the similarities in skeletal structure among these groups. Image © Morphant Creation.

Figure 1.2 One of the usual misrepresentations and wrong portrayals of evolution in general and human evolution in particular. Image © Williammpark.

Figure 1.1 shows the similarities in skeletal structure between humans and the other primates. Since that time, several human fossils have been found (for an overview, see Tattersall, 1998; Wood, 2005). As Darwin had hypothesized, it now seems clear that humans originated in Africa (Tattersall, 2009) and new evidence continuously contributes to a better understanding of human evolution (e.g., White *et al.*, 2009; Berger *et al.*, 2010). However, the idea of evolution in general and of human evolution in particular is usually misrepresented in the public sphere, with illustrations such as the one in Figure 1.2. There are two main problems with this representation of human evolution. First, it portrays evolution as a linear process where each of the species changes into another one. As will be explained in Chapters 4 and 5, evolution is more accurately represented as a branching and not a linear process. Second, this representation shows humans evolving from apes. This is misleading too, because a species cannot evolve from another contemporary species. What is actually happening is that humans and apes share common ancestors, from which they have evolved independently, like branches

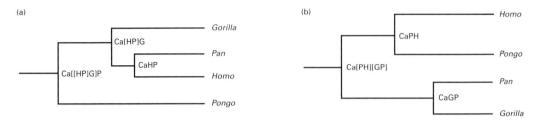

Figure 1.3 (a) Humans and chimpanzees are depicted as the most closely related genera because their common ancestor (CaHP) is closer to the present. These also share a common ancestor with the gorillas (Ca[HP]G), while the orangutans are less related to humans since they share the oldest among the common ancestors (Ca[[HP]G]P) (adapted from Fabre *et al.*, 2009) (*Homo*: humans; *Pan*: chimpanzees; *Pongo*: orangutans). (b) Chimpanzees and gorillas are depicted as the most related genera, sharing a relatively recent common ancestor (CaGP). Humans are depicted closer to orangutans, having diverged from their common ancestor (CaPH) at earlier times, compared to chimpanzees and gorillas. Finally, the two pairs share a common ancestor (Ca[PH][GP]) from which each genus evolved (adapted from Grehan and Schwartz, 2009) (*Homo*: humans; *Pan*: chimpanzees and bonobos; *Pongo*: orangutans). How evolutionary trees are constructed and what kinds of information they provide is discussed in detail in Chapter 5.

starting from a common shoot. Common ancestry and evolutionary change, or descent with modification as Darwin put it, are explained in Chapters 5 and 6, respectively.

Recent advances, such as comparative **genomics** and DNA sequence expression analyses, have contributed to a better understanding of human evolution (Carroll, 2003). Molecular evidence supports the conclusion based on fossils that humans and apes are closely related. A molecular analysis that focused on 27 (from a total of 43 nuclear and 15 mitochondrial) DNA coding sequences, and allowed sampling of 73% to 85% of primate species (Fabre *et al.*, 2009), has concluded that humans are more closely related to chimpanzees (genus *Pan*) than the latter are to gorillas (genus *Gorilla*) (Figure 1.3a). Another line of evidence based on structural, behavioral, and physiological characters, probably not of equivalent status with molecular phylogeny, suggests that humans and orangutans (genus *Pongo*) share a common ancestor not shared by the extant African apes (Grehan and Schwartz, 2009) (Figure 1.3b). Many details on how human evolution actually took place are certainly still missing. Currently we have several scattered pieces of the whole puzzle (Figure 1.4). Nevertheless, the close relatedness between humans and the primates is consistently supported by several different kinds of evidence currently available.

The second line of evidence is a consequence of the ability of evolutionary biologists to make predictions based on existing evidence and test them against it. They might look for particular fossils of particular organisms at particular places, or for particular similarities between specific DNA sequences of certain organisms. Both types of predictions not only have been repeatedly confirmed so far, but have also yielded new evidence of the same kind. You could probably do something like this for your favorite series. You might predict that the producer of the series or a member of the cast would have copies of the old episodes or a copy of the script, and you might look for that person and request these copies. Or you

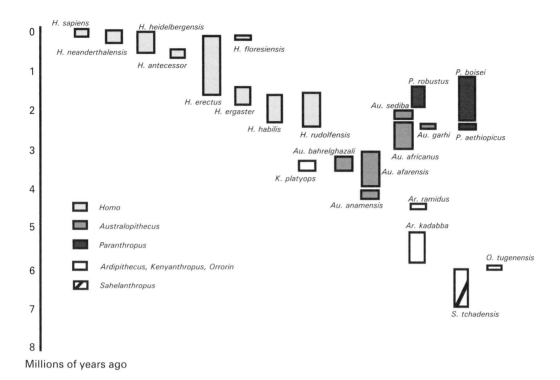

Millions of years ago

Figure 1.4 This is not an evolutionary tree such as the ones depicted in Figure 1.3, because species are not connected with lines. We only have fragmented data about human evolution; much is still missing. The various boxes have different lengths which correspond to the time length (millions of years) during which scientists have found fossils of these species. The various species are not connected with lines because scientists do not know the exact evolutionary relationships (adapted from Wood, 2010). Missing details notwithstanding, we still have a good sense of how our evolution took place. In this figure two words are used to indicate the name of each species; the first refers to the genus and the second to the species. Our species is described as *Homo sapiens*: the word *Homo* indicates the genus and *sapiens* the species.

might predict that some fans of the series would possess what you want and so you could look for their websites or blogs. You might also post a request on your own webpage. Of course, evolutionary biologists cannot find evidence by sending out calls like "fossils of this and that kind wanted." They have to go and look for these themselves. Nevertheless, they often know quite well where to look for evidence and they have been quite successful in finding it. In some cases their predictions would be more successful than your own on finding out what happened previously in the TV series you are watching, because they can have a more solid basis for making predictions.

Although the evolution of tetrapods (four-limbed vertebrates) from sarcopterygian (lobe-finned) fish was generally accepted, there existed few fossils that might suggest how this evolutionary transition might have taken place. The discovery of *Tiktaalik* in Canada has contributed enormously to current **knowledge** of the transition from fish to tetrapods (Figure 1.5). Its skeleton represents a shift from the structure of primitive sarcopterygian fish, toward the structure of tetrapods (Daeschler *et al.*, 2006; Shubin

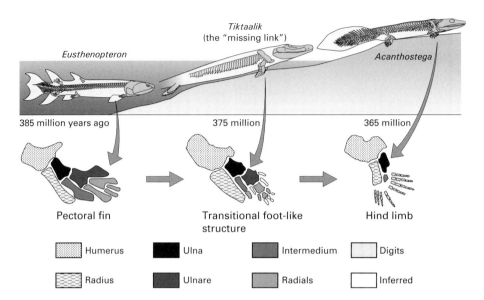

Figure 1.5 The fossils of *Tiktaalik* were found where they were predicted to be and provide evidence about how the transition from fish to tetrapods could have taken place (based on Daeschler *et al.*, 2006; Shubin *et al.*, 2006). Note that this figure does not present the actual transition, but only how it could have been possible. *Tiktaalik* is not the intermediate form or the "missing link," but one that resembles that. Image © Simon Tegg.

et al., 2006). But what is most interesting is why and how these scientists decided to look for the particular fossils at the particular site they did. In his personal account of the discovery, Neil Shubin (2008, pp. 4–5) wrote that: "Most people do not know that finding fossils is something we can often do with surprising precision and predictability. [. . .] Of course, we are not successful 100 percent of the time, but we strike it rich often enough to make things interesting." Shubin then describes how he and his colleagues took into account previous discoveries and decided where to look for fossils of organisms which would be intermediate forms between fish and tetrapods. They had to find rocks of the right age, of a type in which fossils would have been preserved and exposed at the surface. They were aware that amphibian fossils had been recovered from rocks about 365 million years old and that fish fossils had been recovered from rocks about 385 million years old. Consequently, they should look for transitional forms in rocks aged 365–385 million years old. In addition, knowing that sedimentary rocks usually preserve fossils, they had to look for rocks formed in oceans, lakes, or streams, ruling out volcanic and metamorphic rocks in which fish fossils would not likely be found. Finally, they wanted to find areas that were not inhabited and where fossils might be exposed on the surface of rocks. Shubin and his colleagues concluded that the Canadian Arctic was of the right age, type, and exposure, as well as unknown to vertebrate paleontologists. It therefore fulfilled all their criteria. And it was there, at the Fram Formation in Nunavut Territory, Canada, where *Tiktaalik* was eventually found, as they had predicted (Shubin, 2008, pp. 4–27). This discovery, of course, took much time, money, and effort. What is important is that it was based on valid scientific predictions.

The third line of evidence is even more characteristic of science. Contrary to your favorite series, the story of which was the product of human fiction, the history of life on Earth is the product of actual events that are based on natural causal processes such as mutation/recombination, migration, drift, and selection. Under particular circumstances, these processes can cause evolution of a population. For instance, mutation/recombination can produce new DNA sequences and perhaps new characters in a population. In the subsequent generations the population will be different from the initial one, so evolution will have occurred. In the case of migration, some individuals might migrate to new areas, giving rise to a new population which could be different from the old one if some types of individuals but not others from the initial population migrated. Drift results from the random sampling of individuals independently of the characters they possess and of whether these provide them with a particular advantage or not. Some individuals but not others might reproduce, and so the structure of the population might change; the smaller the population, the more significant the effect would be. Finally, during the process of selection some individuals manage to survive and reproduce because they possess characters which contribute to this, whereas others who do not have them fail to survive or reproduce. These processes are discussed in more detail in Chapter 6.

Scientists can make predictions for future outcomes based on their understanding of how these processes take place.[8] Let me give an example. Imagine: a population consisting of green beetles and brown beetles, of the same species, exists in a forest; their color is an inherited character, the **allele**[9] for brown color is dominant[10] and **heterozygotes**[11] exhibit brown color; birds can spot the green beetles on the ground and on the trunks of trees more easily than the brown ones; birds can also spot the brown beetles on the leaves and on the green parts of the plants more easily than the green ones; under these conditions both types of beetles exist in a particular ratio (25% green, 75% brown) in the particular region. It can be predicted that under particular environmental conditions such a population may evolve.

If a new predator is introduced, which lives on the ground and is unable to spot the brown beetles and thus feeds only on green ones, after a number of generations the total number of brown beetles will probably rise. Brown beetles have an advantage because they are concealed in the soil, whereas the green ones are more prone to becoming prey for the new predator on the ground. Consequently, one can make the prediction that

---

[8] Whether these processes are based on laws or law-like (nomological) principles is a discussion that goes beyond the scope of this book (see Sober, 1997; McShea and Brandon, 2010).

[9] An allele is one of several variants of a particular DNA sequence that "encodes" a particular protein or RNA molecule and thus affects a particular biological process. Alleles are identified with particular parts of chromosomes which are described as loci (sing. locus).

[10] Dominance is a concept you probably heard of in your high-school genetics courses: a dominant allele is the one that is "expressed" and the recessive is the one that is not "expressed" when carried together by the same (heterozygous) organism. This concept is problematic as it actually refers to a minority rather than a majority of cases (see Allchin, 2005; Jamieson and Radick, 2013). However, for the purpose of comprehensiveness I will occasionally use the typical terminology of Mendelian genetics taught in high-school biology as most readers of this book will probably be familiar with it.

[11] An individual that carries two different alleles is called a heterozygote. An individual that carries the same allele on both homologous chromosomes is called a **homozygote**.

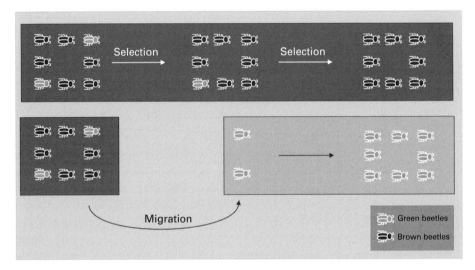

Figure 1.6 Selection and migration; in each case ratios rather than actual numbers of organisms of each type are depicted (see text for the details of the processes).

after a number of generations the population will probably change to one consisting mostly of brown beetles. This will be due to the fact that the brown beetles (and/or the DNA sequence involved in the production of brown color) will be selected. When there is genetic and consequently phenotypic[12] variation in a population (the green and brown colors are inherited characters, i.e., are produced through the expression of particular DNA sequences), natural selection may occur. Not all organisms are equally able to survive and reproduce in a particular environment; some will, others will not. The former are those which are said to be selected. Of course, there is no external agent doing any kind of selection, but one might think that the environment drives the (unconscious) selection of some organisms while others die out. Given this, we can predict that the green beetles in this area will at some point die and the initial population will evolve to one consisting exclusively of brown beetles (Figure 1.6).

Now, consider again the initial population that consisted of 75% brown beetles and 25% green beetles. Imagine that some green beetles only, but not a single brown one, happen to migrate to another area, where they can survive and reproduce without any significant selection pressure. Although brown beetles were greater in number in

---

[12] Which alleles an individual possesses is its genotype. The outcome of the expression of these alleles is described as its phenotype. Alleles may interact in various ways in producing the phenotype. A homozygous individual usually has a particular phenotype, which is determined by its alleles. According to Mendelian genetics usually taught in high-school biology, in a heterozygote one allele may be expressed (dominant) while the other is not (recessive) or in other cases both alleles may contribute to the phenotype observed (co-dominant). It should be noted, though, that how alleles influence phenotype is much more complicated than this simple description because the effect of an allele at one locus may hide the effect of an allele at another locus (**epistasis**) or affect multiple phenomena within the organism (pleiotropy) when, e.g., a protein performs multiple distinct functions or is expressed in multiple tissues (see Stern, 2011 for details).

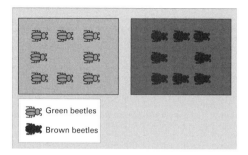

Figure 1.7 Two populations of beetles living in two different areas. This is the outcome of an evolutionary process, and an evolutionary biologist would ask how closely related these populations are, i.e., whether they belong in the same species or not, and how (and when) they diverged from their common ancestor.

the initial population, only green beetles migrated and neither they nor their offspring carry the allele for the brown color (if they did, they would be brown). The consequence of this process is that there will be a deviation from the original frequency of green beetles (25%) in the old area to a new one (100%) in the area to which they migrated. Given this, one can make the prediction that in the new area the migrating beetles will probably give rise to a new population consisting exclusively of green beetles (Figure 1.6). Now, over enough time this population and the population in the original area undergoing selection might independently evolve to two different species of beetles, one with brown and one with green color, which will be distinct from each other and from the initial one (see Figures 1.7 and 1.8 for an example of such a process). This, of course, requires several inherited changes to accumulate to each of the two initial populations before they diverge significantly enough to form distinct species. The details of speciation are discussed in Chapter 6.

From all the above it is clear that evolutionary biologists often make conclusions based on indirect evidence, which has to do with temporal scale unobservability.[13] However, it is possible to observe directly some evolutionary processes. Perhaps the most well known, well studied, and characteristic one is the evolution of the Galápagos finches. Their evolution illustrates how the processes of drift and selection have actually operated in nature. All species of finches currently living at the Galápagos Islands of Ecuador have been derived from a common ancestral species that arrived at the islands

---

[13] Temporal scale unobservability refers to our inability to directly observe particular events and processes which occur on a time scale much greater than a human lifetime. Whereas it is possible to observe and study microevolutionary processes, which involve populations belonging to the same species, it is usually not possible to observe and study macroevolutionary processes like those leading to the emergence of new species from pre-existing ones (although it has been possible to observe this in the lab; this is described as experimental evolution). Some microevolutionary processes can take place within short time spans (e.g., weeks in the case of bacteria or years in the case of finches – see text for details). In contrast, macroevolutionary processes usually span thousands or millions of years and so it is impossible for humans to observe them. The selection of brown beetles or the migration of the green ones described in the example are microevolutionary processes; the evolution of two new species from those initial populations is a macroevolutionary process (macroevolution and speciation are discussed in Chapter 6).

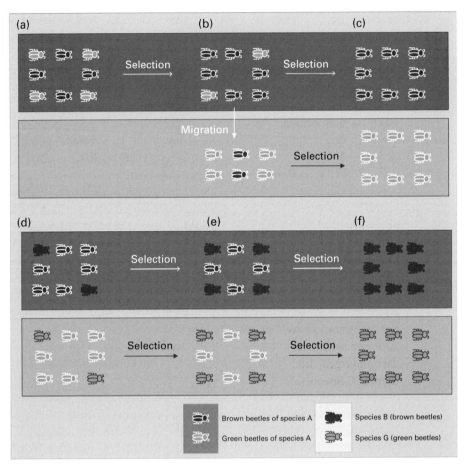

Figure 1.8 Hypothetical evolutionary scenario for the evolution of the two beetle species (green and brown) from a common ancestor. Evolutionary biologists develop such scenarios and then test them against evidence. In each case ratios rather than actual numbers of organisms of each type are depicted. The initial population consists of brown and green beetles. At (a) a new predator causes selection and more green than brown beetles die. In (b) some beetles migrate to a new environment where green beetles have an advantage over brown beetles. Because of different kinds of selection in both environments brown beetles become predominant in the old area and green ones become predominant in the new area (c). The population may diverge further (d–e) and eventually two new species (B and G) evolve (f) which are distinct both from each other and from the initial one (A).

from the American continent (Sato *et al.*, 2001). These species still live in the same environment in which they evolved, and over the years they have been observed to evolve changes in body size and beak shape (Grant and Grant, 2008). Such short-scale studies can provide important insights about how evolution takes place. Although the Grants themselves accepted that in their case they could not have predicted the particular long-term evolutionary outcomes in every detail (mean body size and beak shape at the end of the study) from the beginning, they could certainly make other short-term predictions (Grant and Grant, 2002). The Galápagos finches are a **monophy-letic group** as they have evolved from a single species that arrived from the American continent. Not all species that could possibly live at Galápagos currently live there, only those that are descended from the initial species (common descent will be discussed in Chapter 5). As time went by and new varieties of finches evolved, they dispersed throughout the various islands and, depending on the environmental conditions, some survived and others did not. Through natural selection, the population of finches (and not the individual finches) adapted, which means that the characters of those individuals which could survive and reproduce became prevalent, while other individuals died out (adaptation will be discussed in detail in Chapter 6).

Some moral or social principles that might guide the behavior of the characters of your TV series certainly exist. However, these cannot always be used to make **infer-ences** about what happened in previous episodes, because human behavior can of course violate such principles – e.g., one can behave in an immoral or antisocial way. In contrast, natural processes are based on principles which allow for testable predic-tions, many of which are found to be accurate, as in the case of the *Tiktaalik*, despite the fact that some of them, such as natural selection, have a probabilistic dimension. Based on their understanding of these principles, through the study of fossils and genomes, as well of natural populations in the wild, evolutionary biologists have concluded that all extant and extinct species are related to one another and have diverged from a common ancestor through entirely natural processes of change. Consequently, organisms exhibit both similarities and differences which are due to this fact, and the unity and the diversity of life can be sufficiently explained by evolutionary theory.

Questions answered by evolutionary biology

At this point it is important to note that evolutionary biology, like all science, answers particular types of questions.[14] The classic account is that proposed by Ernst Mayr (1961), arguing that evolutionary biology provides answers to "Why?" questions (e.g., why a character exists or why it has evolved). The respective explanations are based on ultimate causes, which are related to the evolutionary history of a species. These are distinct from the so-called proximate explanations which refer to proximate causes, i.e., causes within individuals, related to their physiology and to how characters actually develop in individuals. The ultimate/proximate distinction has been considered as a major contribution to the philosophy of biology (Beatty, 1994).

Mayr's account has been reconstructed to include a broader conception of evolu-tionary and developmental processes. Thus, proximate explanations are dynamical explanations for individual-level causal events, including not just the decoding of a genetic program but also causal interactions between genes, extra-cellular mechanisms

---

[14] I address the questions that evolutionary biology cannot answer in my Concluding remarks.

and environmental conditions. Accordingly, evolutionary explanations are (causal) statistical explanations that refer to population-level events including not only natural selection but also migration, mutation, genetic recombination, and drift (Ariew, 2003). Recently, the validity of this distinction has been questioned. Given what we currently know about how developmental and evolutionary processes interact and influence one another, it may make no sense to distinguish between evolutionary and proximate causes and explanations (Laland *et al.*, 2011, 2012).

If one pays attention to how evolutionary biologists actually work, one will realize that, depending on their focus, they may have quite different explanatory aims. This means that they may be looking for different kinds of evidence, using different methods, at different places, having different types of questions in mind. There is a diversity of questions that can be addressed by evolutionary biologists. This is due to the fact that virtually every biological character or phenomenon can be studied under an evolutionary perspective. When one asks why any of these exists or how it has come to existence – in other words, when one starts wondering about their origin – the only legitimate answers are found in understanding the patterns and dynamics of evolutionary processes. Evolutionary biologists currently conducting research on evolution around the globe have diverse explanatory aims, but rely on basically the same principles exactly because evolution is the only rational and legitimate answer to every question regarding the origin of biological characters or phenomena. This being said, I do not mean to imply that evolution *can* answer every question about origins; as I have already described for the origins of humans, several open questions remain. What I mean is that if such questions *are* or *can be* answered, evolutionary theory is always part of the answer. As will be explained in detail in Chapter 6, the origin of biological characters can be found in the evolutionary history of the respective species.

What about the alternatives? There is no rational, legitimate, or valid alternative to evolution. Creationists and ID proponents claim that species were created by an intelligent agent who designed all their characters. I will not get into this discussion here because several excellent books provide powerful arguments and ample evidence against this (e.g., Eldredge, 2000; Pennock, 2000; Pigliucci, 2002; Ayala, 2006; Prothero, 2007; Sarkar, 2007; Coyne, 2009; Dawkins, 2009; Avise, 2010). I will only note that organisms are not artifacts – in other words, they are not intentionally designed.[15] As a result, they exhibit many fundamental characters which are useless or even disadvantageous and which are better explained as outcomes of evolution rather than design. This is not to deny that **artifacts** can have useless or badly designed features, but these are not usually fundamental for their **function**. If they are, they were not properly designed. In contrast, organisms can have fundamental characters that are useless or disadvantageous exactly because they were not designed, but because they are products of evolution (the differences between organisms and artifacts will be discussed in

[15] I suggest that artifact or machine metaphors should not be used for organisms (Pigliucci and Boudry, 2011; Brigandt, 2013); if there is a rational for doing so (Becthel, 2013), this should be done with extreme caution.

detail in Chapter 3). Of course, evolutionary biologists are not aware of every detail of the evolutionary history of species, but they nevertheless agree that there is one. Nor should the controversies between evolutionary biologists be taken as evidence of an explanatory insufficiency of science. Evolutionary biologists agree about the fact of evolution and may disagree only on the details. For example, whether humans are more related to orangutans or chimpanzees does not make anyone question the evolutionary history of *Homo*.

Evolutionary biologists focus either on observation, experiments, or both in order to obtain evidence. Some observe species in nature, whereas others conduct experiments in laboratories. Of course, it is possible for an evolutionary biologist to do both. Depending on the research question, evolutionary biologists may employ one or the other strategy. They may study different phenomena or different aspects of the same phenomenon. Their conclusions contribute to the same general framework, and disagreements or different conclusions are part of the game. What is interesting, and not always explicit, is that scientists may draw different kinds of conclusions and thus there may be different kinds of disagreements. Thus, different conclusions may have implications for scientific knowledge only – e.g., which of one or another factor has been more or less important in evolution – but others can have wider implications – e.g., for how we understand human nature and our behavior. Two such examples are given below, in an attempt to make this difference clear. The important point is that scientists are humans and thus there are different, and sometimes idiosyncratic, ways to understand nature. Science is a human activity, after all.

An example of the first kind of disagreement is evident in the conclusions of Jerry Coyne, from the University of Chicago, and Sean B. Carroll, from the University of Wisconsin-Madison. Coyne's major focus is the study of **speciation** – in other words, the process by which new species emerge during evolution (see Chapter 6) – and he is the co-author of the standard academic book in the field (Coyne and Orr, 2004). Carroll's research has focused on DNA sequences which affect animal body patterns; he is one of the most important contributors to the new field of evolutionary developmental biology (often dubbed as evo-devo; see Chapter 5), and he is the author of one of the first books for general readers on this topic (Carroll, 2005a). Carroll, among several others, has advanced the view that it is genetic changes not in protein-coding but in regulatory DNA sequences,[16] often called genetic switches, that have driven morphological evolution – in other words, the evolution of animal form (Carroll, 2005b). Coyne and Hopi Hoekstra, from Harvard University, have questioned Carroll's view, and they have argued that even if this were true it is too early to reach such a conclusion. In addition, they have argued that evolution proceeds via mutations in both protein-coding and **regulatory DNA sequences**, and evidence shows the former to be more

---

[16] Regulatory sequences are DNA sequences which are not transcribed to mRNA like protein-coding sequences, but which affect the expression of these protein-coding sequences. This happens because particular molecules, such as transcription factors, can bind on regulatory sequences and influence transcription of protein-coding sequences. In other words, regulatory sequences act as switches that regulate the transcription of protein-coding sequences and eventually protein synthesis (see Carroll, 2005a; Stern, 2011 for details).

important than the latter (Hoekstra and Coyne, 2007). In a recent book, David Stern (2011) has explained that one needs to take both changes in populations and changes in development into account in order to understand evolution. This disagreement is perhaps technical, although it has significant implications about what the research focus of evolutionary biology should be.

However, different approaches may lead to disagreements with wider implications. Two major contributors to evolutionary thinking are Harvard scientists Richard Lewontin and Edward O. Wilson. Both have worked as scientists, although they have also contributed to the philosophy and the public understanding of biology. Lewontin worked for years on population genetics, focusing both on *Drosophila* and humans. He worked out mathematical models, studied genetic diversity, and argued for a dialectic relationship between organisms and their environment. Wilson was more of what we would call a naturalist, and he focused on social insects, mostly studying them in their natural environments. Lewontin's focus was on the genetic constitution of populations, whereas Wilson's focus was on the social behavior of their members. Consequently, they employed different research strategies, addressing particular questions, and eventually studied evolution from different perspectives. This notwithstanding, their conclusions contributed to the same general framework. However, Wilson and Lewontin have entirely disagreed on the philosophical implications of their scientific conclusions. Wilson is well known for his *Sociobiology* (1975), in which he argued for the genetic basis of (human) behavior. Lewontin has been among its fiercest critics as, for example, a co-author of *Not in Our Genes* (1984) and other books providing counter-arguments (for a brief and informative account of these differences, see Ruse, 1999, pp. 153–193).

The foregoing examples demonstrate the diversity of questions asked by evolutionary biologists: the genetic structure of populations, the social interactions of their members, the emergence of new species, the evolution of animal form, and many more. Let us now see how such questions can be addressed. Returning to the beetles of the previous example, let us assume that an evolutionary biologist conducting field research observes the following: a population of green beetles lives on grassland and its color facilitates its concealment in the particular environment; a population of brown beetles lives in a neighboring, isolated rocky place where brown is the dominant color and where thus its color facilitates its concealment. Let us assume that the beetles do not differ in any other morphological characters (e.g., shape and size) (Figure 1.7).

Two questions that would be of interest to an evolutionary biologist studying these populations are: Are these two distinct species, or two populations of the same species? If they are two distinct species, when and how did they diverge from their common ancestor? To answer these questions, the biologist can make hypotheses and test them against evidence. This is very characteristic of how scientists work. In order to answer particular questions, scientists can develop hypotheses and look for evidence in order to support them or reject them.[17] But this does not mean that scientists

---

[17] This particular, and for some peculiar, way of working is conventionally, but mistakenly, known as the "scientific method." This term is misleading because there is no single method that all scientists use. However, there are particular modes of thought and research strategies which are very common.

cannot work without a hypothesis in mind. Quite the contrary, scientists can formulate research questions and then seek evidence in order to provide answers to these questions. Nevertheless, they often have anticipated outcomes in mind and so they may develop hypotheses based on these. In case they do not, they simply formulate research questions based on existing data, knowledge, or concepts. In all cases, scientists do not start research from nothing; they always look for answers based on previous scientific knowledge and theories. These are taken into account and form the basis for the development of particular research questions or hypotheses which they try to answer or test on the basis of empirical evidence.[18] Of course, depending on the discipline there can be several different approaches or combinations of approaches. Nevertheless, we might summarize the process of answering a scientific question with the following scheme:

---

question → (hypothesis →) empirical data → results → conclusions

---

Let's assume that, in order to answer the foregoing questions, the evolutionary biologist develops a hypothesis for each one of them. These are two different kinds of questions that require two distinct kinds of hypotheses. Whether the two populations belong or not to the same species is a contemporary question that can be answered in a direct manner. Given their morphology, the evolutionary biologist might initially make the hypothesis that these are two distinct but closely related species. Then he might examine their morphologies at a finer level of detail. He might also observe them for some time in their natural habitats in order to see how they behave, mate, feed, and interact. Then, they might be transferred to the laboratory for a more detailed examination of their anatomy, physiology, and behavior. There they might also be studied genetically, and compared to each other in terms of their DNA sequences.[19] This morphological and molecular data would probably suffice for concluding whether the two populations are of the same species or not.

Assuming that support was found for the initial hypothesis that these are two closely related but distinct beetle species, the evolutionary biologist could make another hypothesis about how they have evolved from their most recent common ancestor. But this is a different kind of question: it is not a contemporary but a historical one. This question cannot be answered in a direct but only in an indirect manner. The evolutionary biologist can only make a historical hypothesis about how the two species could have evolved from a common ancestor and test it. This is neither simple, nor easy. Let us assume that from the morphological and the molecular data, as well as from

---

[18] Scientists may ask questions or develop hypotheses. In the second case, it should be noted that they do not always find support for them. However, even rejected hypotheses can lead to valid scientific conclusions, or even open new areas of inquiry and motivate scientists to ask new questions.

[19] These different types of studies are actually more complicated than is described here and usually require several scientists with different areas of expertise in order to be conducted.

comparisons with other extant beetle species, the conclusion is made that the "green" species evolved from a founding population that migrated to the grassy area from the rocky one. According to this, there must have been some color variation within the initial population: the majority of beetles might have been brown, but some green (or green-like) beetles could have existed as well. From an initial population consisting of beetles living in the rocky area, a small part could have migrated to the grassy area. Both populations might have consisted of both brown and green (or green-like) beetles. However, in each area only those beetles that could conceal themselves could survive and reproduce in the long run and thus after a long time these populations may have evolved so that only brown beetles survived and reproduced in the rocky area and only green beetles survived and reproduced in the grassy area. These beetle populations initially differed only in their color, but being isolated from each other for a long time they eventually evolved in different ways as any new characters were confined to the population in which they arose, due to isolation. Thus, the two initial sub-populations of the same species eventually evolved to two distinct species (see Figure 1.8).[20]

How can such a hypothesis be further supported or eventually rejected? There are various ways. The evolutionary biologist might look for other populations of beetles living in the area. There, he might find other beetles living between the "brown" and the "green" species and exhibiting intermediate morphological characters. Or he might find such intermediate forms in fossils of extinct beetle species. Molecular comparative DNA analyses of the "green" and "brown" species and other species living in the area might also yield valuable information. All this might help support or reject the historical hypothesis. It should be noted that such historical hypotheses may be rejected more easily than they can be supported. If one body of evidence strongly contradicted the hypothesis (e.g., beetles living in the intermediate area did not exhibit intermediate characteristics but were very different from both the "green" and "brown" species), then the evolutionary biologist could reject it and make a new one. On the other hand, if all evidence gathered supported his hypothesis, he would have accepted it as the most probable one but this would not be the end of the line. New evidence is always possible which might overthrow even a well-established hypothesis; the history of science is full of such episodes (see many examples in Bowler and Morus, 2005).

Like all science, evolutionary biology does not provide definitive answers. Evolutionary biologists usually base their conclusions on what is called *inference to the best explanation* (IBE). This means that they rely on any available evidence, and develop explanations that explain this evidence in the best possible way (for details on IBE, see Chapter 6). What they actually do is compare potential, often competing, explanations and eventually accept the one that fits best with the available evidence. Of course, over time the details may change or novel explanations may be developed as new evidence becomes available. But there are at least two distinctive parts of evolutionary explanations that will never change. The characters and properties of these two beetle species

[20] This is a very simple, if not a simplistic, scenario. I am only using it for illustrative purposes to indicate the difference between answering questions about the present (are the two populations of the same or different species?) and the past (how have these two different species evolved?).

are either derived from their common ancestor or are the outcome of their evolution in the particular environments. Common characters derived from a common ancestor are described as homologies. However, different species may exhibit common characters that have been selected under the same environmental conditions, and these characters are described as homoplasies (see Chapter 5). Finally, particular characters that have become prevalent in a particular environment, usually through natural selection, and confer an advantage to their possessors are usually described as adaptations (see Chapter 6). Common ancestry and evolution of populations through their interaction with their environment can sufficiently account for the origin of biological characters and are core concepts in any question answered by evolutionary biology.

Consequently, evolutionary biologists may have diverse explanatory aims but they all rely on evidence and some core ideas such as common descent and evolution through the interaction with the environment in order to make conclusions. Why they may end up with different conclusions is another story. This depends on their explanatory aims, the questions asked, and the data obtained. It also depends on scientists' perspectives. Nevertheless, evolutionary biology can in fact provide answers to several other questions, not necessarily asked by evolutionary biologists, but by other scientists. What is more interesting is that some of these questions have a direct impact on human life. It is to these questions that we now turn. In the next two sections I discuss domestication and epidemic infectious disease as two case studies.[21]

Domestication

Humans have had an effect on the evolution of particular species through domestication. Domestication is the process of controlled breeding of a species in a way that makes it useful to humans. In doing so, a species is modified from its wild ancestors under a conscious process of artificial selection performed by humans. In other words, domestication involves a process of evolution by means of artificial selection performed by humans and not by means of selection taking place in nature. In the case of animals, domestication is distinct from taming. In taming, wild-born animals come to live close to humans (e.g., a young tiger, found in the wild, grows up in a zoo), whereas in domestication animals are born and grow up in captivity, belonging in a population that has not lived in the wild for several generations (e.g., dogs). Nevertheless, and despite the strong human influence, domesticated species do not evolve in an entirely directed way because there are particular characteristics that cannot be influenced by humans. For example, domestication took place at particular places in the world where the respective wild species were abundant, even if the initial environmental conditions were not entirely favorable. In addition, humans may cause selection but they do not drive the whole process. Furthermore, humans can never be certain in advance that

---

[21] It should be noted that in this chapter I describe some representative examples and focus on evolutionary mechanisms. Mindell (2007) is an excellent resource on examples of evolution in everyday life, and Poiani (2012) another excellent resource on applications of evolutionary theory.

whatever they intend to "select" will eventually be selected. Interestingly enough, only a small minority of wild species has been domesticated (Diamond, 1997, 2002). Overall, domestication involves an evolutionary process of change in which selection is artificial, but natural selection may actually take place as well.

Perhaps the most familiar domesticated species are dogs (*Canis lupus familiaris*).[22] Dogs are a phenotypically diverse species that consists of at least 400 genetically distinct breeds. What is most interesting is that dogs are not a different species from wolves (*Canis lupus lupus*), as indicated by their names, suggesting that they belong to the same species, *Canis lupus*.[23] However, the origin of dogs is not completely understood. It is widely accepted that dogs are descended from wild wolves through domestication. However, there are open questions regarding the time and place of the first domestication, as well as whether there was a single or more than one domesti-cation event (Wayne *et al.*, 2006). There are different views on these: Some scientists suggest the Middle East as the place of origin of modern dogs, about 12 000 years ago, although there are potential sources of variation from Europe and East Asia (von Holdt *et al.*, 2010). Others suggest Eurasia as the place of origin of modern dogs and also an earlier domestication date of around 30 000 years ago (Skoglund *et al.*, 2011). It seems that dogs have a rich evolutionary history and current attempts focus on relating genomic information to phenotypic variation (Akey *et al.*, 2010). The evolution of dogs is particularly interesting because it shows signs not only of artificial selection, as anticipated because of the domestication process, but of natural selection as well.

Accepting that dogs are descended from wild wolves one might infer that dogs have evolved from wild wolves that were artificially selected by humans, e.g., because they were friendly or because they simply did not cause them any trouble. Under this assumption, some wild wolves came to live close to humans and gradually evolved to dogs as humans selected particular characteristics. Wild wolves first learned to be tame, then they were trained to do whatever humans wanted them to do, and eventually they evolved to domesticated dogs. There are two problems with this view. First, wolves do not easily become tame, and even if they do by growing up among humans, they revert to their wild condition in adulthood. Second, wolves and dogs exhibit significant differences in some characters (e.g., dogs have smaller brains than wolves) that cannot be explained by artificial selection. It has been suggested that the actual process of domestication is different, and that it consists of both natural and artificial selection (Coppinger and Coppinger, 2001, pp. 39–67).

The Coppingers provide evidence that even today it is not easy to tame wolves, and when this is achieved they are still dangerous. Thus, it is hard to imagine how humans living 10 000 or more years ago might have achieved this. Then, they also argue that

---

[22] As already mentioned above, using our species as an example, we use a binomial nomenclature (e.g., *Homo sapiens*) to identify species. In this case, the sub-species is also mentioned. Thus, dogs belong to the sub-species *Canis lupus familiaris* of the species *Canis lupus* of the genus *Canis*.

[23] There is evidence that wolves and dogs have interbred extensively, which may also be a cause of the extraordinary phenotypic diversity of dogs (Vilà *et al.*, 1997).

wolves are not easily trained because they are not obedient and this is why we do not see them in shows with lions, tigers, and other animals. But even if our ancestors managed to tame and train wild wolves some 10 000 years ago, a question remains: How were the abilities to be tamed and to be trained passed on to the descendents of those wild wolves? The answer is that this could not have been possible without some genetic basis. But then, there might have been some genetically tame and obedient wolves that were selected by humans and eventually passed on their characters to their offspring. However, the Coppingers argue that tame and trained wolves do not seem to pass on these characters to their offspring (Coppinger and Coppinger, 2001, pp. 42–50). In addition, dogs are able to understand human signals in a way that wolves are not. Interestingly enough, wolves raised by humans do not show such skills, whereas domestic dogs only a few weeks old do – even if they have had little contact with humans. These observations suggest that the particular cognitive skills of dogs are the outcome of evolution under domestication, with humans selecting those dogs with which they could communicate more effectively (Hare *et al.*, 2002).

The Coppingers then move on to argue that, compared to wolves, dogs exhibit some characters like smaller brains and smaller teeth. Assuming that wolves with smaller brains and teeth once existed, why would our ancestors, who used dogs for hunting or for protection, have selected individuals with these characters? In fact, and although brain size is not everything, wolves seem to be "smarter" than dogs, or at least are better at solving problems, by watching not only humans but other animals as well. Dogs, on the contrary, are generally not able to simply learn through observation as wolves do and they need repetition in order to learn a task (Coppinger and Coppinger, 2001, pp. 46–49). The answer that the Coppingers provide to the question of why humans selected individuals with these characters is that they simply did not make such a selection. They argue that some of these characters were already there, as inherited characters that became prevalent through natural selection. They suggest that some wolves first became self-domesticated by natural selection. Then the tamer and the more trainable ones were selected by humans (Coppinger and Coppinger, 2001, p. 57). Here is a plausible scenario: humans started living in settlements where food was abundant so that waste food existed. Some wolves who were genetically predisposed to be less afraid of humans started having access to this new food resource. A measure of this can be whether they flee from humans or not and whether they bite them or let them handle them – for simplicity we can make the distinction between more "friendly" and "wild" wolves (see Trut, 1999, p. 163). The initial wild wolf population had already evolved to one consisting of scavenger dogs with smaller brains and teeth (large brains and teeth are a waste of energy for a scavenger dog and they might have simply been eliminated by natural selection) (Coppinger and Coppinger, 2001, pp. 58–62). Among these, the tamer ones were selected over the less tame ones (which might have simply been killed by humans). The whole evolutionary process as suggested by the Coppingers is illustrated in Figure 1.9.

That the ability to be tamed is inherited and that selection for such characters can take place during domestication has been experimentally shown by Dmitri Belyaev and his colleagues (Belyaev, 1979; Trut, 1999). During an experiment that lasted for more than 40 years, they managed to produce a domestic population of silver foxes

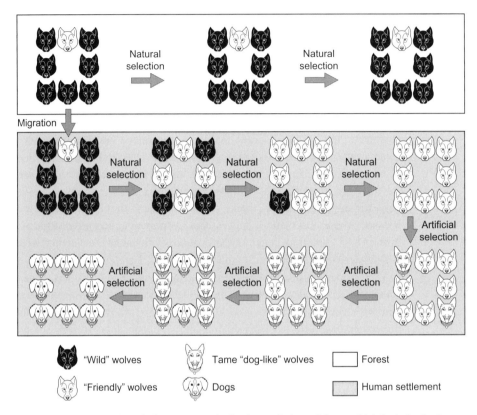

Figure 1.9 Hypothetical evolutionary scenario for the evolution of dogs, which includes both natural and artificial selection (see main text for details). Image © Simon Tegg.

(*Vulpes vulpes*). This is a species which is taxonomically very close to dogs and which had never been domesticated in the past. Foxes were first evaluated for tameness. Starting from one month old and performing the test every month until they were sexually mature (7–8 months), foxes were given food from the hand of an experimenter who also tried to handle them. At the end of this period, the foxes were ranked as Class III (flee from humans or bite when handled), Class II (let themselves be handled but do not seem to be friendly to humans), Class I (friendly toward humans), and Class IE the "domesticated elite" (try to establish human contact and attraction, like dogs). Each time they let the most tame foxes reproduce. In 1999, 40 years and 45 000 foxes after Belyaev started the experiment, there was a unique population of 100 domesticated foxes who differed significantly from the wild ones: they responded to sounds and opened their eyes earlier, and they showed fear response for the first time at an older age, which was also linked to changes in the levels of hormones related to stress. Other changes included loss of pigment in certain areas of coat color, new characters such as floppy ears and rolled tails. Belyaev's initial conclusion, which seems to have been confirmed by subsequent work after his death, was that domestication caused selection of behavioral characters: Selection of individuals carrying DNA sequences which have an influence on the nervous and endocrine systems and thus affect not only behavior (on the basis of which individuals were selected), but their development as well (Trut, 1999). Interestingly enough, such DNA sequences may have recently been identified (Albert *et al.*, 2009). This is a very important conclusion to which I return in Chapter 5: changes in development can have an important influence on evolution.

If you consider the whole process depicted in Figure 1.8 carefully, you will realize that it is a familiar one. It is a repetition of cycles of the two-step process described earlier in this chapter. This process consists of two distinct steps: variation and selection.[24] In the fox experiment that stands as evidence of how dogs may have evolved from wild wolves, small differences in behavior among individual foxes (the element of variation) were (artificially) selected and eventually gave rise to a quite different population of tame foxes which had evolved from wild ones. Evolutionary theory can thus account for the evolution of the numerous breeds of dogs which live among us. They have evolved through a combination of natural and then artificial selection. Why are these so many? Because of us. Living under human protection and care, even the weakest breeds are able to survive. Some dogs of small size might even be threatened by cats if they were left on the street to live on their own. By taking care of them, we help them survive and reproduce. By making them breed, we are also able to produce several different kinds of combinations of dog characters. This is how our so-called "best-friends" have evolved. But evolutionary theory can also account for the evolution of organisms which have perhaps a more direct and significant impact on human life: pathogens. These will be the focus of the next section.

Epidemic infectious disease

One of the topics in which evolution has been crucial to our understanding is that of some major pandemics such as Acquired Immune Deficiency Syndrome (AIDS). By the end of 2011 an estimated 34 million people were living with HIV all over the world, with 1 in every 20 adults being infected in sub-Saharan Africa. An estimated 2.5 million people became infected with HIV in 2011, but there seems to be a steady decrease in this number, which is 20% lower than it was in 2001 (UNAIDS, 2012). There are some major issues here stemming from the evolution of the respective pathogens: the inability of the human immune system to respond effectively to infection by these pathogens; the insufficiency of the produced vaccines; and the fact that pathogens occasionally become resistant to drugs. Evolutionary theory can explain why the human immune system cannot respond effectively, why vaccines and drugs are insufficient, and why drug-resistant pathogens emerge.

AIDS is caused by human immunodeficiency viruses HIV-1 and HIV-2. These viruses are considered as relatively new pathogens since they were introduced in humans during the twentieth century. Both of these have evolved from strains of simian

---

[24] There is, of course, more than that to evolution (see Chapters 5 and 6). The process of evolution is not identified with the process of selection.

immunodeficiency viruses (SIVs) on separate occasions (Heeney *et al.*, 2006).[25] These viruses infect human T-cells, which are among the leucocytes (white blood cells) which initiate the immune response against pathogens. In general, when a pathogen infects the human organism it eventually encounters the T-helper cells (usually described as CD4 cells, because of a specific kind of receptor on their cell membrane). As soon as these cells interact with the molecules of the outer surface of the pathogen (antigens) they excrete molecules which initiate an immune response. This consists of various cells and molecules that are produced, but most notably of T-cytotoxic cells (usually described as CD8, because of another specific kind of receptor on their cell membranes). These are white blood cells which destroy the human cells infected by HIV and thus do not permit the reproduction of the virus inside them. In addition, B-cells, another type of white blood cell, produce antibodies, specialized proteins which neutralize the respective antigens. This immune response is most often effective in the sense that the pathogens are destroyed, although the host may die before this happens. In short, we can think of T-cells and B-cells as the police officers of the human organism which arrest and eliminate any intruder. As soon as the latter gets inside the organism, it is recognized and its description is distributed all over the whole organism. Then other police officers (cells) are able to identify the intruders and eliminate them. Normally, any pathogen will at some point be eliminated by the immune system (unless the host dies in the meantime, which is, unfortunately, something natural).

A distinctive characteristic of AIDS is the deficiency of the natural, protective immune response. Why is that? In a sense, HIV viruses manage to "disguise" themselves and avoid being neutralized or destroyed by antibodies. How is this possible? Can these viruses think? Are they so wise that they can avoid the far more complex cells of the immune system? The answer is absolutely not; quite the contrary, the viruses are defective. All viruses can only reproduce themselves in their host cells. Actually, some scientists do not consider them as alive because they exhibit no other property of living systems besides reproduction. In addition, they do not reproduce on their own, but rather through the machinery of their host cell, which they usually destroy.[26] The HIV viruses include an enzyme, called reverse transcriptase, which synthesizes DNA that is complementary to the viral RNA. This is the crucial stage. This enzyme is error-prone and thus several mistakes happen during this reverse transcription[27] process. Eventually, a DNA molecule is produced which is quite different from the one

---

[25] People often wonder how the virus was transmitted from primates to humans. Interestingly enough, it has been found that the primate bush meat sold in African markets is often infected with SIV. Thus, it is not surprising that people working in this market might have become infected (Peeters *et al.*, 2002).

[26] Actually this is more or less what happens when one inserts a compact disk (CD) into a computer and makes 1000 copies of it. The CD does not reproduce itself; it is the computer that makes the new copies. Imagine that, during the reproduction, the new CDs were created in the computer, which was eventually destroyed afterwards. In the case of the reproduction of HIV, the new viruses destroy the host cell as they are released out of it.

[27] Synthesis of an RNA molecule which is complementary to a DNA strand is described as transcription. HIV has an enzyme that synthesizes a DNA strand which is complementary to an RNA molecule. This process is called reverse transcription and the enzyme is called reverse transcriptase.

that should have been produced, were the process accurate. Reverse transcriptase is estimated to make approximately 0.2 errors per genome during each replication cycle, and further errors occur during transcription of RNA from DNA. In addition, HIV has a generation time of approximately 2.5 days and produces $10^{10}-10^{12}$ new viruses each day. This DNA molecule will integrate in the host-cell DNA and, when transcribed, it will produce a new RNA that, when translated, will produce proteins that will be different from the ones that should have been produced. Some of these proteins will be part of the outer surface of the newly produced HIV viruses, which will in turn be new to the white cells of the immune system.

An example of the resulting changes is the so-called glycan shield: There is a constantly changing pattern of glycosylation of the HIV envelope proteins that prevents antibody binding and does not affect the capacity of the virus to infect cells. Consequently, there will be no memory, and no effective immune response, in the human organism against these new viruses. The old ones will eventually be eliminated, but the new ones will have to be identified again, right from the start. In the meantime these viruses will reproduce, and new ones will emerge which will again be different and thus impossible to fight (see Rambaut *et al.*, 2004 for details, and Figure 1.10 for an overview of this process). In the meantime, more T-cells will be destroyed. Thus, the problem with HIV is that it not only kills the police officers (T-cells) which "arrest" it, but also "disguises" itself so that it is not recognizable. Consequently, other police officers (T-cells) are not able to track it, and thus it has to be "arrested" again and again.

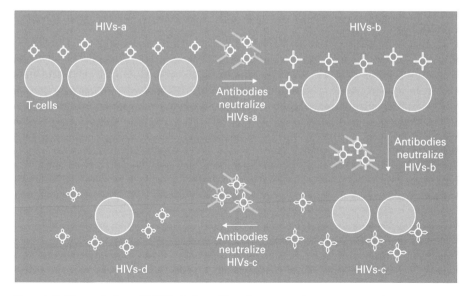

Figure 1.10 The evolution of HIV in its host. Note that new variants are constantly produced while T-cells become fewer with time (the proportion of viruses and cells is not accurate; the figure only shows the viral particles at stable numbers and the numbers of T-cells decreasing).

In the meantime, the available police officers become fewer and fewer. This cycle takes place again and again, and more and more viruses are produced while T-cells become fewer as time goes by. Eventually, the host organism dies because of some other pathogen that infects it because it can no longer fight it. HIV patients die because they no longer have an effective immune system and so they become vulnerable even to pathogens that might have no effect on a healthy individual.

If you consider the whole process depicted in Figure 1.10 carefully, you will realize again that it is the familiar two-step process described earlier in this chapter. The mistakes made during the process of reverse transcription produce genetic variation: New viral DNA and protein molecules and consequently new HIV viruses are made.[28] These are then "selected" in the environment of the host organism, in the sense that they are not eliminated by the cells and molecules of the immune system, whereas those viruses from which they were derived are. The immune response will eliminate those viruses that initially infected the host, but not the recently produced ones, which are new to it, as a recognition process by T- and B-cells is required. Consequently, the prevalence of AIDS in some countries is explained through the evolution of HIV.[29] In fact, there seems to be a close connection between the evolution of a particular pathogen and the spread of the disease it causes: The evolution of the pathogen drives the spread of the disease and then the spread of the disease may affect the evolution of the pathogen.[30] At this point it should be noted that the actual within-host evolutionary process is more complex than is presented here. When a new type of HIV virus emerges within a host, it does not necessarily follow that it will replace the initial type that infected the host. However, in the case of AIDS many new types can emerge and this leads to a complex process of successive replacements of different HIV types within a single host. Which of these will eventually be transmitted is not easy to predict. It should also be noted that natural selection is one of the mechanisms by which HIV evolves. Drift seems to play some major role as well (Alizon et al., 2011).

The evolution of HIV has major implications for the production of effective vaccines. Evidence so far suggests that generating an immune response similar to the one generated after infection is not enough to immunize effectively against HIV. In addition, and strange as it may seem, several fundamental questions are still unanswered: Which antigens should be included in the vaccines? How should we deal with the tremendous variability produced during the reproduction of the virus? Should vaccination focus on protection from infection or from the onset of the disease? What seems to be a crucial

---

[28] The error-prone HIV reverse trancriptase is actually an advantage that drives its evolution. However, new HIV viruses also emerge because of the recombination of segments of HIV genomes in hosts that have been infected by several different viral particles (Rambaut et al., 2004).

[29] This is, of course, part of the explanation. AIDS is a pandemic because in particular regions (mostly in Africa and Asia) the public is not well informed. HIV viruses are neither transmitted easily, nor do they live for long outside their host organisms compared to other viruses like, for example, the H1N1 virus that causes swine flu. HIV viruses are only transmitted through unprotected sexual intercourse or through blood, and only live for a few minutes outside their host. This notwithstanding, millions of people are living with AIDS today and this is due to social rather than biological factors (see Whiteside, 2008 for an overview).

[30] Such studies have given rise to the field of evolutionary epidemiology (Restif, 2009).

point is that HIV initially infects a small number of cells. Thus, vaccination might aim at preventing or clearing this initial infection, or at keeping the virus at low levels so that no AIDS will occur. Of enormous interest is also the fact that exposure to HIV is not always followed by infection. Research has shown that T-helper cells (CD4) have very important roles during immune response, whereas T-cytotoxic cells (CD8) are important but less useful for vaccination strategies. It is also of interest that despite the large diversity of the HIV surface glycoprotein gp120, there may not be a need for a respectively large diversity of antibodies. All these require further research (see Virgin and Walker, 2010 for further details).

Another distinctive feature of AIDS is its co-incidence with tuberculosis, which is actually a case of a co-epidemic. In 2011 there were an estimated 8.7 million new cases of tuberculosis globally, 13% of which were co-infected with HIV; 1.4 million people died from tuberculosis, including 430 000 HIV-positive people (WHO, 2012). Tuberculosis is caused by *Mycobacterium tuberculosis*, actually a family of strains with significant variation in how quickly they spread through populations, or how capable they are of causing active tuberculosis. Although most diseases that affect people with AIDS are not a threat to others, tuberculosis is an exception. Being HIV-positive is the most important risk factor for susceptibility to *M. tuberculosis* infection and progression to active disease (Dye *et al.*, 2010). It seems that each disease facilitates the progression of the other: HIV-positive people are more likely to develop active tuberculosis, which in turn shortens their expected life span. In addition, diagnosis of tuberculosis is more difficult in HIV-positive people. Several studies provide evidence of how this may be possible (e.g., Reuter *et al.*, 2010; Diedrich and Flynn, 2011).

In one case, from a population of 1539 patients, 542 were positive for *M. tuberculosis* and 221 of them had multi-drug resistance (MDR) to tuberculosis. Of these, 53 patients had extensive drug resistance (XDR) to tuberculosis.[31] These came from different regions, and their only possible contact could have been made in the hospital where they received health care. In addition, none of them had a family member with tuberculosis. What is interesting is that all 44 of these 53 patients with XDR tuberculosis who were tested for HIV were found to be HIV-positive. Their median T-helper cells count was 63 cells per mm^3 – in other words, lower than the limit of 200 cells per mm^3 that indicates immune deficiency. Fifty-two of these patients eventually died (Gandhi *et al.*, 2006). How can one explain the fact that all 44 patients with XDR tuberculosis were HIV-positive? Is it possible that XDR strains were not found in patients not infected with HIV? Evolution could provide a hypothesis that could be tested: HIV-positive patients were more vulnerable and thus were at high risk of illness when they were infected by drug-resistant strains. In other words, within a population of patients infected by *M. tuberculosis*, there may have been selection for the XDR strains in the HIV-positive patients. For some reason, the XDR strains might not proliferate in patients not infected with HIV, perhaps due to their effective immune response. But although protective immunity suppresses the proliferation of XDR strains,

---

[31] Resistance at least to isoniazid and rifampicin is described as MDR tuberculosis; additional resistance to any fluoroquinolone, and one of capreomycin, kanamycin, or amikacin is described as XDR tuberculosis.

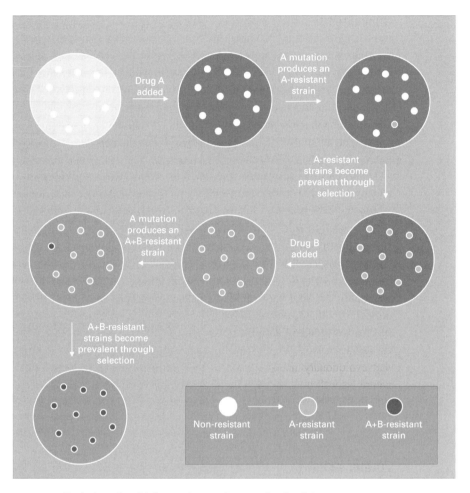

Figure 1.11 Evolution of multi-drug resistance (see text for details).

in patients with immune deficiency these strains may become prevalent. This is, of course, a hypothesis that should be tested against evidence.

What is even more interesting, and again explained in evolutionary terms, is how these drug-resistant strains arise. Resistance of *M. tuberculosis* strains to drugs is the result of new mutations and not of DNA transfer from resistant to non-resistant bacteria. Such mutations have a possibility of 0.01–1 in one million. Thus, populations of resistant bacteria arise in patients and gradually become prevalent as the others are eliminated by drugs. In other words, in the presence of drugs, drug-resistant bacteria multiply whereas non-resistant ones are killed. Once again, this is the familiar two-step process described earlier. Mutations in bacterial genomes produce variation and then there is selection among the different bacterial strains, with the resistant ones being selected. Through this process *M. tuberculosis* strains that are resistant to various drugs become prevalent, producing what has been called MDR to tuberculosis (see Figure 1.11 and Gandhi *et al.*, 2010 for details).

Conclusions

In all the examples of evolution described above, a simple process always takes place: selection from within an available set of variants. This, of course, is not the only mechanism by which evolution proceeds, but it is an important one and certainly the most widely discussed. Given the knowledge we currently have about how evolution takes place in the everyday world, as illustrated in this chapter with just a few examples, one wonders why people reject it. My view is that the excellent books already available provide numerous examples of, and evidence for, evolution, but do not explicitly address human emotions and intuitions. In the next two chapters I argue that people reject evolution because they do not understand it, and this is due to two unconscious mistakes: First, they do not clearly distinguish between what they know and what they believe; second, they hold preconceptions resulting from deep intuitions which make scientific explanations seem counter-intuitive. In Chapter 4 I describe the process of Charles Darwin's own conceptual shift, as an exemplar case of what it takes to overcome obstacles and understand natural processes. I also describe how his theory was disregarded for some time during the last decades of the nineteenth century, not only due to religious reactions but also because of sound criticism on scientific grounds. In Chapters 5 and 6 I describe some core evolutionary concepts, taking into account the earlier discussion of conceptual obstacles. In my concluding remarks I conclude that evolutionary theory is important for understanding our world and our place in nature, as well as that religion and morality are not threatened by it. Thus, let us now turn to the serious and important conceptual issues behind evolution.

Further reading

There exist numerous books which present the evidence for evolution as well as the main processes. A nice book to start with is Jerry Coyne's *Why Evolution is True*, which provides an authoritative overview of evidence and processes. Another book with several examples and useful information is *The Greatest Show on Earth: The Evidence for Evolution*, by Richard Dawkins. A similar, brief book is Alan Rogers' *The Evidence for Evolution*. A must-read is David Mindell's *The Evolving World: Evolution in Everyday Life*, which is full of interesting examples of the importance of evolution for everyday life. A similar, but more technical, book is *Pragmatic Evolution: Applications of Evolutionary Theory*, edited by Aldo Poiani. For those interested in the fossil evidence for evolution a great resource is Donald Prothero's *Evolution: What the Fossils Say and Why it Matters*. The current advances and the future directions of evolutionary theory are nicely presented in *Evolution: The Extended Synthesis,* edited by Gerd Müller and Massimo Pigliucci. Two books discussing the implications of evolutionary theory are *Darwin's Legacy: What Evolution Means Today*, by John Dupré, and *The Evolutionary World: How Adaptation Explains Everything from Seashells to Civilization*, by Geerat Vermeij. For those seeking an encyclopedia-style book, *Evolution: The First Four Billion Years*, edited by Michael Ruse and Joseph Travis, provides more than 500 pages of alphabetical entries on topics and personalities relevant to evolution, while it also includes recent essays on some of the most important evolutionary topics. Beyond evolution, an impressive short book about what science is and how it is done is Stuart Firestein's *Ignorance: How it Drives Science*. Ronald Giere's *Scientific Perspectivism* provides a very interesting, but more philosophical, account of how science is done. Another interesting and philosophical account of the nature of science is *Systematicity: The Nature of Science* by Paul Hoyningen-Huene.

2 Religious resistance to accepting evolution

The idea of evolution has been widely, and sometimes fiercely, debated in the public sphere. Various polls in the United States, Europe, and other countries have shown that there is a low public acceptance of evolution (see, for example, Miller *et al.*, 2006).[1] This low public acceptance is usually related to **Creationism** (**Intelligent Design** or ID is often considered as its most recent version, although there are some differences between them – see Numbers, 2006, and especially Numbers, 2011). Although Creationism is prevalent in the United States, it is by no means restricted to there; it seems to be emerging in other countries too (Numbers, 2009a). In general, Creationism is the **belief** that God created the universe, including the Earth and humans, through a series of miracles. Young-Earth Creationists perceive the world to have been created in six days of 24 hours each, sometime within the last 10 000 years, whereas Old-Earth Creationists accept the scientific account of the age of the Earth but still believe that the creation of life and organisms took place through a series of miraculous interventions. ID proponents provide a seemingly more sophisticated criticism of evolutionary theory, claiming that it cannot explain everything about life. They consider, e.g., the vertebrate eye or the bacterial flagellum, as **irreducibly complex** systems – i.e., systems that become non-functional if a part is removed – , which cannot therefore have evolved simply through evolution by natural selection. In their view, such systems can only have been created for their current roles by an intelligent agent or Creator God, and so stand as evidence for ID. In addition, they claim that evolution by natural selection is explanatorily insufficient since it has a **low probability** to occur. However, such arguments are fallacious because many events can have a very small probability to occur and yet are possible (e.g., the outcome of repeated tosses of a coin resulting in heads five consequent times has a low probability, but it is possible). Most importantly, an event (e.g., the origin of complex biological structures) having a small probability does not imply the improbability of the theory assigning this probability (e.g., evolutionary theory) (see Brigandt, 2013 for a recent overview and critique).

Such claims have a widespread appeal, in part because superficially they seem to be correct. However, Creationism and ID have been criticized as non-scientific,

---

[1] It is important to note at this point that attention should always be paid to what kind of questions/items, and with what kind of content, are included in surveys about the public acceptance of evolution, as well as to whether they might cause discomfort for respondents or lead to biased or otherwise distorted representations of their beliefs (see, for example, Rughinis, 2011).

religiously founded approaches to such an extent that no further discussion is required here (see, for example, Eldredge, 2000; Pennock, 2000; Pigliucci, 2002; Sarkar, 2007; Avise, 2010). Starting with a misrepresentation of how evolution occurs, their proponents claim that something more than natural mechanisms is required in order to explain the origin of species and their characters. This additional factor is God, whose existence and power are the ultimate explanation for everything. In this view, whenever there are no scientific explanations about something, God's divine intervention stands as the always-sufficient alternative, an approach that has been characterized as the "God of the gaps." This means that whenever there is a "gap" in the explanatory potential of science, this is filled by assuming that God intervened and so His intervention stands as the explanation for whatever cannot be explained otherwise. However, the argument "there is no scientific explanation for *X*, therefore *X* can only be explained by assuming divine intervention" is ambiguous. There is an important difference between the propositions "*X* is *unexplained* by science," which means it has not been explained by science *yet*, and "X is *unexplainable* by science," which means that science cannot *in principle* explain *X*. The former proposition refers to questions that science has not answered *yet* but which it can in principle answer, whereas the latter proposition refers to questions that fall outside the realm of science because it cannot explain everything.

Consequently, the "**God of the gaps**" **argument** does not work either way. That there is no scientific explanation for something because, e.g., no relevant evidence has been found yet, does not entail that such evidence will never be found. The classic example here is the quest for transitional fossils. If someone had argued that divine intervention was necessary for the special creation of marine and land organisms since there was no conclusive evidence about the transition from life in the sea to life on land, what would that person have to say about *Tiktaalik*, the discovery of which is described in Chapter 1? There also exist important questions about the nature, the meaning, or the purpose of life. But such questions fall outside the realm of science, which by definition has an empirical character, and they rather belong in the realm of philosophy and metaphysics. Questions such as whether God exists or not cannot, in principle, be answered by science because our epistemic access is limited to the empirical data available in the natural world. Without any epistemic access to the supernatural, the existence of God or any such question cannot be considered a scientific question (see Pennock, 2000, pp. 163–172 for a detailed discussion; see my Concluding remarks about questions which cannot be answered by science).

Another reason that Creationism and ID have a wide appeal is that evolutionary theory has been identified with a form of materialism, which is often perceived as amoral, if not immoral. A usual criticism is that evolutionary theory deprives human life of moral values and principles as it presents humans as just one animal species among all the others. In this view, if humans accept that they are just animals, as evolutionary theory suggests, they may start behaving like them: compete, kill, mate promiscuously, etc. Interestingly enough, proponents of this view overlook the possibility that morality may itself be a consequence of evolution (e.g., Ayala, 2010b). It seems that arguments against evolution persist and what makes things even worse is the outbreak of what has been called *militant modern atheism*. The latter provides a critique of religious fundamentalism, but also often stands as an attack against all religion. Although the former is quite accurate, the latter is perceived by many religious and irreligious people as mistaken (Ruse, 2010; Kitcher, 2011). Richard Dawkins is unquestionably the most famous proponent of this view. When the criticism of religious fundamentalism is extended to all religious attitudes, and when acceptance of evolution is promoted as the only rational alternative, a conflict arises. A consequence is that evolutionary theory is identified with atheism or materialism and thus many people become afraid of it and its consequences before they even have the chance to understand what it is about.

Evolutionary science, as all science, documents characteristics of the natural world. In what follows I start from the argument that the natural world was designed by a wise and competent Creator, whose existence was confirmed by the complexity and the design observed in nature, articulated in most detail by William Paley. Then, I explain why the notion of design is so widely accepted and why the idea of evolution seems to clash with religious worldviews (for two excellent and complementary accounts of how evolutionary theory relates to religion, see Alexander, 2013; Ayala, 2013). Then, I argue that the real problem is epistemological, a matter of what one actually knows and of whether this is distinguished from what one believes. The supposed conflict between science and religion is a very complicated issue and its features have changed over time (Brooke, 1991). People with similar backgrounds and knowledge may have very different beliefs, and a rich diversity of views has always existed among theologians and scientists. To illustrate this, I focus on three evolutionary biologists and show how differently they perceive the implications that evolutionary biology has for religion. I conclude by putting emphasis on the distinction between what one knows and what one believes.

Creation and design in nature

Contemporary evolutionary theory emerged historically in a Christian religious context. Thus, I will focus on the contrast of evolutionary theory with notions of creation and design in nature within the Christian worldview.[2] Predominant in this case has been the ***argument from design***.[3] According to this argument, if nature seems to exhibit design, it is because it is God's creation. Therefore, this design stands as evidence for His existence. The strategy of arguing for the existence and attributes of God from features of nature – **Natural Theology** – uses **the argument from design** to support this worldview. Perhaps the most well-known proponent of Natural Theology, who actually had a major influence on Darwin, was William Paley. Paley believed that the complexity and perfection of the natural world, documented by its empirical study,

---

[2] An excellent resource about the relation between science and various religions is Brooke and Numbers (2010).

[3] For the history of the argument from design, see Ruse (2004).

were the most powerful arguments for the existence of God. According to him, all patterns and laws found in nature reflected the thoughts of God. Consequently, the study of nature was a way to prove His existence. Paley lived at the end of a long period of production of new knowledge and of expression of heretical ideas. Empiricist philosophers, such as David Hume, had expressed skeptical arguments about the authoritarian status of religion, suggesting that it was based only on faith and not on factual data (Ruse, 2004). For instance, Hume's criticism included the following objections: (1) evil, imperfections, and wasted material could be found in the world, (2) more than one designer could exist, and (3) the argument as it was originally stated could explain some but not all types of complexity found in the world (see Oppy, 1996). Hume's criticism resulted from his empiricist approach to the study of nature and from his intention to protect philosophy from the anti-empiricist approach of theology, responding to natural theologians who thought they could build a theology wholly on the study of nature:

In a word, CLEANTHES, a man, who follows your hypothesis, is able, perhaps, to assert, or conjecture, that the universe, sometime, arose from something like design: But beyond that position he cannot ascertain one single circumstance, and is left afterwards to fix every point of his theology, by the utmost licence of fancy and hypothesis. This world, for aught he knows, is very faulty and imperfect, compared to a superior standard; and was only the first rude essay of some infant Deity, who afterwards abandoned it, ashamed of his lame performance; it is the work only of some dependent, inferior Deity; and is the object of derision to his superiors: it is the production of old age and dotage in some superannuated Deity; and ever since his death, has run on at adventures, from the first impulse and active force, which it received from him. [. . .] You justly give signs of horror, DEMEA, at these strange suppositions: But these, and a thousand more of the same kind, are CLEANTHES's suppositions, not mine. From the moment the attributes of the Deity are supposed finite, all these have place. And I cannot, for my part, think, that so wild and unsettled a system of theology is, in any respect, preferable to none at all. (Hume, 1993/1779, p. 71)

Paley tried to show that the existence of God could be confirmed with rational arguments based on data from the study of nature. In particular, Paley used the metaphor of the organism as a watch and of God as a watchmaker, according to which a complex structure, like a watch, could not have emerged accidentally but required the existence of a designer-watchmaker. Let us consider Paley's argument. Right from the start of his book entitled *Natural Theology* (Paley, 2006/1802), he suggested that:

In crossing a heath, suppose I pitched my foot against a *stone*, and were asked how the stone came to be there, I might possibly answer, that, for any thing I knew to the contrary, it had lain there for ever: nor would it perhaps be very easy to shew the absurdity of this answer. But suppose I had found a *watch* upon the ground, and it should be enquired how the watch happened to be in that place, I should hardly think of the answer which I had before given, that, for any thing I knew, the watch might have always been there. Yet why should not this answer serve for the watch as well as for the stone? Why is it not as admissible in the second case, as in the first? For this reason, and for no other, viz. that, when we come to inspect the watch, we perceive (what we could not discover in the stone) that its several parts are framed and put together for a purpose, e.g., that they are so formed and adjusted as to produce motion, and that motion so regulated as to point out the hour of the day; that, if the different parts had been differently shaped from what they are, of a different size from what they are, or placed after any other manner, or in any other order, than that in which they are placed, either no motion at all would have been carried on in the machine, or none which would have answered the use, that is now served by it. (Paley, 2006/1802, p. 7)

This is a teleological[4] argument, but one that is based on intention and design. The parts of the watch were formed and adjusted *in order to* produce motion; and the whole system was regulated in such a way *in order to* show what time it is. If the parts of the watch had a different shape or size, or if they were placed in another manner or order, the watch would not be functional. This particular argument implies the existence of intelligent and **intentional design**, and consequently of a watchmaker who designed the watch in a particular way so that it would be functional. His purpose was to make an artifact that tells the time and it was fulfilled by the implementation of a particular design that involved a specific arrangement of its parts. This design in turn stands as evidence for the existence of a designer. The stone has no such features; it does not (seem to) consist of parts properly shaped and adjusted for some purpose. Consequently, there is no inference of it being designed and thus about the existence of a designer. Paley concluded that:

This mechanism [of the watch] being observed [. . .], the inference, we think, is inevitable; that the watch must have had a maker; that there must have existed, at some time and at some place or other, an artificer or artificers who formed it for the purpose which we find it actually to answer; who comprehended its construction, and designed its use. (Paley, 2006/1802, p. 8)

Leaving (philosophical) technical details aside, the structure of Paley's argument could be presented as follows:

Paley's argument

The stone does not consist of parts formed and adjusted for some purpose.

Therefore there is no reason to infer that the stone was designed by anyone.

The watch consists of parts formed and adjusted for some purpose which would not be fulfilled had they been formed and adjusted differently.

Therefore there is reason to infer that the watch was designed by an intelligent and competent watchmaker.

Paley wrote that the "inference is inevitable" (this is a case of inference to the best explanation or IBE).[5] That no one ever observed the process of making the watch, or that it might go wrong sometimes, or that we might not understand how some of its parts were contributing to its overall function, would not, according to Paley, allow any questioning of the fact that it was designed. And he continued:

Suppose, in the next place, that the person who found the watch, should, after some time, discover, that, in addition to all the properties which he had hitherto observed in it, it possessed the unexpected property of producing, in the course of its movement, another watch like itself; (the thing is conceivable;) that it contained within it a mechanism, a system of parts, a mould

---

[4] **Teleology** is a mode of explanation in which the existence of a feature is explained on the basis of its contribution to some end. Teleology does not necessarily entail intentional design and **teleological explanations can be natural** (see Lennox and Kampourakis, 2013; see also Chapter 3).

[5] This topic is discussed in detail in Chapter 6; for Paley and IBE, see Ruse (2004).

for instance, or a complex adjustment of laths, files, and other tools, evidently and separately calculated for this purpose; let us enquire, what effect ought such a discovery to have upon his former conclusion? I. The first effect would be to increase his admiration of the contrivance, and his conviction of the consummate skill of the contriver. (Paley, 2006/1802, p. 11)

Here Paley noted that even if one found that the watch could reproduce, and could thus explain that it came into being by a pre-existing watch, this would not undercut the need to explain how the first watch and its design emerged. The first consequence of such an observation would be that the watchmaker would be even more capable than we thought before. This observation would, of course, stand as additional proof for the existence and attributes of a designer.

If that construction *without* this property, or, which is the same thing, before this property had been noticed, proved intention and art to have been employed about it; still more strong would the proof appear, when he came to the knowledge of this further property, the crown and perfection of all the rest. (Paley, 2006/1802, p. 11)

Paley argued that the property to produce another watch should be ultimately attributed to the designer who created the first watch and not to the first watch itself. A complex property like reproduction could not simply exist without having been designed. Paley actually noted that if we attributed the existence of a watch to another watch and of that to another watch, and so on, we would never get a definite answer and we would still need a designing mind:

There cannot be design without a designer; contrivance without a contriver; order without choice; arrangement, without any thing capable of arranging; subserviency and relation to a purpose, without that which could intend a purpose; means suitable to an end, and executing their office, in accomplishing that end, without the end ever having been contemplated, or the means accommodated to it. Arrangement, disposition of parts, subserviency of means to an end, relation of instruments to an use, imply the presence of intelligence and mind. (Paley, 2006/1802, p. 12)

Paley then made the move from the watch to organisms:

This is atheism: for every indication of contrivance, every manifestation of design, which existed in the watch, exists in the works of nature; with the difference, on the side of nature, of being greater and more, and that in a degree which exceeds all computation. I mean that the contrivances of nature surpass the contrivances of art, in the complexity, subtlety, and curiosity of the mechanism; and still more, if possible, do they go beyond them in number and variety: yet, in a multitude of cases, are not less evidently mechanical, not less evidently contrivances, not less evidently accommodated to their end, or suited to their office, than are the most perfect productions of human ingenuity. (Paley, 2006/1802, p. 16)

Paley argued that organisms not only seem as designed as instruments and artifacts do, but they are also more complex than any human-made object. He then compared the eye with the telescope and argued that they are both instruments operating under the same principles. Before getting into the details of the comparison, Paley concluded that "there is precisely the same proof that the eye was made for vision, as there is that the telescope was made for assisting it" (p. 16). He then made a detailed comparison between the structures of the two and concluded that:

How is it possible, under circumstances of such close affinity, and under the operation of equal evidence, to exclude contrivance from the one, yet to acknowledge the proof of contrivance having been employed, as the plainest and clearest of all propositions, in the other? (Paley, 2006/1802, p. 18)

In other words, given the similarities, we cannot conclude that the telescope is designed and the eye is not.

The evidence we have for their close affinity in terms of structure and function do not allow us to question the conclusion that they are both designed, and hence that there must be some designer for each of them. But then Paley goes on to notice the superiority of the eye over the telescope:

But further; there are other points, not so much perhaps of strict resemblance between the two, as of superiority of the eye over the telescope; yet, of a superiority, which being founded in the laws that regulate both, may furnish topics of fair and just comparison. Two things were wanted to the eye, which were not wanted, at least in the same degree, to the telescope; and these were, the adaptation of the organ, first, to different degrees of light; and, secondly, to the vast diversity of distance at which objects are viewed by the naked eye, viz. from a few inches to as many miles. These difficulties present not themselves to the maker of the telescope. (Paley, 2006/1802, p. 18)

This adds more to the argument. A "natural" instrument is superior to a human-made one because there exist several different versions of it adapted to different conditions. If we replace the watch with the telescope, given that they are both instruments designed each for a particular purpose, the argument takes a new form:

Paley's argument

The stone does not consist of parts formed and adjusted for some purpose.

Therefore there is no reason to infer that the stone was designed by anyone.

The telescope consists of parts formed and adjusted for some purpose which would not be fulfilled had they been formed and adjusted differently.

Therefore there is reason to infer that the telescope was designed by a competent telescope-maker.

The eye is superior to the telescope.

Therefore the eye-maker is superior to the telescope-maker.

The eye is adapted to different degrees of light and to observing objects at a variety of distances. Paley notes that these "difficulties" were not faced by the creator of the telescope; in other words, the creator of the eye had to overcome more difficulties. What is the inference, then? If he did overcome these difficulties, the creator of the eye must be more competent than the creator of the telescope. Not surprisingly, this is the case. But there is more. The creator of the eye not only overcame these difficulties, but he did so in many ways, depending on the way of life of the various organisms. For example, Paley notes that "by different species of animals the faculty we are describing [seeing objects at different distances] is possessed, in degrees suited to the different range of vision which their mode of life, and of procuring their food, requires" (Paley, 2006/ 1802, p. 21). He then describes several examples of eyes in various organisms and concludes that: "Thus, in comparing the eyes of different kinds of animals, we see, in their resemblances and distinction, one general plan laid down, and that plan varied with the varying exigencies to which it is to be applied" (Paley, 2006/1802, p. 22).

After providing more details about how the eyes of different organisms operate, Paley asks the final question:

One question may possibly have dwelt in the reader's mind during the perusal of these observations, namely, Why should not the Deity have given to the animal the faculty of vision *at once?* Why this circuitous perception; the ministry of so many means? an element provided for the purpose; reflected from opaque; substances, refracted through transparent ones; and both according to precise laws: then, a complex organ, an intricate and artificial apparatus, in order, by the operation of this element, and in conformity with the restrictions of these laws, to produce an image upon a membrane communicating with the brain? Wherefore all this? Why make the difficulty in order only to surmount it? If to perceive objects by some other mode than that of touch, or objects which lay out of the reach of that sense, were the thing proposed, could not a simple volition of the Creator have communicated the capacity? Why resort to contrivance, where power is omnipotent? (Paley, 2006/1802, p. 26)

And he immediately provides the answer:

Contrivance, by its very definition and nature, is the refuge of imperfection. To have recourse to expedients, implies difficulty, impediment, restraint, defect of power. This question belongs to the other senses, as well as to sight; to the general functions of animal life, as nutrition, secretion, respiration; to the oeconomy of vegetables; and indeed to almost all the operations of nature. The question therefore is of very wide extent; and, amongst other answers which may be given to it, beside reasons of which probably we are ignorant, one answer is this. It is only by the display of contrivance, that the existence, the agency, the wisdom of the Deity, *could* be testified to his rational creatures. [...] Whatever is done, God could have done, without the intervention of instruments or means: but it is in the construction of instruments, in the choice and adaptation of means, that a creative intelligence is seen. It is this which constitutes the order and beauty of the universe. God, therefore, has been pleased to prescribe limits to his own power, and to work his end within those limits. (Paley, 2006/1802, pp. 26–27)

Although the Creator was capable of creating the eye once and for all, He did it many times in different ways, taking into account the peculiarities of life of each organism. And He did this on purpose because He wanted to make His competence evident to humans. Thus, He was not simply capable of creating the eye; He was capable of creating many different eyes, operating under different circumstances. He did not simply create but also posed limitations to His creative process which He was then able to overcome. This is further evidence of His competence.

This gives to the argument its complete form:

Paley's argument

The stone does not consist of parts formed and adjusted for some purpose.

Therefore there is no reason to infer that the stone was designed by anyone.

The telescope consists of parts formed and adjusted for some purpose which would not be fulfilled had they been formed and adjusted differently.

Therefore there is reason to infer that the telescope was designed by a competent telescope-maker.

The eye is superior to the telescope.

Therefore the eye-maker is superior to and eventually more competent than the telescope-maker.

The eye-maker did not simply create the eye, but set limits to his creative process which he was eventually able to overcome.

Therefore the eye-maker is not only competent but also made sure that his competence is evident to us.

Therefore God is the eye-maker.

Based on this, we can summarize Paley's argument as follows:

Paley's argument

Complexity indicates design and hence the existence of a designer.

The more complex an instrument, the more competent is its designer.

Animal organs are superior and much more complex than human-made instruments.

Therefore the creator of animal organs is superior and more competent than any human creator.

Therefore the creator of animal organs is God.

Why did Paley make the inference from complexity to design? It seems that humans tend to intuitively link complexity to design and to think that any complex structure is more likely to have been designed rather than to have emerged through some other process. Paley thought so, which could be due to his religious beliefs. But is religiosity the cause of such intuitive thinking? Is it due to religiosity that we look for purpose and design in nature? It seems that intuitions about purpose and design come first and religiosity could either be a consequence of this intuitive thinking or it may simply happen to fit nicely with these intuitions. This is, in my view, a potential major issue in the public debates about evolution. We humans intuitively perceive purpose and design in nature. Religion is in accordance with these intuitions and thus seems intuitive; evolution is not in accordance with these intuitions, in fact it suggests that imperfections and cruelty exist in nature, and eventually seems counter-intuitive. Let us now examine some evidence from conceptual development research that supports this view.

In one study in the United States it was investigated whether children from Christian fundamentalist school communities expressed more Creationist views than children coming from non-fundamentalist school communities on the issue of the origin of species. The conclusions of this study were particularly interesting. The participants were divided into three groups based on their age: 5–8 years old, 8–10 years old, and 10–13 years old. Most students from fundamentalist school communities provided Creationist explanations to all tasks, independently of their age. On the contrary, students from non-fundamentalist school communities provided explanations that were different in different age groups. However, 8–10-year-old students of this background provided mostly Creationist explanations, which is a very interesting finding that shows that it is not only the religious background of the family that may have an influence on students' beliefs. All participants seemed to endorse mixed beliefs, with evolution mostly applied to organisms besides man, for whom creation was preferred instead (Evans, 2001). In general, in many instances the social-religious background of the students was found to have an influence on their explanations of the origin of species; however, 8–10-year-old children were found to exhibit a bias for endorsing intentional Creationist accounts of how species originated, regardless of the religiosity of their background. This finding suggests than there is more than religion involved in human intuitions about purpose and design in nature.

Similar findings were reported by another study that involved American and British students. Based on the assumption that America and Britain share many cultural characteristics but differ in religiosity, with the British being less religious than Americans, the study aimed at investigating whether there was a difference in the preferences of American and British elementary school students, aged 7–10, for teleological explanations. Despite some differences in details, the explanations of students of both groups were quite similar; they generally preferred teleological explanations both for organisms and for **non-living natural objects**. Although British children were considered less likely to be exposed to influences about intention or design in nature, they nevertheless provided teleological explanations. Thus, it may be that children require a minor or no influence at all from their environment to provide teleological explanations (Kelemen, 2003). These results suggest that children are naturally inclined to prefer explanations of nature as an intentionally designed artifact, and that this tendency is not necessarily the result of the religiosity of their social background. Such a preference for intentional explanation may then be what leads children, in the absence of other knowledge, to a generalized view of objects as intentionally created by someone for a purpose, and perceive nature as an artifact of non-human design (Kelemen, 2004).

To investigate this further, it was examined whether children's tendency to reason about natural phenomena in terms of a purpose (Kelemen, 1999a) and their intuitions about ID in nature, whether or not they came from fundamentalist religious backgrounds (Evans, 2001), were related in any systematic way. British elementary school children (aged 6–10 years old) were given tasks which might document their intuitions about purpose and ID in the context of their explanations about the origins of natural phenomena. It was found that children were most likely to provide teleo-**functional explanations** for artifacts as well as for artifact-like natural objects and animals, but not for natural events, to open-ended questions. Moreover, children's teleological and ID intuitions were found to be interconnected. The results suggested that there was a systematic connection between children's teleo-functional explanations and their intuitions about the non-human intelligent design of nature. Children who provided purpose-based explanations of nature also endorsed the existence of a creator agent, in a manner that might be informed by their understanding of artifacts. However it was not clear how robust this connection was and if it existed at the pre-school age (Kelemen and DiYanni, 2005).

The conclusions of these studies suggest that cultural factors other than religion, such as our understanding of artifacts, may be why we tend to perceive purpose and design in nature. This is precisely what Paley seems to do in his book. Even if he was convinced that an intelligent Creator exists and he was thus looking for evidence to confirm His existence, his argumentation is also informed by his understanding of machines in particular and of artifacts in general. What I suggest is that it may be that purpose and design come first, due to a bias to perceive nature as an artifact, and it is then that there is a need to look for a designer, and so such an inference is made. Children in the studies reported above may have intuitively perceived nature as an artifact and then may have looked for a potential artificer. This may be called an artifact-thinking argument and has the following structure:

Artifact-thinking argument

Artifacts are designed by competent designers and this is why they have complex structures.

We observe enormous complexity in nature, in organisms in particular, which is larger than the complexity of artifacts.

Therefore the designer of organisms is more competent than the (human) designers of artifacts, and this could only be God.

Paley's argument, outlined above, has more or less the same structure:

Paley's argument

Complexity indicates design and hence the existence of a designer.

The more complex an instrument is, the more competent is its designer.

Animal organs are superior and much more complex than human-made instruments.

Therefore the creator of animal organs is superior and more competent than any human creator.

Therefore the creator of animal organs is God.

Thus, complexity in nature comes first; it reminds us of artifacts with which we are familiar and it is only then that we start looking for a designer. The complexity in nature is so enormous that a very competent designer is required and this can only be God. There are more studies that provide evidence that we intuitively tend to think in terms of purpose and design, but this is a discussion left for Chapter 3. For now, it is enough to note that it may not be our religious beliefs (only or at all) that make us perceive purpose and design in nature, but that we are otherwise prone to do so; then, our religious beliefs only make us infer that God is the designer of the complexity we observe.

Deborah Kelemen has made the suggestion that because we grow up surrounded by artifacts and we become familiar with their intentional use from very early in our lives, we may come to the conclusion that everything around us is an artifact made for a purpose. Thus, we extend our artifact thinking to nature and so come to intuitively believe that organisms, and even non-living natural objects, have also been intentionally designed for some purpose (see Kelemen, 1999a). This view echoes the suggestion of Jean Piaget, that children are artificialists (see Piaget, 1960/1929, 2013/1947). However, it seems that what is actually the case is not very clear. In Chapter 3 I will present different conclusions from different bodies of research; one of them suggests that children are able to clearly distinguish between organisms and artifacts, the other that they do not. Whatever the case, the important issue for our discussion here is that thinking of organisms as artifacts is something that seems intuitive for many people. This could be a major conceptual obstacle to understanding evolution. This will be the topic of Chapter 3, in which the differences between organisms and artifacts will be discussed in detail. Let us now return to how evolutionary theory seems to be in conflict with worldviews.

Evolution and worldviews: perceived conflicts

Richard Dawkins famously described Darwin's natural selection as the **blind watch-maker**, the natural equivalent and eventually the alternative to Paley's divine designer. The argument is simple and shared by many biologists nowadays. **There is design in nature, but it is natural**; it is neither purposeful nor intentional. There is a designer in nature, but it is a blind and unconscious one:

Paley's argument is made with passionate sincerity and is informed by the best biological scholarship of his day, but it is wrong, gloriously and utterly wrong. The analogy between telescope and eye, between watch and living organism, is false. All appearances to the contrary, the only watchmaker in nature is the blind forces of physics, albeit deployed in a very special way. A true watchmaker has foresight: he designs his cogs and springs, and plans their interconnections, with a future purpose in his mind's eye. Natural selection, the blind, unconscious, automatic process which Darwin discovered, and which we now know is the explanation for the existence and apparently purposeful form of all life, has no purpose in mind. It has no mind and no mind's eye. It does not plan for the future. It has no vision, no foresight, no sight at all. If it can be said to play the role of watchmaker in nature, it is the *blind* watchmaker. (Dawkins, 2006a, p. 5)

Based on this, we can summarize Dawkins' argument:

Dawkins' argument

Artifacts are designed by competent human designers and this is why they have complex structures.

We observe enormous complexity in nature, in organisms in particular, which is larger than the complexity of artifacts.

The complexity we observe in nature, and in organisms in particular, is the outcome of natural processes, which are unconscious and automatic.

Therefore, if we need to identify a "designer" of organisms that is more competent than the (human) designers of artifacts, this is natural, mindless, and sightless; it can only be natural selection.

Now where is the problem with this? Why do many people not accept natural selection as an alternative to a divine designer? I am going to use an analogy in order to compare the two explanations. Imagine a class of 20 students finishing elementary school and getting ready to enter middle school. Imagine that these students have poor grades so far – their average is 10 out of 20. Imagine also that after six years, when the class is finishing high school, the grades of the students have improved so that the class average has reached 19 out of 20. How is this possible? There can be two competing explanations, one according to Paley's argument and one according to Dawkins' argument. The former presupposes an external agent (like Paley's divine designer) who intervenes, whereas the latter does not require one and does not rely on any intention or purpose but results out of a process of unconscious selection.

Here is a possible explanation. The director of the middle school, which has high standards, decides that such poor grades are unacceptable for his school. Therefore he makes appropriate changes in the curriculum and assigns his teachers to small groups of students so that each student has his or her own mentor. He also provides students with extra courses and extracurricular material so that they study and learn more. In this case, the director acts like Paley's designer. He has a particular purpose, to improve the average grade of this group of students, and in order to achieve this he implements design. He thus designs a whole teaching sequence which lasts for six years and which focuses on each individual student. At the end of this process, or rather throughout it, each student improves and gradually gets higher and higher grades. At the end of the six-year process the whole class has improved significantly and eventually they achieve the high average of 19 out 20. The director's purpose has then been fulfilled, and this was due to the implementation of his design.

Let us now consider an alternative explanation, that of an unconscious and unintentional process that does not assume any intervention. The students with poor grades are enrolled in the middle school which, however, belongs to an educational institution with high standards that expels students who keep getting low grades. As a result, it is difficult for many of the students to achieve what the school requires. As a result, through the six years a process of differential attainment takes place. Those students from the initial group, who try hard enough and learn more, improve and eventually make it to the next grade; in contrast, the rest of the students from the initial group who do not try hard enough will not make it to the next grade and will be expelled from school. Because the school has set high standards, any new student who joins the class in the intermediate grades has to pass the same exams and thus has to be a high attainer as well. Consequently, over the years the average grade of the class improves as students with low attainment are expelled and only high achievers from other schools are allowed to take their place. Thus, without any designed intervention but only through a process of differential attainment, the average grade of the group improves.

I am aware that this analogy has an implicit assumption for the second case: that the high standards of the school are not due to the intention of anyone (director, school board, etc.) but that they simply exist. If we can overlook this, what we get are two cases that represent the differences between divine creation and evolution by natural processes. In the first case the director (who is the analogue of the divine creator) implements design to fulfill his purpose: students manage to attain high grades and stay at this particular school (which is the analogue of the organisms' managing to survive in their environment). The fact that students improve and are not expelled is due to the implementation of this design (which is the analogue of the divine design of a benevolent God who acts for the good of organisms and who designs their adaptations). In the second case no such design exists; rather, a process of differential attainment takes place (the analogue of the process of differential survival) and as a result the constitution of the class changes, perhaps dramatically, over the years (which is the analogue of evolution by natural selection – a process of change through differential survival). In short, the first example involves design and benevolence for the graduation (survival in nature) of all students from the initial group; the second process exhibits no design and no benevolence, but only a purposeless process of attainment or failure (survival or death in nature).

Apparently, evolution in nature takes place according to the second case. And of course, organisms in nature have more difficulties to overcome. Dying is, of course, worse than being expelled from school; in the latter case there are other options (other schools to go to), whereas in the first case there is no other option. Here is where a moral problem arises: it seems that it is intuitively more acceptable (or moral) for people to think in terms of the first case rather than in terms of the second. It makes more sense to accept that someone will take care of those who do not make it rather than accept that they will simply be eliminated. However, elimination is what often happens during evolution and this is perhaps why it seems to be counter-intuitive. I think that the following quote by Peter Bowler summarizes the problem appropriately:

> this view [that any mechanism for maintaining adaptation is compatible with design] is difficult to sustain in the case of natural selection, where the element of chance variation seems to make the outcome unpredictable and where death is the driving force of change. (Bowler, 2009a, p. 205)

If we intuitively think of organisms as designed in order to be adapted[6] (or adaptable), it is difficult to accept that adaptations originate from chance variations that produce non-designed and thus unpredictable characters which might later become prevalent due to natural selection. It is even more difficult to accept that evolutionary change in a population results from death. In nature, it is not individuals that evolve by undergoing particular changes; it is populations that evolve because some individuals die and some others manage to survive in particular environments. How can one accept this?

There are two issues here. The first is pragmatic, the other is logical. Most of us detest death, especially if we have experienced the death of a beloved person. And if we have not, we still feel that death is a bad thing, and most would agree that it is not moral to cause death. Indeed, the moral values of most human cultures prescribe that we should not cause death, neither to humans nor to other organisms. I do not think I need to explain why we try to protect human life. In many cases we try to protect animal life as well; we even take animals to our homes as pets in order to protect them, or we protest for their rights. The problem, in my view, is that this is happening in a discriminative, and consequently inconsistent, manner. We may protect and love our favorite dog or cat (or whatever – some people make very strange choices of pets), while we domesticate cattle, sheep, goats, pigs, and other animals in order to eat their muscles (mostly), which are full of proteins. Why do we accept their deaths as necessary?[7] To make things worse, we consider someone who kills several people within a single day as a murderer (no question that he is), but we tend to consider someone who is doing the same during a war as a hero. Are human lives less important if there is a war going on? I do not really want to get into the metaphysics of these issues. I only want to show that sometimes we (discriminatively and questionably) consider death as necessary or natural, and sometimes we do not. We make inconsistent distinctions for pragmatic purposes. Although death is a consequence

---

[6] I am referring to the term adaptation in the wider, everyday use of the term. The term adaptation has a more specific meaning in science, and scientists and educators should be very careful in their discussions about it (see Chapter 6 and Kampourakis, 2013b).

[7] The truth is that not all of us accept the death of these animals as necessary. Some people protest against killing animals even for food, and they feed themselves only on plants. But, biologically speaking, plants may also die when we eat them. And even if plants are not as sentient as animals are, they nevertheless are exactly as alive!

of life and in fact it is the only predictable outcome once we are born, it seems that it is very difficult for us to get along well with it.

This might be a consequence of not appropriately distinguishing between what we know and what we believe. We know that death is a fact of life. We do not know what happens after that, but we are ready to believe in life after death. We observe coincidental death happening around us and yet we are ready to believe that it has some deep or transcendent purpose. Even when there is no rational explanation of why someone and not someone else dies, we are ready to explain in fatalistic terms that there was some reason that someone died and someone else did not. Here the problem of inconsistency arises: Why would God allow someone to die but not someone else? Why do some people live short and miserable lives, like children born in various regions of sub-Saharan Africa who will die of famine before they grow up enough to die of AIDS, whereas other people live long and wealthy lives in Western countries? A well-known reply is that this may be God's will. He may have some reason that some people die younger than others; that some people live a happy and long life whereas others live miserably and die young. This is called *theodicy*: God delivers justice as He wishes. But then this assumes that there exist at least two kinds of people: those who have to suffer or die young, and those who do not. Why is that? (For a relevant discussion, see Kitcher, 2007, pp. 120–131.)

An important question about this was elegantly asked by David Hume:

Is he [God] willing to prevent evil, but not able?
Then is he impotent.
Is he able, but not willing?
Then is he malevolent.
Is he both able and willing?
Whence then is evil? (Hume, 1993 [1779/1777], p. 100)

According to our moral values, we could consider any death caused by any cause other than natural causes (e.g., old age) as death caused by some evil power. For our purposes here, let us call this kind of death unnatural. Then, in paraphrasing Hume, we might ask:

Is God willing to prevent unnatural death, but not able?
Then is he impotent.
Is he able, but not willing?
Then is he malevolent.
Is he both able and willing?
Whence then is unnatural death?

The problem here is that any attempt to answer these questions would rather be based on belief, and not on the empirical evidence on which science relies as no human seems to have any privileged access to His mind. Many people think they *know* the answers to these questions; however, it could be the case that they simply *believe* they do. We do not know why humans and other organisms die unnaturally; we may of course believe that there is a divine cause for this or that life is purposeless or even pointless. This is one of the major obstacles in really understanding what is going on around us. In the following section, I am going to show how scientists may extend scientific knowledge beyond the realm of science to answer questions about which we actually have no knowledge. Then in the last section I will focus on the crucial distinction between what ones knows and what one believes.

Evolution and religion: scientists' views

There is a widespread view that a conflict exists between science and religion. However, history makes clear that the interaction between the two has been extremely complicated and context-dependent. John Hedley Brooke (1991), and more since then, have shown this. There are exemplar cases of biologists who are also devout believers and who find no conflict between evolutionary theory and religion, such as Francisco Ayala and Ken Miller. In his 2007 book entitled *Darwin's Gift to Science and Religion*, Ayala argues that evolutionary theory provides a solution to the problem of evil discussed in the previous section (pp. x–xi):

I assert that scientific knowledge, the theory of evolution in particular, is consistent with a religious belief in God, whereas Creationism and Intelligent Design are not. This point depends on a particular view of God – shared by many people of faith – as omniscient, omnipotent, and benevolent. This point also depends on our knowledge of the natural world, and, particularly, of the living world. The natural world abounds in catastrophes, disasters, imperfections, dysfunctions, suffering and cruelty. [...] I shudder in terror at the thought that some people of faith would implicitly attribute this calamity to the Creator's faulty design. I rather see it as a consequence of the clumsy ways of the evolutionary process. The God of revelation and faith is a God of love and mercy, and of wisdom. Darwin's theory of evolution is a gift to science, and to religion.

Similarly, Ken Miller (2007, p. 291) writes that:

Those who ask from science a final argument, an ultimate proof, an unassailable position from which the issue of God may be decided, will always be disappointed. As a scientist I claim no new proofs, no revolutionary data, no stunning insight into nature that can tip the balance in one direction or another. But I do claim that to a believer, even in the most traditional sense, evolutionary biology is not at all the obstacle we often believe it to be. In many respects, evolution is the key to understanding our relationship with God. God's physical intervention in our lives is not direct. But His care and love are constants, and the strength He gives, while the stuff of miracle, is a miracle of hope, faith, and inspiration.

These views notwithstanding, not all scientists hold the same religious views. This is one case where generalizations would be wrong. Not all scientists are devout believers like Miller and Ayala. But also not all scientists are atheists or irreligious, as is Dawkins, whose views will soon be discussed in detail. Thus, being a scientist does not necessarily entail anything about one's religious views. In a recent book (Ecklund, 2010) presenting the conclusions of a systematic study about what scientists actually think about religion, it was found that approximately 50% of scientists consider themselves to be religious. The author conducted research with a sample of about 1646 natural and social scientists from 21 elite[8] universities in the United States, interviewing 275 of

---

[8] Under the assumption that "elites are more likely to have an impact on the pursuit of knowledge in American society." It should be noted, as the author acknowledges, that the sample did not include professors from the middle and southern United States, where more people are highly religious, but mostly from the northeast and west coast (Ecklund, 2010, p. 158).

them. Approximately 53% of the scientists surveyed stated that they had no religious affiliation, whereas approximately 47% of them claimed they had one (Ecklund, 2010, p. 33). Ecklund's detailed discussion presents the variety of views and attitudes toward religion among scientists. What is important is not whether most scientists are religious or irreligious, but that there is a variety of views.

To illustrate this further, in this section I will briefly describe the views of three evolutionary biologists: Richard Dawkins from Oxford University, Simon Conway Morris from the University of Cambridge, and the late Stephen Jay Gould who was at Harvard University. All three of them are well known for different reasons, with different kinds of contributions to science. While all of them are proponents of evolution, they disagree on how evolution actually proceeds, but most importantly they disagree on the implications that evolutionary theory has for our understanding of life and nature. Interestingly enough, all three of them have written books in which they make these views explicit (Gould, 1999; Conway Morris, 2003; Dawkins, 2006b). In this section I will briefly describe these views and I will attempt an analysis of these in order to show how even scientists can conflate what they actually know with what they believe. I have selected these scientists as representatives of three distinct views: atheism on one side (Dawkins), religiosity on the other (Conway Morris), and agnosticism as an intermediate position (Gould).[9]

Richard Dawkins is a well-known atheist. Right from the start of his 2006 book *The God Delusion* he notes that "Being an atheist is nothing to be apologetic about. On the contrary, it is something to be proud of, standing tall to face the far horizon, for atheism nearly always indicates a healthy independence of mind and, indeed, a healthy mind" (Dawkins, 2006b, p. 3). Dawkins actually considers religious devotion as evidence of unhealthiness:

You say you have experienced God directly? Well, some people have experienced a pink elephant, but that probably doesn't impress you. Peter Sutcliffe, the Yorkshire Ripper, distinctly heard the voice of Jesus telling him to kill women, and he was locked up for life. George W. Bush says that God told him to invade Iraq (a pity God didn't vouchsafe him a revelation that there were no weapons of mass destruction). Individuals in asylums think they are Napoleon or Charlie Chaplin, or that the entire world is conspiring against them, or that they can broadcast their thoughts into other people's heads. We humour them but don't take their internally revealed beliefs seriously, mostly because not many people share them. Religious experiences are different only in that the people who claim them are numerous. (Dawkins, 2006b, p. 88)

Thus, according to Dawkins there is a subjective attitude toward religion. Although the claims of religious people are as irrational as those of mad people, we tolerate religious beliefs because too many people share them.

Dawkins suggests that we are prone to accept the illusion of design in nature as true and this is why we turn to religion: "Maybe you think it is obvious that God must exist,

---

[9] This categorization is somehow abstract, but also quite representative. There in fact exists a continuum of different views from Dawkins to Conway Morris, and even further. In my discussion I will not describe how these scientists have criticized each other's view.

for how else could the world have come into being? How else could there be life, in all its rich diversity, with every species looking uncannily as though it had been 'designed'?"[10] Dawkins suggests that natural selection is a plausible alternative explanation for the design perceived in nature: "Far from pointing to a designer, the illusion of design in the living world is explained with far greater economy and with devastating elegance by Darwinian natural selection" (Dawkins, 2006b, p. 2). Having argued about the power of natural selection, Dawkins argues "why there almost certainly is no God." At the end of this chapter he summarizes what he considers as the central argument of his book:

(1) One of the greatest challenges to the human intellect, over the centuries, has been to explain how the complex, improbable appearance of design in the universe arises.
(2) The natural temptation is to attribute the appearance of design to actual design itself. In the case of a man-made artefact such as a watch, the designer really was an intelligent engineer. It is tempting to apply the same logic to an eye or a wing, a spider or a person.
(3) The temptation is a false one, because the designer hypothesis immediately raises the larger problem of who designed the designer. The whole problem we started out with was the problem of explaining statistical improbability. It is obviously no solution to postulate something even more improbable. We need a 'crane,' not a 'skyhook,' for only a crane can do the business of working up gradually and plausibly from simplicity to otherwise improbable complexity.
(4) The most ingenious and powerful crane so far discovered is Darwinian evolution by natural selection. Darwin and his successors have shown how living creatures, with their spectacular statistical improbability and appearance of design, have evolved by slow, gradual degrees from simple beginnings. We can now safely say that the illusion of design in living creatures is just that – an illusion.
(5) We don't yet have an equivalent crane for physics. Some kind of multiverse theory could in principle do for physics the same explanatory work as Darwinism does for biology. This kind of explanation is superficially less satisfying than the biological version of Darwinism, because it makes heavier demands on luck. But the anthropic principle entitles us to postulate far more luck than our limited human intuition is comfortable with.
(6) We should not give up hope of a better crane arising in physics, something as powerful as Darwinism is for biology. But even in the absence of a strongly satisfying crane to match the biological one, the relatively weak cranes we have at present are, when abetted by the anthropic principle, self-evidently better than the self-defeating skyhook hypothesis of an intelligent designer. (Dawkins, 2006b, pp. 157–158)

According to Dawkins, it is an illusion to see design in nature and it is a delusion to attribute this design to God, as He is improbable. Natural selection, on the other hand, is a more probable and thus a more plausible alternative. Dawkins considers the idea of an intelligent designer as a self-defeating one, because he seems to be sure that eventually humans will come up with a cosmological alternative as good as the biological one developed by Darwin. What is most crucial is that Dawkins believes that the question of

---

[10] It should be noted that Dawkins asks how the world came into being, but then leaves this question behind. It is important to keep in mind that the primary aim of evolutionary theory is not to explain the **origin of life** on Earth, although some plausible explanations are available, but to explain the unity and the diversity of life thereafter.

the existence of God is a scientific one, which implies it is a question that can be answered based on empirical data. He writes:

The view that I shall defend is very different: agnosticism about the existence of God belongs firmly in the temporary or TAP [Temporary Agnosticism in Practice] category. Either he exists or he doesn't. It is a scientific question; one day we may know the answer, and meanwhile we can say something pretty strong about the probability. (Dawkins, 2006b, p. 48)

Dawkins considers the question about the existence of God as one about which there really exists a definite answer, which we are not able to reach yet (because we lack the necessary evidence or simply because we do not understand it). However, and despite the lack of evidence, Dawkins is convinced that the existence of God is something quite improbable.

   Simon Conway Morris is at the other extreme, but without being as explicit as Dawkins. His response to such arguments[11] is that:

It seldom seems to strike the ultra-Darwinists[12] that theology might have its own richness and subtleties, and might – strange thought – actually tell us things about the world that are not only to our real advantage, but will never be revealed by science. [...] But to assume that science itself can produce or verify the truths upon which it depends is, as many have pointed out, simply circular. (Conway Morris, 2003, p. 316)

First, we need to recall the limits of science. It is no bad thing to remind ourselves of our finitude, and of those things we might never know. [...] At its simplest it is a precautionary principle, and more significantly a belated acknowledgement that the architecture of the Universe need not be simply physical. [...] Second, for all its objectivity science, by definition, is a human construct, and offers no promise of final answers. We should, however, remind ourselves that we live in a Universe that seems strangely well suited for us. [...] The idea of a universe suitable for us is, of course, encapsulated in the various anthropic principles. These come in several flavours, but they all remind us that the physical world has many properties necessary for the emergence of life. (Conway Morris, 2003, pp. 326–327)

Conway Morris suggests that there are questions which cannot be answered by science, and theology might give appropriate insights instead. He implies that there are no empirical grounds on which we can seek answers to the questions about the existence of God. This is one major difference from Dawkins. The other major difference is that Conway Morris seems to be convinced that God exists, and that the following "facts of evolution are congruent with a **Creation**":

(1) its underlying simplicity, relying on a handful of building blocks;
(2) the existence of an immense universe of possibilities, but a way of navigating to that minutest of fractions which actually work;
(3) the sensitivity of the process and the product, whereby nearly all alternatives are disastrously maladaptive;
(4) the inherency of life whereby complexity emerges as much by the rearrangement and co-option of pre-existing building blocks as against relying on novelties per se;

---

[11] The book by Conway Morris which I am quoting was published in 2003, before Dawkins' 2006 book. However, Dawkins' views were already widely known, even if they were not elaborated in the detail they are in his 2006 book.
[12] Dawkins is of course one of them, Daniel Dennett is another.

(5) the exuberance of biological diversity, but the ubiquity of evolutionary **convergence**;

(6) the inevitability of the emergence of sentience, and the likelihood that among animals it is far more prevalent than we are willing to admit. (Conway Morris, 2003, p. 329)

In short, Conway Morris argues that there are particular features of life on Earth that indicate the influence of more than just natural processes: the emergence of workable and adaptive options, the emergence of complexity through the re-use of extant matter, the convergence of features despite the enormous diversity, and the inevitability of sentient life. He implies that these characteristics of life are not accidental and point to factors beyond nature (and thus outside the realm of science).[13] Thus, he suggests that:

given that evolution has produced sentient species with a sense of purpose, it is reasonable to take the claims of theology seriously. In recent years there has been a resurgence of interest in the connections that might serve to reunify the scientific world view with the religious instinct. [...] In my opinion it will be our lifeline. (Conway Morris, 2003, p. 328)

Conway Morris suggests that despite its achievements, science alone cannot guide us in our quest to explore nature and understand life. Theology has much to add to this quest, and thus we should aim at unifying these two approaches. What we observe around us cannot be explained solely in terms of scientific inquiry; there is more to it. And although it does not prove the existence of God, we must nevertheless have our eyes open:

the complexity and beauty of "Life's Solution" can never cease to astound. None of it presupposes, let alone proves, the existence of God, but all is congruent. For some it will remain as the pointless activity of the Blind Watchmaker,[14] but others may prefer to remove their dark glasses. The choice of course, is yours. (Conway Morris, 2003, p. 330)

Conway Morris concludes that those who perceive life as pointless and do not realize that there is something more out there are short-sighted. Others may choose not to do so, but seek answers beyond science. It is up to us to decide.

There is an interesting contrast so far. For Dawkins, religiosity is evidence of irrationality, if not stupidity. For Conway Morris it is evidence of open-mindedness and thoughtfulness. For Dawkins it is almost certain that God does not exist, and this will eventually be proved by science. For Conway Morris it is almost certain that God exists, but there is no need to prove it; there is abundant evidence around us that should make us undertake scientific and theological quests simultaneously. The late Stephen Jay Gould expressed an intermediate view:

I do not see how science and religion could be unified, or even synthesized, under any common scheme of explanation or analysis; but I also do not understand why the two enterprises should experience any conflict. Science tries to document the factual character of the natural

---

[13] Evolution can be compatible with religion in different ways. **Deism** is the idea that God created the physical laws needed for the universe to exist and function, but otherwise does not interact with the universe. In this view, God may have created the world, set natural selection as the main mechanism of evolution, and then let it evolve. **Theism** is the idea of a personal creator God who interacts with His creation and answers prayer. In this case, evolution cannot take place only through natural processes, but is guided by God.

[14] No question that this refers to Dawkins and to the title of his 2006 book.

world. [...] Religion on the other hand, operates in the equally important, but utterly different, realm of human purposes, meanings and values. (Gould, 1999, p. 4)

For Gould, religion is valuable and can co-exist with science. It does not add to the aims of science, but nevertheless has much to contribute to human life. Whether God exists or not is irrelevant, and actually is a question that we cannot answer; but religion is important, this notwithstanding:

I am not a believer. I am an agnostic in the wise sense of T.H. Huxley, who coined the word in identifying such open-minded skepticism as the only rational position because, truly, one cannot know. Nonetheless [...] I have great respect for religion. (Gould, 1999, pp. 8–9)

Gould notes that science and religion occupy two distinct non-overlapping **domains**, or magisteria.[15] He suggests:

that these two domains hold equal worth and necessary status for any complete human life; and second that they remain logically distinct and fully separate in styles of inquiry, however much and however tightly we must integrate the insights of both magisteria to build the rich and full view of life traditionally designated as wisdom. (Gould, 1999, pp. 58–59)

For Gould, science and religion both contribute independently to the fullness of life:

NOMA honors the sharp differences in logic between scientific and religious arguments. NOMA seeks no false fusion, but urges two distinct sides to stay on their own turf, develop their best solutions to designated parts of life's totality, and, above all, to keep talking to each other in mutual respect, and with an optimistic forecast about the value of reciprocal enlightenment. (Gould, 1999, pp. 210–211)

In this view science and religion are entirely distinct and non-overlapping domains that address different questions by providing answers that independently contribute to our understanding of the world. Science and religion can neither be unified nor be in conflict, because they are so different that there is no point in even trying to compare them to each other. They nevertheless can communicate, but without falling into each other's realm.

The views of Dawkins, Gould, and Conway Morris are summarized in Table 2.1. What is very important and clear so far is that these three scientists have entirely different views about religion. The widespread idea of a conflict between science and religion has been challenged by many historians of science (Brooke, 1991; see also chapters in Numbers, 2009b; Dixon *et al.*, 2011). In order for such a conflict to exist, all scientists should share the same views against religion. Is it possible to talk about conflict when scientists can have such different views as Dawkins, Gould, and Conway Morris do?[16] I think not.

The question which is of interest to us is whether their views are based on actual knowledge or on belief. These three evolutionary biologists all accept the fact of evolution,

---

[15] Hence his NOMA principle (non-overlapping magisteria).

[16] I do not mean to imply that this small sample of three scientists is representative of all scientists. Nor do I want to imply that scientists hold one of these views. Rather, I simply want to argue that scientists may hold very different, and contrasting, views about religion by providing three exemplars that display the range of differences that are extant.

Table 2.1 An overview of the views of Dawkins, Gould, and Conway Morris

	Dawkins (atheism)	Gould (agnosticism)	Conway Morris (religiosity)
God's existence	Improbable	Unknowable	Probable
Question about God's existence	Scientific	Cannot be answered	Theological
Status of religion for humans	Unimportant and unnecessary, if not harmful	Important	Necessary
Relation between religion and science	Conflict	Co-existence	Unification

although they have disagreed about its details. Disagreement among scientists is plausible because there may be different ways to interpret scientific data or to draw different conclusions from them. But in these cases there are quite objective grounds, raw data obtained from the study of the natural world, on which conclusions are based. Thus, other scientists can offer constructive criticism or just comment on the different conclusions by studying the available data themselves. But which are the grounds for comparing views about God and religion? I do not think there are any objective grounds; these views are highly idiosyncratic and subjective. The quotes above indicate this, as none of the three scientists provides any evidence for his claims. For example, Dawkins argues that the existence of God is improbable, Conway Morris that it is very probable, and Gould that it is something we cannot really know. But none of them actually develops any scientific argument to support his view: provide evidence and draw conclusions from it. They simply express their personal beliefs. This is a major issue in the debates about evolution: People often do not distinguish between what they know and what they believe. This will be the topic of the next and final section of this chapter.

Distinguishing between knowing and believing

It is quite easy to confuse what we know with what we believe, as we may use the two verbs as synonyms. For example, I may say that I *believe* that my friends love me, while in fact I *know* they do because of their attitude (they visit me often, they always seem happy to spend time with me, and they do this quite often). In contrast, I may say that I *know* that my friends love me while in fact I simply *believe* this because I have no reason to think otherwise (they may dislike me, but nobody has ever told me that, and I am not concerned that they do not visit me often). In another example, I may say that I *know* that my wife is cooking my favorite food while in fact I *believe* she is because I just smelled it (but she may not be cooking; someone else may be doing this). Or I may say that I *believe* that my wife is cooking my favorite food when I smell it, although I actually *know* that she is because I just saw her doing it. The distinction I want to make is between when I believe and when I know. I believe that my wife is cooking my favorite food simply because at that point I have no justification/reasons/evidence that it is her (and not someone else, e.g., her mother who lives downstairs) who is cooking it. In this case, what I know is that my food is being cooked by someone, but I have no evidence of who is cooking it and so I simply believe it is my wife who is doing it (because neither me nor our children cook). This implies that *knowing* requires something more than *believing*. To resolve the confusion we need more than just a superficial treatment, and this requires feedback from **epistemology**.

It is said that knowledge is a justified belief. We use our senses (vision, hearing, smelling, tasting) to perceive the world around us. When we describe what we experience, we also describe what we believe. When I smell my favorite food being cooked, I believe this is happening without any need to observe it. I know that someone is cooking it because I can smell it. Based on my previous experience, I can also imagine how it will look on my plate or what its taste will be, as well as that it is my wife who is cooking it because she does all the cooking at home. In this case I *justifiedly believe* what I believe because of my past experience. But what does it mean to *justifiedly believe* something? It means that what I believe is probably, but not necessarily, true. Smelling my favorite food makes me think my wife is cooking it, and this is what is probably happening because neither our kids nor I ever cook. So, I justifiedly believe that my wife is cooking my favorite food based on past experience. This is different from *being justified in believing* something. In sensing the smell of my favorite food I am *justified in believing* that it is my wife who is cooking, although I may have no evidence that she is (e.g., I have not seen her) cooking it. There could be an alternative explanation, such as that it is her mother doing the cooking.

I should note at this point that to *justifiedly believe* something presupposes that I am *justified in believing* it. To believe something I must have some reason to do so. However, the opposite is not necessary; I do not have to *justifiedly believe* anything I am *justified in believing*. Now, although justified belief is necessary for knowledge, they are not the same. Knowing something means it is true, whereas I may *justifiedly believe* something which is false. An example will help here: I may be *justified in believing* that my wife is cooking my favorite food because of past experience. She does all the cooking at home. I would be justified in believing she is cooking my favorite food even if I later found out that she was out shopping and that it was her mother downstairs who was cooking. In contrast, I would *justifiedly believe* that my wife was cooking my favorite food if I had heard her in the kitchen doing something. In this case I would be justified in believing that she was cooking the food (because she is cooking), and I would also justifiedly believe that she actually does so because I heard her in the kitchen. That she generally does the cooking does not entail that she is cooking right now. I am justified in believing this, but it may not be true. That I heard her in the kitchen makes me justifiedly believe that she is cooking; this is most probable but not necessarily true.

Here is another example in order to distinguish between *justifiedly believing* and *being justified in believing* something. I may have good reasons to anticipate that I will soon be promoted because I believe my boss thinks very highly of me. Over the years I have worked hard and I have contributed a lot to the success of my company. So, I have very good reasons for anticipating that I will be promoted, because I believe that my boss appreciates what I am doing and that he will reward me for that. However, I may also believe this not because of such good reasons, but because of really poor reasons, e.g., because I read my horoscope and found out that my boss is fond of me and that I will soon have professional success, related to a higher salary and a promotion. In this case I am justified in believing that my boss thinks very highly of me, but I do not justifiedly believe so (my belief isn't justified) (for other examples and details, see Audi, 2011). The important point made here is that there can be different kinds of justification, so that not all justified beliefs are equivalent to knowledge. Knowledge requires well-grounded beliefs.

Audi (2011) distinguishes between three kinds of ways in which beliefs are grounded: causal, justificational, and epistemic. Each of these is related to a particular kind of question, as presented in Table 2.2. We can distinguish between these grounds by using the example of my wife cooking my favorite food. Sensing the smell of my favorite food would be a causal grounding for believing that my wife is cooking it. She might well be doing this. However, it might also be her mother downstairs doing the cooking. If my wife had told me that morning that she was going to cook my favorite food, I would justifiedly believe that she was cooking it. Again, she might well be doing this. But I might be wrong again, because in the meantime she might have asked her mother to cook the food because something else came up. However, if I went to the kitchen and saw my wife cooking my favorite food, then I would know that she was indeed doing this. Therefore, there may be causal and justificational grounds for a belief, but some independent kind of evidence is required in order to claim that we know something. In the example I used, I actually *knew* when I both smelled the food and saw my wife cooking it. What I am suggesting here, following Audi (2011), is that a correct answer to the question "How do you know that?" *must* cite more than a single causal grounding for the belief.

Actual epistemological issues are complicated and difficult, and certainly fall outside the scope of this book. Here I will focus on scientific knowledge only. Audi (2011) provides a thoughtful discussion of scientific and religious knowledge in chapter 12 of his book. What I will attempt here is a simplified (but hopefully not simplistic) discussion of the relevant issues. Both religious knowledge and scientific knowledge can be considered as grounded causally and justificationally. But are they both well grounded in epistemic terms? Let us consider science first. One might claim that South

Table 2.2 Grounding of beliefs and related questions

Grounding of beliefs	Question	Example with wife and food
Causal	"Why do you believe that?"	"I smell my favorite food and I believe that my wife is cooking it"
Justificational	"What is your justification for believing that?"	"I smell my favorite food and my wife told me this morning that she would do it"
Epistemic	"How do you know that?"	"I smell my favorite food and I saw my wife in the kitchen cooking it"

America and Africa were once parts of a single continent. What is the causal grounding for this belief? It could be that the two continents seem to have complementary shapes. What is the justificational grounding for this belief? It could be that the same types of stones are found on both continents, and this would justify the belief that the two continents once used to be one.[17] However, the epistemic grounding for this belief requires even more; a plausible explanation for what is observed. We now know enough about plate tectonics to explain continental drift.[18] Thus, one can claim that one knows that Africa and South America are parts of what was once a single continent. In this case, different kinds of evidence were brought together to form an inference to what is currently considered as the best explanation for the initial observation.

Now, let us turn to religion. One might claim that God sent him a message that something bad would happen via a sign, e.g., a depiction of a falling angel was formed by moisture on a window while it was cold outside. This person's causal grounding for the belief that God sent him a message was the sign he saw. His justification for getting this message was the particular depiction he saw (he did not see a dinosaur but a falling angel) and therefore considered it as a message from God. Because the angel was a falling one, he considered the message as indicating that something bad would happen. But what are the epistemic grounds for this belief? Unfortunately, there are no epistemic grounds unless other people have seen the same sign and then something bad happened to them. In fact, in order to have real epistemic grounds for this belief it must be the case that the majority of people who have seen this sign in the past have eventually experienced something bad. But even if this is the case, there is no actual evidence that the sign is a message from God. It might be, but we cannot know for sure whether it is. There may be a secret cause–effect relation in nature of which we are unaware.

---

[17] This is actually how Alfred Wegener started thinking about his theory of continental drift. Alfred Wegener was not the first to observe that there was an apparent fit between the coastlines of Africa and South America, but he was the first to use this insight to develop a theory that aimed at explaining a wide range of phenomena. He initially considered the idea of continental drift as improbable, but he changed his mind as soon as he came across a report that discussed the paleontological similarities between the strata of Africa and Brazil. For Wegener, the contraction mechanism of mountain building, accepted at that time, was insufficient. Continents might be invaded by shallow seas but they could never form a deep ocean bed. Wegener thought that horizontal movements of the continents could provide an alternative explanation. In 1915 he published a book in which he had concentrated all the evidence that was against the old theory of mountain building and in which he proposed his own as an alternative (Gribbin, 2003, pp. 444–448; Bowler and Morus, 2005, pp. 238–242).

[18] It was Harry Hess who proposed that ocean ridges occurred at sites where molten rock welled up from the interior of the Earth. According to his "sea-floor spreading" model the hot mantle material spread out, pushing the continents on either side of the ridge apart. The youngest rocks were solidified next to the ridges, whereas the oldest rocks were found further away from the ridges as they were pushed to make room for the new material. This idea was supported by the patterns of magnetism that had been revealed on the seabed, particularly the existence of parallel stripes of normal and reversed magnetism alongside the mid-ocean ridges, which led Fred Vine and Drummond Matthews to set the foundations for the theory of plate tectonics. They suggested that as new rock welled up, it was imprinted by the current direction of the magnetic field of the Earth. When this field reversed, a new strip of reverse-magnetized rock would begin to form that would push the initial strip away from the ridge (Gribbin, 2003, pp. 457–459; Bowler and Morus, 2005, pp. 247–249).

And even if one claims that it was God who set this up, we can never be absolutely certain that it was Him who did so.

Let me give another example, because that of the falling angel as an example of religious belief may have connotations of superstition or intellectual naivety.[19] Let us assume that someone believes that how he leads his life matters to God. Someone who leads a moral life may end up living a very happy and peaceful life. He might thus consider this as a reward from God as He acknowledges that this person never behaved immorally. This could be a sufficient causal grounding for such a belief. If he also knew that all other people leading moral lives were also living happily and peacefully, he would even have justificational grounding for such a belief (although I am inclined to think that anyone reading these lines has already thought that there exist moral people who nevertheless live miserable lives or immoral people who live very happy lives). Even in such a case, it would be very difficult to actually know that how one leads one's life matters to God, because there is no way to test this belief – unless of course someone has some kind of privileged access to His mind.

With the above I do not mean to imply that we can know through science and that we cannot know through religion. The distinction I want to make is just that between testable and non-testable beliefs. So, since we can test our beliefs about the natural world, can we also claim that science provides us with actual knowledge? My answer is yes and no. Science is perhaps the most objective means we have to come to know anything about the world around us. But, in fact, there is nothing in science about which we can be absolutely certain. There are, of course, numerous claims we can make that no one would question, such as that organisms consist of one or more cells (viruses are not usually considered as organisms) or that gravity affects the movements of all natural objects that we know (from stones on the surface of Earth to the larger celestial bodies of our solar system). But generally speaking, scientists do not claim to have absolute knowledge and they are aware that they are based on approximations. Scientists build on the available evidence to develop the best possible explanations, through asking questions and trying to provide plausible and legitimate answers to them. This, in turn, involves human imagination and creativity, which are necessary in developing and testing hypotheses. Science is a purely human activity which depends on human perception. I think that Giere's *Scientific Perspectivism* (2006) provides an accurate view of how science is actually done. I endorse his view that:

The inescapable, even if banal, fact is that scientific instruments and theories are human creations. We simply cannot transcend our human perspective, however much some may aspire to a God's eye view of the universe. Of course, no one denies that doing science is a human activity. What needs to be shown in detail is how the actual practice of science limits the claims scientists can legitimately make about the universe. [. . .] By claiming too much authority for science, objective realists misrepresent science as a rival source of absolute truths, thus inviting the charge that science is just another religion, another faith. A proper understanding of the nature of scientific

---

[19] I must admit, though, that except from those who are scholars or well educated, many other religious people whom I have encountered have exhibited signs of superstition or intellectual naivety. I do not know if religiosity causes one to be superstitious, but it certainly allows one to be so.

investigation supports the rejection of all claims to absolute truths. The proper stance, I maintain, is a **methodological naturalism** that supports scientific investigation as indeed the best means humans have devised for understanding both the natural world and themselves as part of that world. That, I think, is a more secure ground on which to combat all pretenses to absolute knowledge, including those based on religion, political theory, or, in some cases, science itself. (Giere, 2006, pp. 15–16)

Science asks particular questions about the world, and develops tentative explanations that can be empirically tested. As Stuart Firenstein (2012) nicely puts it, it is ignorance and not knowledge that drives science. It is what we still do not know that opens new areas of inquiry and it is by asking questions and seeking their answers that we advance what we know. Empirical testing is crucial for this purpose, because this is how we can find sufficient epistemic grounding for our beliefs. But to have something tested empirically means that it can be perceived by our senses, or – as I prefer to call this – it "fits into our heads." There is too much around us that does not "fit into our heads," and this is where metaphysics, in the form of philosophy or religion, comes in. Science, on the other hand, attempts to study the natural world from the human perspective, which is based on human perceptions and senses. But can't we perceive God? So many people have claimed they have seen God, the Holy Mother, or Jesus Christ. Are they lying? In most cases they are not; they really believe they have seen them. But why don't we question their claims as we do for those of people who have seen UFOs? I think Dawkins is right that we consider people who have experienced God differently than people who have seen pink elephants or anything else. I am not implying that people who claim to have experienced God are mad (at least, most of them are not). People, whether religious or not, often see what they need to see because they are desperate or sad.[20] I am also skeptical about what those people who have claimed to have experienced God have actually experienced. Could it be the case that they have not really experienced anything but just say so because they know that such claims have an impact on other people? Can we test this? Here is, in my view, where the main problem lies.

A main difference between knowledge and justified belief lies on testability. Simply put, a justified belief may become actual knowledge not only when it is accidentally

---

[20] I should note at this point that my personal views notwithstanding, I certainly understand the need for religion. I entirely agree with Michael Ruse's and Philip Kitcher's views on this. These two prominent (and irreligious, if not atheist) philosophers have written two book-length treatises about how there is always place for religion in the rational world of science. Instead of describing their views in detail, I prefer to quote their conclusions with which I could not agree more:

Can a Darwinian be a Christian? Absolutely! Is it always easy for a Darwinian to be a Christian? No, but whoever said that the worthwhile things in life are easy? Is the Darwinian obligated to be a Christian? No, but try to be understanding of those who are. Is the Christian obligated to be a Darwinian? No, but realize how much you are going to foreswear if you do not make the effort, and ask yourself seriously (if you reject all forms of evolutionism) whether you are using your God-given talents to the full. (Ruse, 2001, p. 217)

There is truth in Marx's dictum that religion [. . .] is the opium of the people, but the consumption should be seen as medical rather than recreational. [. . .] Genuine medicine is needed, and the proper treatment consists of showing how lives can matter. [. . .] In addressing these issues we may discover that the deliverances of reason can be honored without ignoring the most important human needs – and going beyond supernaturalism, that we can live with Darwin, after all. (Kitcher, 2007, pp. 165–166)

repeated, but when it is deliberately and independently tested and eventually confirmed. Let us return to the previous examples. Whether Africa and South America were once parts of the same continent is something that can be independently tested in various ways and this is why it can turn into knowledge. What happened in this case is that several hypotheses were put to the test and it is from such tests that conclusions were made. Whereas the claim that God sent a message through the sign of the falling angel on the window, or that how we lead our lives matters to Him, cannot be tested no matter how many any relevant instances may occur in life. We cannot cause the appearance of the sign of the falling angel on the window and we cannot access God's mind. Consequently, we cannot put such beliefs to the test. In contrast, all the evidence we need to test the hypothesis about Africa and South America once being one continent are out there. This does not mean we can have knowledge of anything, e.g., our own unique introspective states. For example, assume that I am currently having a very particular sort of pain sensation in my left leg, and I think I know that as surely as I could know anything. However, this doesn't seem to be something that can be deliberately and independently tested. How could someone independently test whether I am having this very particular sort of pain sensation? This would be difficult to do. However, scientific knowledge most of the time does not have to do with our unique introspective states, but with factual evidence which can be used to test hypotheses or assumptions.

Let me give another example here. Imagine that someone claims to have achieved human reproductive cloning,[21] and is ready to take advantage of this development. Whether he actually achieved it or not can quite easily and efficiently be put to the test. Experts can extract DNA from the cells of the clone and from the donor of the original cell. The DNA molecules of the clone and the donor can be compared in their entirety, base by base. If the DNA sequence of the two molecules is almost 100% identical then we can claim that we know that cloning has probably[22] been achieved. If, on the other hand, the DNA sequences are not 100%, but are 99% identical, we will know that the supposed clone is not actually one.[23] Any justified belief that can be tested and confirmed in such a way can become knowledge. Testing can provide the independent causal grounding for the initial belief. Unfortunately, there is no such way to test religious experience. And I think that Dawkins and Conway Morris are both wrong, in a way. We currently have no reason to believe that the question for the existence of

---

[21] Human reproductive cloning, or the production of a clone of an adult, can be achieved by somatic cell nuclear transfer. It is this process that resulted in the birth of Dolly the sheep. Ian Wilmut and his colleagues at the Roslin Institute took cells from an adult ewe's mammary glands and starved them of nutrients to arrest further cellular development and to restore them to a totipotent state. They then transplanted the nuclei of those cells into enucleated sheep oocytes and applied an electric current to fuse and activate them. Out of 277 enucleated eggs, Dolly was born on July 5, 1996 (Wilmut *et al.*, 1997).

[22] We can never be absolutely certain about this due to the high degree of similarity of the DNA sequences of closely related organisms. Also, even the DNA in the cells of the same organism is not 100% identical. As Peter Godfrey-Smith shows, after 40 cell divisions any two cells of an adult human would, on average, differ by about 144 point mutations: "Although most of these mutations would have no phenotypic effects, it is important to recognize that most human cells within a human body are not genetically identical" (Godfrey-Smith, 2009, pp. 82–83).

[23] In this case we can be absolutely sure that there was a failure or fraud.

God will one day be tested scientifically. Of course, this does not disprove the existence of God in any way. On the other hand, the fact that we observe evolutionary convergence is not necessarily evidence for the existence of God. Evolutionary developmental biology suggests that much of what we perceive as evolutionary convergence could be the outcome of evolutionary conserved, and not converged, developmental processes (see Sober, 2003; see also Chapter 5). Does this mean that Gould is right? No. He writes that one cannot really know. But how do we know that one day we will not be able to know? We may, we may not. Who knows?

My suggestion is that we should stick with what we actually and currently know, and not speculate about what we may one day know or what we may never know. Dawkins suggests that one day we will know that God does not exist; Conway Morris implies that one day we will know that God exists; and Gould implies that we will never know if God exists. None of these beliefs can currently be put to the test, and so there is no need to worry about such questions. We are free to believe whatever we want. But we need to be responsible enough to distinguish between what we believe and what we actually know. And in terms of scientific knowledge we have everything that is required in order to be able to be aware of what we know. This does not mean that we can know everything; it only means that we can be quite sure about what we know and what we do not know. And although there is much in science that we know, there also exists too much that we do not know (and interestingly we know that we do not know). For example, we know the cause of the thalassemias, but we do not know the cause of all cancers. Science communication should be more effective in this respect. I strongly believe that scientists should explain to the public not only what we know, but also what we do not know.

Realizing what scientists know, do not know, and cannot know depends on understanding the nature of questions asked by scientists and the kinds of answers it is possible to provide. An important distinction which is relevant to that has to do with how one approaches nature, already mentioned in Giere's excerpt above. It is important to clearly distinguish between two types of naturalism: *metaphysical* naturalism and *methodological* naturalism. **Metaphysical naturalism**, also called philosophical or ontological naturalism, suggests that only natural entities exist and thus denies the existence of anything supernatural. Methodological naturalism does not deny the existence of supernatural entities, but nevertheless recognizes that one cannot study them and consequently that there is no reason to be concerned about them. Science is a practice of methodological naturalism: Whether a realm of the supernatural exists or not, it cannot be studied by the rational tools of science. Science does not deny the supernatural, but accepts that it has nothing to say about it. Science is a method of studying nature, hence methodological naturalism. Science is also concerned with the metaphysics of nature (causes of natural phenomena, etc.), but does not always have something to say about them.

So what should we do? I think that rationality suggests three steps: First, we need to ask if our beliefs are justified; second, we need to question them and put them to tests; third, we need to accept that some of them constitute knowledge and some do not, and that we hence need to distinguish between knowing and believing. Scientific knowledge is never absolute, but it is constantly put to the test. The outcome of this process is valuable because, as in the case of evolution, it confirms what we know. And if one day it fails to do so, there is no problem. Because then we will again know that something we thought we knew was not entirely accurate. Knowledge is important, even if it simply debunks what we thought we knew. How about the rest? In my Concluding remarks I address those questions which cannot be answered by science and thus about which we should not invoke science in order to support our own answers and beliefs.

Conclusions

In this chapter I explained why the idea of design in nature may stem from our understanding of artifacts and not from our religious beliefs. It may not be the case that people believe in God and then make the inference that He has designed organisms; it can very well be that people see complexity and design in nature and it is from this that the inference to a designer-God is made. Then, I explained why this idea seems to be in conflict with what evolutionary theory suggests. Evolutionary theory invokes death and suffering to explain how evolution takes place, and these may not seem to be compatible with a benevolent God. The idea that organisms are designed by a benevolent God is more intuitive, perhaps due to our awareness of why and how artifacts are constructed, than the idea that organisms have evolved through natural processes in which cruelty and death persist. This apparent incompatibility depends on one's worldviews, and there is no point in talking about a clash between science and religion since, for example, scientists who are proponents of evolution may hold very different religious views. To illustrate this, I described the views of three evolutionary biologists about religion. There is no single consensus view about religion among scientists, and the different views held by different scientists point to an important distinction between what one knows and what one believes. This is in my view where the problem of religious resistance to the acceptance of evolution lies, in part. People do not easily distinguish between what they know and what they believe. I do not mean to underestimate the religious and emotional issues. However, in my view the conceptual issues are equally important and perhaps have not been given the required attention so far. In the next chapter I argue that the low acceptance of evolution could be due to the fact that people do not really understand it because it is a counter-intuitive idea. Humans have deep intuitions that form conceptual obstacles to understanding evolution. Therefore, I suggest that the widespread discussion about evolution and how it relates to religious belief is somehow misleading because it focuses on religious issues, which are one part of the problem, but overlooks the conceptual issues which are the other important part. In order to be able to distinguish between what we know and what we believe, we first need to properly understand what we know. But this is difficult in the case of evolution due to particular conceptual obstacles that will be the focus of the next chapter.

Further reading

Myths about the relationship between science and religion are quite widespread, so a good book to start with should be *Galileo Goes to Jail and Other Myths About Science and Religion*, edited by Ron Numbers. For a historical overview, a great book is

John Hedley Brooke's *Science and Religion: Some Historical Perspectives*. It is also important to remember that science and religion are different in different parts of the world. Two great books on this topic are *Science and Religion Around the World*, edited by Brooke and Numbers, as well as David Livingstone's *Putting Science in its Place: Geographies of Scientific Knowledge*. Elaine Ecklund's *Science vs. Religion: What Scientists Really Think* presents interesting findings about scientists' religious views. There are many books about religion written by scientists, some of which were cited in this chapter (Richard Dawkins, Stephen J. Gould, Simon Conway Morris, Francisco Ayala, Ken Miller, and others). These present their personal views and are, of course, of great interest. Philosophers of science have also provided detailed accounts of the impact of Darwinism and evolutionary theory on religiosity, some of which are quite objective. A good one is Philip Kitcher's *Living with Darwin*. Michael Ruse has also written much on this topic; among his books one could suggest *Can a Darwinian be a Christian? The Relationship between Science and Religion* and, most recently, *Science and Spirituality: Making Room for Faith in the Age of Science*. There exist even more books about Creationism and ID. A nice book to start with is *Denying Evolution: Creationism, Scientism, and the Nature of Science*, by Massimo Pigliucci. Another book with an interesting approach to the topic is Robert Pennock's *The Tower of Babel: The Evidence Against the New Creationism*. A very detailed account of the history of Creationism and ID is given by Ron Numbers in his book *The Creationists: From Scientific Creationism to Intelligent Design*. Finally, a very clear and informative introduction to the difficult domain of epistemology is given by Robert Audi in *Epistemology: A Contemporary Introduction to the Theory of Knowledge*.

3 Conceptual obstacles to understanding evolution

In the previous chapter I argued that religious resistance to evolution is only part of the problem of its low public acceptance. I also argued that intuitions about purpose and design in nature may stem from our knowledge of artifacts, and not (exclusively at least) from our religious beliefs and worldviews. Remember that in two studies 8–10-year-old children generally preferred Creationist explanations, independently of the religiosity of their families. So there must be something more in play than religious instinct. Thagard and Findlay (2010) have suggested that there are two major types of obstacles to accepting evolution: cognitive obstacles and emotional obstacles. Religion is certainly related to emotional obstacles and there is no question that these are important. However, this book focuses on conceptual difficulties and obstacles to understanding evolution. I must note that I do not underestimate the importance of emotional obstacles; however, I think that the importance of conceptual obstacles has been underestimated so far in the public debates about evolution. Psychologists write about cognitive biases that give rise to intuitions/preconceptions and eventually to misconceptions. In this book I will be using a rather different terminology which I find more comprehensive to non-psychologists. Thus I will explain how intuitions – a term referring to what comes spontaneously to mind or to how we intuitively tend to think – give rise to preconceptions about the natural world, which in turn can change to misconceptions when new knowledge is inappropriately added.

As explained in the previous chapter, it is important (but difficult) to distinguish between knowledge and belief. In this chapter I will focus on what constitutes knowledge and on the concepts that are associated with it. I will explain why particular human intuitions make evolutionary explanations seem counter-intuitive, or at least less intuitive than explanations such as those based on artifact thinking and intentional design. This chapter includes the core argument of this book: Our intuitions make us think of organisms as artifacts and it is because of this that we may then seek an artificer/creator. I should note at this point that I do not claim that belief in a god or deity of whatever kind, and consequently religion, stems from our intuitions. The origin of religious belief is a question I will not deal with here.[1] My suggestion, already outlined in the previous chapter, is that human intuitions make evolutionary theory difficult to understand. With this I do not deny that religious beliefs and worldviews have an effect

---

[1] It is interesting, however, that it might be explained in evolutionary terms – see Boyer, 2001; Wilson, 2002; Atran, 2004; Dennett, 2006.

on the low public acceptance of evolution. I rather argue that there is another important issue: Evolutionary theory does not contradict our religious belief about the existence of God – indeed many people claim that evolution and religious belief are in several ways compatible – but our intuitive ways of perceiving the world around us. My suggestion, therefore, is that the main problem with evolutionary theory is not its perceived conflict with religious belief, but its conceptual conflict with human intuitive ways of thinking.

Our knowledge and understanding of the world is formulated in terms of concepts that are mental representations of the world. **Scientific concepts** provide systematic representations through which explanations of and predictions about phenomena are possible (Nersessian, 2008, p. 186). Concepts should be distinguished from conceptions, the latter being the different meanings of, or meanings associated with, particular concepts. From our early childhood we experientially formulate conceptions of the world which are described as preconceptions. As we grow up, we often assimilate knowledge which further modifies our preconceptions, occasionally turning them into more complex but incorrect conceptions which are described as misconceptions (Vosniadou, 2012). The distinction between preconceptions and misconceptions is important because although people may restructure or reorganize their conceptions when they acquire new knowledge, this may be done in the wrong way. In other words, new knowledge (e.g., through schooling) does not guarantee a correct understanding of concepts if the old conceptions are not properly restructured. This is what **conceptual change** is about: the change of our conceptions with development and learning. For conceptual change to occur, existing misconceptions must be properly addressed so that individuals understand that their prior conceptions are wrong or explanatorily insufficient. Ideally, the outcome of a conceptual change process is the replacement of preconceptions or misconceptions with accurate concepts (see chapters in Vosniadou, 2008 for relevant details).[2] Consequently, in this book conceptual change is roughly defined as the change of conceptions in the wider sense (including the change in the meaning of concepts or the change in the relationship among concepts within an explanatory scheme or model) as a result of conceptual conflict, i.e., the realization that prior conceptions are wrong or explanatorily insufficient.

Achieving conceptual change is not an easy task because preconceptions build on intuitions, which are often strongly held and form the bases for misconceptions which may even persist into adulthood (Bloom and Weisberg, 2007). It is important to note that these intuitions seem to be deeply rooted and strongly held, so that they are not completely overwritten even by expert knowledge. To give such an example about a fundamental distinction, that of living/non-living entities, I will briefly present the results of an interesting study. In the first experiment of this study it was examined whether adults (undergraduate students) had any difficulty classifying as living

---

[2] There are several aspects of conceptual change (Chi, 2008; Keil and Newman, 2008; Duit and Treagust, 2012). The literature on conceptual change research cannot be reviewed here, nor could one discuss all the contemporary issues in conceptual change research. An excellent resource on conceptual change for the interested reader is Vosniadou (2008). Another interesting collection of papers on this topic is Thagard (2012). An interesting account of conceptual change based on a cognitive-historical approach is provided by Nersessian (2008). For a very interesting philosophical exchange, see Lennox (2013a, 2013b) and Burian (2013).

particular items that children often find difficult to classify correctly. The items included animals, plants, non-moving artifacts (e.g., a towel), non-moving natural objects (e.g., a stone), moving artifacts (e.g., a truck), and moving natural objects (e.g., the water in a river). Participants were asked to state as quickly as possible whether each item was living or non-living (a procedure described as rapid categorical processing). Results suggested that undergraduate students had difficulties, similar to those that young children have, in correctly classifying the items presented to them. This shows that childhood intuitions persist into adulthood and are expressed when people are asked to provide answers spontaneously. The same study was repeated with university biology professors in order to see whether expertise in a particular domain affects intuitions. Interestingly enough, even biology professors had the same difficulties that undergraduates had. Professors did better than undergraduates in classifying both animals and plants as living, but they did not perform better in their answers for artifacts and non-living natural objects. What this study shows is that there are intuitions relevant to biological knowledge which have developmental "roots" and cannot be completely overwritten despite expert knowledge (Goldberg and Thompson-Schill, 2009).

Consequently, intuitions that generate persistent misconceptions may serve as **conceptual obstacles**: conceptions which are strongly held, which are resistant to change, and which thus impede understanding and acquisition of correct concepts. These, in turn, make scientific theories like evolutionary theory difficult to understand.[3] In this chapter I first describe what I consider as conceptual change in science: the change in explanatory schemes or models as a result of change of prior conceptions due to conceptual conflict. Then I analyze two conceptual obstacles which are impediments to our understanding of evolution: **design teleology** and **psychological essentialism** – which, as conceptual development research suggests, develop early in childhood. In this chapter I focus on studies on pre-school and young elementary children and, unless otherwise specified, I use the word *children* to refer to this age range (approximately 4–8 years old). Children of these ages have received little formal training and so their answers often reflect their intuitions. It seems that formal education largely leaves these intuitions unchallenged and so they may persist into adulthood. Finally, I describe what conceptual change in evolution consists of, in the light of these conceptual obstacles.

Conceptual change in science

A careful study of the history of science reveals interesting cases of conceptual change. The work of Thomas Kuhn (1996/1962)[4] used to be the exemplar here and has also served as the basis for the classical approach to conceptual change in science education

---

[3] Conceptual obstacles to understanding a scientific theory constitute one important aspect of how science is perceived by people. For a different, interesting aspect that has to do with what is considered as rational, see Wilkins (2011).

[4] For an interesting overview of how the problem of conceptual change was addressed by philosophers of science during the twentieth century (including pre- and post-Kuhnian developments), see Arabatzis and Kindi (2008).

(Posner *et al.*, 1982). However, it seems that a solely historical approach to the understanding of conceptual change in science is not adequate. Rather, a cognitive-historical approach seems more appropriate because cognitive science can help us better understand scientific practice (Arabatzis and Kindi, 2008; Nersessian, 2008). The history of science is not simply a narrative of events, and its understanding requires more than a superficial reading. In addition, conceptual change in (the history of) science is not simply a process of replacement of concepts by new ones, and so a careful study of history and historiography can provide an understanding of conceptual change at a deeper level. A thorough understanding of how past scientists created new concepts or redefined existing ones or restructured the relations between them reveals how conceptual change occurs. To give an example relevant to evolution, it is interesting to study history in order to understand how Charles Darwin underwent a conceptual change from his initial natural theological assumptions to develop his theory of natural selection. There is no point to talk about a single shift from pre-Darwinian to Darwinian views, because no such shift ever took place in the history of science (Bowler, 2005; Corsi, 2005; Hodge and Radick, 2009b). This topic will be the focus of Chapter 4. In short, the history of science may illuminate conceptual development research, but it does not provide all the clues for how conceptual change in science occurs.[5]

But what exactly is conceptual change in science and in what way does it differ from simple knowledge acquisition or enrichment? A major difference is that conceptual change usually occurs because of conceptual conflict. Conceptual change is more than simply acquiring new knowledge or enriching a pre-existing, incomplete body of knowledge. Conceptual change occurs when someone already possesses some prior idea/belief and realizes that it is internally inconsistent or contradictory (Chi, 2008; Carey, 2009, p. 415). In this case a prior conception may be replaced by some new one or be restructured to properly accommodate elements of the new one. When this happens, conceptual change has occurred. What is more important is that conceptual change does not take place immediately when one realizes the conflict between the prior and the new knowledge, but often takes time to accomplish. This is usually due to the fact that the initial conceptions (in the form of preconceptions or misconceptions) are strongly held and thus are resistant to change. Conceptual change may involve changes in the internal structure of concepts, or changes, central to their meanings, in the relations of concepts to others (see Keil and Newman, 2008). In this sense conceptual change may also involve changes in the meaning of concepts or introduction of new ones in the structure of explanations, but also the restructuring of explanations and consequently changes in the relations among concepts. We intuitively rely on our conceptions to provide explanations for the phenomena we perceive in the world, and so a significant change in our conceptions (conceptual change) is directly related to significant changes in the explanatory schemes we use. As already mentioned above, I consider conceptual change as the change in the structure of the explanatory schemes or models we use.

---

[5] It should also be noted that individual conceptual change during development does not simply parallel conceptual change in the history of science. This is an important point for educational purposes (see Levine, 2000; Greiffenhagen and Sherman, 2008; Van Dijk and Reydon, 2010; Kampourakis and Nehm, 2014).

But let us first examine what a scientific explanation is. In general an explanation consists of an explanandum (whatever is being explained) and an explanans (whatever is doing the explaining). For example, if one asks "Why *X*?" and the answer is "Because *Y*," then *X* is the explanandum and *Y* is the explanans (see Godfrey-Smith, 2003, p. 191; Rosenberg, 2005, p. 26 for comprehensive accounts).[6] It seems that there is agreement among many philosophers that the concept of cause is central to the process of scientific explanation (Kitcher, 1989; Salmon, 1989; Okasha, 2002, p. 49; Godfrey-Smith, 2003, pp. 196–197; Woodward, 2003; Rosenberg, 2005, p. 27; Strevens, 2009).[7] Cognitive scientists also suggest that causes are perceived by humans as central in the process of explanation (Keil and Wilson, 2000; Lombrozo and Carey, 2006). Scientists are often able to identify the cause or causes of a particular phenomenon and so provide an explanation for its occurrence. In other cases, scientists develop models which can be used in order to explain what they observe (Woodward, 2011; Frigg and Hartmann, 2012). Thus, conceptual change could take the form of a restructuring of an explanatory scheme: change in causal relations or change in the structure of causal models used to explain a phenomenon. To illustrate how this is possible, we now turn to a particular case from the history of science.

When we wake up in the morning, one of our first questions usually is: "Has the Sun risen?" During the day we can observe the Sun rising and until midday we can even navigate using its position (since we accept that the Sun rises from the east, we know where the east is as the Sun seems to come up from there). Then, in the afternoon, we can also see the Sun going down, until it eventually disappears. And so on. Why do we think that the Sun rises from the east and goes down toward the west? Because this is what we observe, and the same pattern is repeated every single day. However, anyone who has attended a high-school course on astronomy or who has read any relevant, even non-technical, book is aware that the previous account for the motion of the Sun is not scientifically accurate. The Sun neither rises, nor goes down. It is Earth, our planet, which orbits the Sun, as do all the other planets of our solar system. The rotation of Earth around its axis and the revolution[8] of Earth around the Sun produce the motion of the Sun relatively to our position on Earth that we observe. Astronomers have

[6] In the philosophy of science there are several accounts of scientific explanation (see Woodward, 2008 for an overview). It has been suggested that to explain something is: (1) to show how it is derived in a logical argument (**deductive**-nomological account [Hempel and Oppenheim, 1948]; **inductive**-statistical account [Hempel, 1965, pp. 376–412]), (2) to give information about how it was caused (Salmon, 1984, 1989; see also Scriven, 1959, 1969), (3) to connect a diverse set of facts by subsuming them under a set of basic patterns and principles (Kitcher, 1981, 1989), or (4) to analyze (deconstruct and reconstruct) the responsible mechanism (mechanistic account [Machamer et al., 2000; Bechtel and Abrahamsen, 2005; Bechtel, 2013]).

[7] Several theories of causality have been developed. Causes are generally considered as difference-makers for their effects, i.e., causes make a difference to whether or not the effect occurs (see Hitchcock, 2008). For example, in the case of a forest fire, oxygen, combustible materials, and a lighted cigarette are possible necessary conditions. In this case, the lighted cigarette is the cause of the forest fire. Oxygen and combustible materials are there, and there could be a fire for other reasons, but it is the lighted cigarette that makes the difference.

[8] It should be noted that the word revolution initially described (and still refers to) a periodically recurring cycle. However, after it became known that it is the Earth which revolves around the Sun, the word revolution was given a new meaning: that of radical and irreversible change (Shapin, 1996, p. 3).

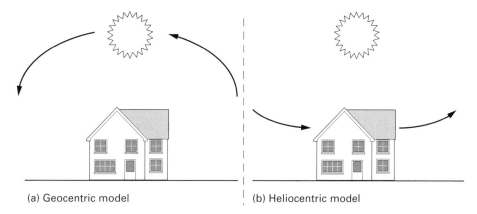

(a) Geocentric model | (b) Heliocentric model

Figure 3.1 Different perceptions of reality. In both cases, the Sun and Earth move relatively to each other. According to the geocentric model, Earth is static and the Sun moves around it (a), whereas according to the heliocentric model, the Sun is static and Earth moves (b). Although we intuitively think that (a) is what is happening, (b) can describe the relative movement as well. Based on this evidence only, one cannot conclude what is actually happening. Image © Simon Tegg.

explained that what we intuitively think about the relative movement of Earth and the Sun is wrong. Contrary to what we perceive, it is our planet, Earth, which orbits the Sun, not vice versa.

This suggestion was made by Nicholas Copernicus,[9] who questioned the geocentric (Earth-centered) model of Claudius Ptolemy, according to which the planets encircled a static Earth that was placed in the middle of the universe. Copernicus attempted to present a new and simpler way of explaining the motions of the planets. What he suggested was that it would be possible to explain curious features of the planetary motions (retrograde motion) if it was assumed that the Sun and not Earth was placed at the center of the universe. Copernicus' model was published in 1543 in a book entitled *De Revolutionibus Orbium Celestium* (*On the Revolutions of the Celestial Spheres*). In developing his heliocentric (Sun-centered) model, Copernicus did not introduce any new concepts, but maintained Ptolemy's central concepts. However, Copernicus' heliocentric model required a substantial reorganization of the geocentric model (Thagard, 1992, pp. 195–199; Shapin, 1996, pp. 20–22; Gingerich, 2005). What Copernicus actually did was to restructure an explanatory scheme and produce a new one that accounted better for the observed phenomena. But what made Copernicus conceive of such a model that contradicts our everyday perception? What made him come up with the counter-intuitive idea that the Sun does not move around Earth, as we observe every day?

Figure 3.1 includes very simple representations of a geocentric and a heliocentric model. Both representations include Earth and the Sun, as well their relative movements according to the two models. In the case of the geocentric model (Figure 3.1a), the Sun is shown to move around Earth, which is static. In the case of the heliocentric model

---

[9] And before him, by Aristarchus of Samos (see Gingerich, 1985).

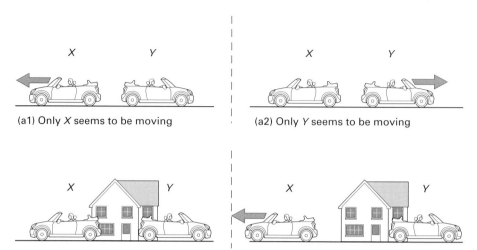

(a1) Only *X* seems to be moving   (a2) Only *Y* seems to be moving

(b) When the house is included in the system, it is concluded that it is *X* and not *Y* who is moving.

Figure 3.2 Relative movement. Both (a1) and (a2) can explain the relative movement, so based on this observation only, one cannot conclude what is actually happening. In order to understand which of the elements of a system is moving and which is not (persons *X* and *Y*), we must examine their movement relatively to a third element which we consider static (e.g., a house). In a similar but much more complicated manner, Copernicus not only examined the motion of the Sun relatively to Earth, but carefully studied the system that included the other planets. He concluded that a heliocentric model provided a better explanation for the motions of planets compared to a Ptolemaic/geocentric model. Image © Simon Tegg.

(Figure 3.1b), the Sun is static and it is Earth that moves around the Sun. On the basis of what we perceive, the explanations provided by the two models are equivalent (but most people would intuitively accept the first). Consequently, more information is required about the wider system, not only about Earth and the Sun.

To make this clearer we can use a simple thought experiment from physics. Imagine two people (*X* and *Y*), each of whom drives a car (Figure 3.2a). If they suddenly start moving away from each other, two possible explanations are: either *Y* is static and *X* is moving (Figure 3.2a1), or *Y* is moving and *X* is static (Figure 3.2a2). Both explanations are equally correct. Thus, in order to understand what is happening we need more information about the system. If the system includes a house, and if we examine the movement of *X* and *Y* in relation to it, then we realize that it is, e.g., *X* and not *Y* who is moving because *X* moves away from the house, whereas *Y* stays close to it (Figure 3.2b). The point made here is that scientists go beyond intuitions and simple perception to study phenomena by means of detailed rather than superficial observations, and by taking into account their contexts rather than simply studying them in isolation.[10]

So what was Copernicus' contribution? Ptolemy had suggested that the planets performed uniform circular motions. Each planet moved with uniform speed around a

---

[10] Of course, it may be the case that scientists study some phenomena in isolation, e.g., the effect of factor *F* on process *P*. But even then they are aware of the simplifications made and they do not overlook the wider system in which *F* and *P* belong.

small circle, the epicycle, the center of which also moved uniformly on a larger cycle, the deferent, inside which Earth was located (see Figure 3.3a). Ptolemy developed different models which, when combined, could account for the observed planetary positions. These were the eccentric model (Earth was not located in the center of the deferent but in another point close to it), and several variants of the epicycle-on-deferent model (the planets did not only orbit Earth but also moved around an epicycle, the center of which actually orbited Earth) (Lindberg, 2007, pp. 98–105). Although Copernicus retained the idea of uniform circular motions, he rejected the idea that Earth was static and introduced three motions of Earth. The first motion was the diurnal rotation of Earth around its axis; the second motion was the movement of Earth along with the other planets around a center close to which a static Sun was located; and the third motion was that of Earth's axis in a conical path – to keep it pointed toward the north celestial pole (Ravetz, 1996). This new heliocentric model – despite its own problems – explained more than Ptolemy's geocentric model (Thagard, 1992, pp. 197–199).

What motivated Copernicus to reject the geocentric model and develop a heliocentric one? Copernicus found appealing the principle that the periods of the planets should be longer the farther their orbits were from the center of motion, assuming that their speeds were the same. According to the geocentric model, this principle worked well for Saturn, Jupiter, and Mars, but not for the Sun, Venus, and Mercury. Copernicus gave a correct account of Ptolemy's distances between the planets and rejected them because they did not conform to the distance–period principle. He also rejected all versions of geocentric models that he was aware of and suggested that either a new center of motion should be found or that it should be concluded that there was no principle governing the order of the orbits of the planets. He had already read accounts that described Venus and Mercury as orbiting around the Sun while the latter and the other planets orbited Earth. So he thought it could be the case that all planets orbited the Sun, but he had to calculate the new periods for Venus and Mercury and show that they fitted such a model. Copernicus made these calculations and found that the results fitted with a heliocentric model. He calculated the period of Venus at 225 days and of Mercury at 116 days, thus less than a year – as it was expected.[11] As already described, according to the geocentric model each planet moved uniformly around an epicycle (with radius r), the center of which also moved uniformly on a larger cycle, the deferent (with radius R). Figure 3.3a presents the motion of Venus according to the Ptolemaic/geocentric model. As is obvious in Figure 3.3b, the heliocentric model explains the motion of Venus without the assumption of a large epicycle. The geocentric model assumed that the center of the Venus epicycle was on the line from Earth to the Sun and that the deferent rotated with the same speed as the Sun. Whereas the geocentric model required that Venus move on an epicycle that in turn revolved around Earth, the heliocentric model suggested that there were no epicycles, only orbits of all planets around the Sun. The Sun was at the

---

[11] The periods of Mercury and Venus should be less than the period of the Earth, which is one year or 365 days. All descriptions that follow refer to Mercury and Venus as "inner" planets, i.e., planets between the Earth and the Sun in a heliocentric system.

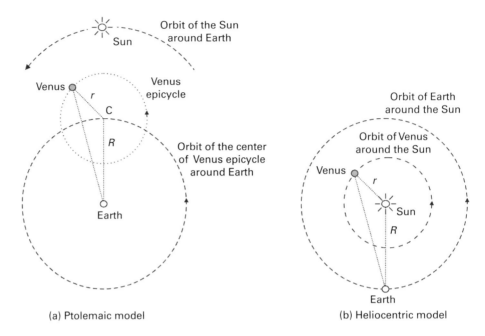

(a) Ptolemaic model   (b) Heliocentric model

Figure 3.3 Comparison of the Ptolemaic/geocentric and the heliocentric model (adapted from Goldstein, 2002, p. 227). In order to explain the relative movement of Venus and Earth, one had to assume that the Venus epicycle as well as the Sun orbited Earth. The heliocentric model was free of such assumptions and the Venus epicycle was transformed to the orbit of Venus around the Sun. In this case, Earth also orbited the Sun, like Venus and any other planet. It should be noted that (a) only shows Ptolemy's epicycle on the deferent model (this does not include the eccentric and the equant models).

center of the orbit of the planets. Note that r became the radius of the orbit of Venus and R became the radius of the orbit of Earth around the Sun. It should be noted that the r/R ratio was not affected by the transformation (Goldstein, 2002).

Copernicus' heliocentric model was initially received as a theoretical scheme rather than a representation of reality. It was neither accepted nor debated immediately, only later, when Galileo entered the stage. The heliocentric model eventually challenged the geocentric view of our world (that Earth is in the center of the universe), but it did not alone cause the wider change in our perception of our place in the universe (see Ravetz, 1996; Shapin, 1996; Danielson, 2009). However, Copernicus had raised questions that were important to answer. One of these was the explanation for the missing phases of Venus. In December 1610 Galileo observed the half-planet phase (phase 5 in Figure 3.4b). This observation provided important support for a model in which the planets move around the Sun (Heilbron, 2010, pp. 164–170).[12] Thus, another point that should be noted is that in this case of conceptual change the new model was not

[12] Heilbron (2010) gives an interesting account of this discovery. It is not clear how exactly Galileo came to think that the phases of Venus might provide answers in support of one system over the others, nor did he immediately accept the Copernican system.

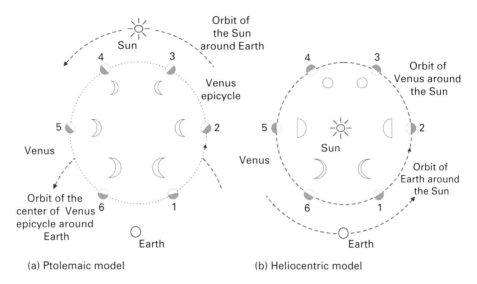

(a) Ptolemaic model   (b) Heliocentric model

Figure 3.4 The anticipated phases of Venus are different between the Ptolemaic and the heliocentric models (adapted from Heilbron, 2010, p. 168). In 1610 Galileo observed phase 5 in (b), which was not a phase consistent with Ptolemy's model.

immediately accepted because further evidence that supported it was required. This is usually the case. Any new model should both adequately capture existing data and allow new predictions. As will be explained in more detail in Chapter 4 on Darwin's theory, even proponents of a new model or theory may question it in their attempt to understand if it is accurate. Constructive criticism and the quest for more evidence is a distinctive feature of science and one that strengthens the conclusions of scientists. Questioning a theory or a model is different from rejecting it. Both rejection and acceptance of a scientific theory require some solid evidential basis, and questioning it is the best way to find the necessary evidence and reach the appropriate conclusions.

What is the general conclusion from Copernicus and his case of conceptual change? It is that there was a change, actually a restructuring, of a model. The initial geocentric model was restructured to produce a heliocentric one. Note that there was no replacement of an old model with an entirely new one. Copernicus did not actually introduce new concepts (Thagard, 1992, pp. 196–199) but restructured the old model by considering new relations between the planets (e.g., that Earth moves like all the other planets) and by rejecting the old ones (that Earth is the center of the planetary system). What is more important is that the intuitive belief, firmly held for thousands of years, that Earth lies at the center of the planetary system, was eventually rejected after the crucial contributions of Kepler, Galileo, and Newton. This example nicely summarizes how scientists work. They go beyond intuitive beliefs to study phenomena which may be imperceptible in everyday life. Scientists accumulate data that become evidence for a theory that explains phenomena more effectively than our intuitive theories. But it is here that the problem of public acceptance of science arises. It is not enough to have scientists gather evidence to support their theories; they also need to make the public understand why scientific theories are valid and why they have more explanatory power than human intuitions. Why should I accept the heliocentric model which suggests that Earth orbits the Sun? I do not feel that I am moving at all; I also feel neither any centripetal nor any centrifugal force acting on me. Rather, what I perceive is that I live on a static Earth and I see the Sun and the Moon moving around it. In order to be convinced that this is not the case, someone must convince me that my intuitions are wrong.

This is exactly the problem with evolution. People tend to intuitively think in particular ways that make the conclusions of scientists about the evolution of life on Earth seem entirely counter-intuitive. As we might intuitively think that we live on a static Earth and that the Sun revolves around it,[13] we might also think that organisms are designed and that they have fixed **essences**. The first intuition about purpose and design in organisms is described as design teleology, whereas the second about essential and unchanging characteristics of organisms is described as psychological essentialism.[14] Conceptual development research suggests that both design teleology and psychological essentialism are deeply rooted and strongly held intuitions that generate misconceptions which arise during early childhood and persist into adulthood. These intuitions eventually make evolutionary theory seem entirely counter-intuitive. In the next two sections I will explain in detail why design teleology and psychological essentialism are conceptual obstacles to understanding evolution.

Design teleology as a conceptual obstacle to understanding evolution

Why do airplanes have wings? A reasonable answer would be: *in order to fly*. Of course, planes usually do not fly because of their wings only; they also need powerful engines in order to take off and maintain flight, although gliders can fly without engines as soon as they take off. How about birds? Why do birds have wings? A reasonable answer would also be: *in order to fly*. Again, birds do not fly because of their wings only; they also need to have relatively light bones and strong muscles. Birds usually have to move their wings up and down in order to fly, but are also able to maintain flight for some time by extending their wings and following air currents. Such "in order to" answers to "why?" questions are described as teleological because they imply the existence of an end or goal (Lennox, 1992; Ariew, 2007; Walsh, 2008; Lennox and Kampourakis, 2013).[15] A character exists in order to fulfill a goal; in the previous example wings exist

---

[13] Apparently, science communication about this issue has been successful and I doubt there are any educated people who would nowadays doubt the fact of the heliocentric system.

[14] This is not the place for a detailed philosophical analysis of teleology or essentialism. The focus of the sections on teleology and essentialism of this chapter is mostly on their impact as conceptual obstacles rather than their content. The interested reader should consult Lennox and Kampourakis (2013) and Wilkins (2013) for detailed philosophical analyses of teleology and essentialism, respectively, as well as for the implications of these philosophical analyses for conceptual development research.

[15] *Telos* is the Greek word for a final end or goal. Consequently, *teleology* is the study of final ends or goals. In the philosophy of science, answers to "why?" questions are considered as explanations. Consequently, an "in order to" answer to a "why?" question is a teleological explanation because it explains the existence of

in order to make flight possible. Thus, we may be able to explain the existence of a character by reference to the goal that it serves. How about explaining wholes? We might ask the following question: Why do airplanes exist? A reasonable explanation would be: *in order to be used by people for travelling*. In order to travel far away, one may use a car, a train, or a ship, but nothing compares to an airplane. A trip that might last for days if one used any other means of transportation will last just a few hours if one decides to travel by airplane. How about birds? Could we ask why birds exist? Certainly. But would it be reasonable to answer that birds exist for a purpose? Perhaps. Birds may contribute something important to the economy of nature; they may have some role. Otherwise, why should they exist at all? So we may also give teleological explanations both to questions about parts and to questions about wholes for artifacts and organisms.

My teleological explanations for birds in the previous paragraph may intuitively seem correct, but they are basically wrong. There are two reasons for this. The first reason is that airplanes are artifacts, whereas birds are organisms; as I will explain, it is not appropriate to provide explanations for artifacts and organisms on the basis of the same assumptions.[16] The second reason is that answers about parts and wholes are also different in many respects from each other, because a part may have been designed to serve a purpose, but a whole may consist of both purposeful and purposeless parts, or some arrangements inside a whole may serve a purpose, whereas others do not. In what follows I explain which teleological explanations are legitimate, distinguishing between teleological explanations for organisms and teleological explanations for artifacts, as well as between teleological explanations for parts and teleological explanations for wholes. Then I present the findings of conceptual development research on teleological explanations and explain why design teleology is a conceptual obstacle to understanding evolution.

Let us return to the wings example. Although it may seem reasonable to suggest that both airplanes and birds have wings *for* flying, there is a major difference between them: artifacts are intentionally designed/created (by humans) for a purpose (Keil, 1989, p. 49; Bloom, 2004, p. 55; Hilpinen, 2011). A sharp object is not an artifact unless it was made sharp in order to cut things or open holes. One may use a sharp branch from a tree as a tool to open holes, but this is not an artifact unless it was intentionally broken in a particular way in order to be sharp.[17] If someone found a branch that was accidentally cut to be sharp and used it to open holes, this would certainly be a tool, but not an artifact. Now, since artifacts are intentionally designed for a purpose, they usually have the appropriate size/shape for the intended use. Airplanes have wings which are proportionally large to their size. A Cessna airplane has smaller wings than an Airbus, and in both cases the wings are long enough to facilitate take-off and flight. No rational

---

something in terms of the goal it serves. Hereafter, I will describe all such answers in terms of goals or purposes as explanations.

[16] For philosophical discussions of organisms and artifacts, see McLaughlin (2001) and Lewens (2004).

[17] Such natural objects have been called *naturefacts*, because they stand between natural objects and artifacts. However, if they are modified further with an intended use in mind, they can become genuine artifacts (Hilpinen, 2011).

aircraft builder would ever design an Airbus with the wings of a Cessna, or vice versa. In addition, there are no airplanes without wings. Other aircraft, of course, exist, such as helicopters, which do not have wings – but we would not consider them as airplanes. To summarize, airplanes are artifacts intentionally designed in order to fly and so not only do they all have wings in order to fulfill this goal, but their wings are always of an appropriate size.

Is this the case for birds? The answer is no. All birds have wings (these are actually their forelimbs), but not all of them use these for flying. It is a vague generalization to suggest that *birds have wings for flying*. Eagles fly *because* they have long wings (and skeletons and muscular systems, etc.) which make flight possible. But what about penguins? Penguins have relatively small wings for their size, so it is impossible for them to fly. However, penguins use their wings for swimming, and they can actually swim very fast underwater. Should we then state that penguins have wings for swimming? Yes, we can. Most interestingly, ostriches also have wings but use them neither for flying nor for swimming. And their wings are small in proportion to their body size, at least compared to eagles. Consequently, all birds have wings, but not all birds use their wings in order to fly because they are not always of the appropriate size. This is the case because birds are not artifacts and their wings were not intentionally designed for flying. Birds, like all organisms, have come to possess their characters through evolution and are not intelligently designed. I will return to the details of evolutionary processes in the next chapters. Consequently, the existence of an intelligent designer of birds (especially if that was perceived to be a divine one) does not seem to be consistent with the characteristics of ostriches. Why would an intelligent designer design a big bird with relatively small wings that do not help it to fly? A human aircraft designer who designed the airplane on the right of Figure 3.5, which is apparently unable to fly, would be a bad designer, and not one we would think of as intelligent.

Let us now turn to the parts/whole question. Teleological explanations for wings are actually explanations about parts. How about wholes? Can we explain the existence of

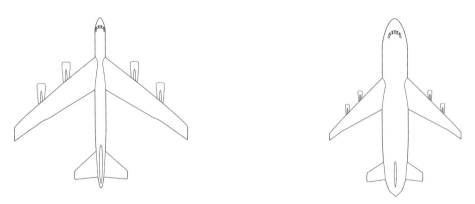

Figure 3.5 The relative sizes and body length/wing length ratio for an eagle (left) and an ostrich (right) in airplane form (with a high degree of approximation); or how eagles and ostriches would look if they were airplanes. Could you imagine the plane on the right being designed for flying and yet be unable to fly? Image © Simon Tegg.

airplanes in teleological terms? The answer is definitely yes. Airplanes are artifacts and are thus designed for a purpose. The existence of (whole) airplanes can be explained in teleological terms insofar as they are used to transfer people from place to place. This is the main use of airplanes and this is what they were designed and constructed for. An airplane may be found in a museum, but this is not what it was designed for – this is an incidental use. If you think carefully, the existence of any whole artifact can be explained in teleological terms: scissors are designed and created in order to cut; nutcrackers are designed and created in order to crack nuts; pencils are designed and created in order to be used for writing; cars are designed and created in order to be used for transportation. And so on.

Is this the case for organisms? The answer is probably no. We know that approximately 99% of all the species that have ever existed on Earth have become extinct (Jablonski, 2004a; see also Chapter 6). Assuming that whole organisms were designed and created for some purpose or role (even if we do not know what this is), the question that is really hard to answer is why would anyone (God or Nature) create so many species for some purpose and then let them go extinct? Assuming that their supposed role in nature was performed by other organisms as soon as they went extinct, why, then, should have they emerged in the first place? If organisms serve a purpose in nature, why would anyone create some species and then let them go extinct and be replaced by others and not create or let the latter emerge right from the start? These are questions that are difficult to answer, and whatever answer one gives in terms of purposes would be more or less speculative. Any explanation seems reasonable, but the explanation provided by evolutionary theory requires fewer assumptions than one in terms of design and purpose. Isn't it more plausible to accept that organisms went extinct due to natural causes rather than that they emerged for a purpose and were then allowed to become extinct?[18] We will return to this question in the chapters that follow.

So far I have argued that there is a major difference between airplanes and birds, and more generally between organisms and artifacts. Teleological explanations for artifacts presuppose design, whereas teleological explanations for organisms presuppose natural processes, i.e., evolution. It makes more sense to think of imperfections as the outcome of evolution, rather than the outcome of intelligent design. Imperfections can be the outcome of design, but of bad and incompetent design which we would probably not characterize as intelligent. Thus, teleological explanations can be given for both organisms and artifacts, but these are of a different kind. Teleological explanations *are* legitimate for organisms, as long as they are based on natural processes, such as natural selection, and not on intentional design. The crucial distinction here is that artifacts have particular features *in order to* perform some role as a consequence of their being designed for this purpose, whereas organisms have particular features (characters) *in order to* perform some role as a consequence of their being selected during evolution.

---

[18] It could certainly be the case that some organisms became adapted to a certain environment through selection and later went extinct because either the environment changed or a better-adapted organism outcompeted them. But there is no purpose in such a process. Selection did not intentionally design organisms the way an artificer would do.

In addition, artifact teleology is external, whereas organism teleology is internal. The wings of airplanes and eventually airplanes themselves serve their human creators and their intentions. In contrast, the wings of birds serve (if they do so) their possessors (and probably their own intentions: find food, avoid predators, etc.). This is an important point for the discussion that will follow. If artifacts possess some character for some purpose, this is a purpose external to them which has been set by their human creators. In contrast, the characters of organisms do not serve any external purpose because they were not intentionally designed; organisms are not artifacts. If organisms possess some characters that seem to serve some purpose, e.g., eagles have wings *for* flying, what is actually happening is that flying is a consequence of having wings and other appropriate body parts that serves the organisms themselves and not some agent external to them. Thus, organism teleology is based on consequences without a presupposition of intentional design and so differs significantly from artifact teleology.

There exists an enormous body of research which suggests that people tend to intuitively provide teleological explanations for organisms and artifacts from very early in childhood. There is some disagreement on whether children provide teleological explanations for organisms and artifacts only, or for non-living natural objects (rocks, clouds) as well. It has been suggested that children provide teleological explanations discriminatively for organisms and artifacts and that they are able to accurately distinguish between them (Keil, 1992, 1994, 1995). Another body of research suggests that children provide teleological explanations in a non-discriminative manner for organisms, artifacts, and non-living natural objects (Kelemen, 1999a, 1999b, 1999c). Recent research suggests that the first view may be more probable for particular ages (Kampourakis *et al.*, 2012a), but a shift from a non-discriminative to a discriminative teleology may also be possible (Kampourakis *et al.*, 2012b). The issue at stake here is how children really perceive objects around them. Kelemen has suggested that children may provide teleological explanations for organisms because they consider them as artifacts (Kelemen, 1999c), whereas Keil has suggested that children may have autonomous biological thinking and thus limit teleological explanations to seemingly advantageous features of animals and artifacts (Keil, 1992).[19]

One question immediately arises: Does knowledge about organisms influence knowledge about artifacts, or is it knowledge about artifacts that influences knowledge about organisms? Kelemen and Carey (2007) suggest the second – a proposal originally made by Piaget (1960/1929) – but they also recognize that there is conflicting evidence (see Keil, 1992; Greif *et al.*, 2006; Keil *et al.*, 2007) and that eventually both may be possible. However, the suggestion that knowledge about artifacts influences knowledge about organisms seems quite plausible. From very early in their lives, most of the objects that children encounter are artifacts and they learn that they exist in order to achieve some goal. Just think how many artifacts were around you when you were an infant; you may have spent most of your first 2–3 years of life inside your home

---

[19] Some of the conclusions from these studies have implications for the origins of biological thought, e.g., whether it is autonomous or whether it stems from other fundamental ways of thinking, but this topic cannot be discussed here (see Carey, 1985; Keil, 1992; Springer, 1999; Inagaki and Hatano, 2002).

surrounded by toys, feeding bottles, a crib, dummies, chairs, tables, sofas, spoons, plates, electronic devices, and numerous other artifacts. How many animals did you encounter as an infant? Did you even pay any attention to non-living objects such as clouds or rocks? Even if you grew up in the countryside and your parents were farmers, the animals and the non-living natural objects that you encountered were probably much less in number and variation compared to the artifacts that you were familiar with at a young age. Thus, it may be the case that in the absence of alternative explanations, children intuitively draw on their understanding and knowledge of artifacts and eventually conclude that organisms, like artifacts, exist in order to be used for some purpose (see Kelemen, 1999a for a comprehensive review). What is also important is that adults provide teleological explanations for organisms and non-living natural objects in a similar manner, under particular conditions (e.g., Kelemen and Rosset, 2009). Let us now examine in more detail the relevant evidence.

In one study, four- and five-year-old children (and adults) were shown photographs of various organisms, artifacts, and non-living natural objects, and were asked to explain what the objects and their parts were *for*. What is important is that participants were explicitly given the option to answer that the objects and/or their parts were not *for* anything. Results indicated that while there was no significant difference between children and adults in providing teleological explanations for the parts of organisms, the parts of artifacts, and whole artifacts, there was a significant difference in that children provided more teleological explanations for whole natural objects, parts of natural objects, and whole organisms compared to adults. In a second study, it was examined whether children really believed that whole organisms, whole artifacts, and whole non-living natural objects were *made for* something or whether they thought they could simply perform or be used for certain roles. Children had to choose between statements suggesting that an object was made for something or that it was not made for anything. In this study there was no difference between children and adults in providing teleological explanations for organisms. In addition, while both children and adults generally provided teleological explanations for artifacts, adults provided significantly more teleological explanations. Finally, there was a significant difference between children and adults in the teleological explanations they provided for non-living natural objects, with twice as many children providing such explanations compared to adults. Interestingly enough, a third study concluded that both children and adults share the same notion of function, as they suggested that organisms' parts and artifacts were *for* some particular role (Kelemen, 1999b).[20]

To investigate further whether children and adults provide teleological explanations for both organisms and natural objects, Kelemen (1999c) attempted to replicate the findings of Keil (1992). In his study Keil had investigated whether children preferred teleological explanations for organisms and artifacts in a similar way. Children (second grade) were given two possible explanations for why plants and emeralds were green; one was that being green helps there to be more of them, the other that they were green

---

[20] The three studies reported in Kelemen (1999b) involved different participants.

because they consisted of tiny green parts. The former explanation was given mostly for plants, whereas the latter explanation was given mostly for emeralds. In short, children preferred teleological explanations for organisms than for non-living natural objects (Keil, 1992, pp. 129–130). In Kelemen's replication study, children and adults were shown pictures of different pairs of organisms and non-living natural objects, and were then asked "why?" questions about their properties. Participants could choose between two answers for each question, one physical and one teleological. For instance, when shown a picture of an extinct aquatic reptile (*Cryptoclidus*) and a pointy rock found in the same area, and asked why the rock was pointy, participants had to choose between the physical explanation "They were pointy because bits of stuff piled up on top of one another for a long time," and either that "They were pointy so that animals wouldn't sit on them and smash them" or that "They were pointy so that animals like *Cryptoclidus* could scratch on them when they got itchy." Results indicated that adults provided teleological explanations for organisms but not for non-living natural objects such as pointy rocks. However, children at all grade levels preferred teleological over physical explanations for the properties of non-living natural objects such as rocks and stones. In a second study, an attempt was made to influence children's explanations about non-living natural objects. Thus, children were not only shown the picture of a cloud, but also additional pictures that presented the stages of cloud formation. This was a hint toward physical explanations; however, results indicated that this had no influence as there were no overall differences from the previous study (Kelemen, 1999c).[21]

This is the evidence on which Kelemen bases her suggestion that children provide teleological explanations for all objects in a non-discriminative manner, drawing on their understanding of artifacts. Assuming that children become familiar with artifacts before they become familiar with organisms and non-living natural objects, it is reasonable to conclude that children extend their intuitive artifact thinking to organisms and to non-living natural objects. However, Greif *et al.* (2006 – Keil being one of the co-authors of this study) have concluded that this may not be the case, as in their study four- and five-year-old children asked different types of questions for animals and artifacts. More children asked questions about the functions of artifacts than about the functions of animals. For instance, the question "What does it do?" occurred more often for artifacts than animals. In addition, children asked what artifacts were designed for or how they worked, but never asked such questions about animals. Such results certainly stand as evidence that children perceive organisms, or at least animals, differently than artifacts.

A relevant important finding is that young children are sensitive to intentionality. In a study, three-year-olds, five-year-olds, and adults were shown representations of objects (drawings and paintings) and actual objects. Both children and adults tended to attribute a function to the item if it was described as intentionally created rather than if it was described as accidentally created. For instance, when shown a knife-like structure made of Plexiglas that looked very much like a knife, more children called it a knife when it

---

[21] The two studies reported in Kelemen (1999c) involved different participants.

was intentionally created rather than if it was accidentally created. Thus, perceptual information was not enough (Gelman and Bloom, 2000; Gelman, 2003 p. 256). In a follow-up study, children were not simply shown the items and told how they were created, but they were shown the processes through which they could have been made (intentional and accidental) without being given any further information. For example, a splotch of yellow paint that looked like the Sun was made once by accident (and the woman who made it seemed to be disappointed by what she had made) and once intentionally (and the woman who made it seemed to be satisfied by what she had made). Children used intentionality mostly to name the objects that were being created (Gelman and Ebeling, 1998; Gelman, 2003, pp. 259–260).

Finally, in another study, three- and four-year-old children were asked questions about novel artifacts. Then, some children were shown a function that plausibly accounted for the structural features of the object, whereas others were shown an implausible function. Children given plausible functions were more satisfied than those given implausible functions, because the latter asked more questions about function. This suggests that children seem to think in terms of intentional design when they think about functions, and that "they understand the true functions of the artifacts to be the design functions" (Asher and Kemler-Nelson, 2008).

Despite the conflicting findings and conclusions of the Keil and Kelemen studies, there are two common findings which should not be left unnoticed. First, in all studies children generally provided teleological explanations for organisms and artifacts. The major disagreement between the Keil and Kelemen studies is on whether teleological explanations are provided for non-living natural objects as well. Second, children may perceive animals as being different from artifacts, but this does not mean they perceive animal parts differently from artifact parts. In the Greif et al. (2006) study, children's questions about function were more frequent for animal parts than for whole animals, and overall the number of questions about parts was similar for organisms and artifacts (p. 458). Kelemen (1999b) has also reported that four- and five-year-olds provided teleological explanations for both animal and artifact parts, while they also realized that parts of organisms are more probable to have some use or function compared to whole organisms. In my view, this is very important. Both whole artifacts and artifact parts may be perceived to have some use (e.g., the airplane is useful to humans for travelling and the wings are useful to the plane for flying). In contrast, whole organisms have no (natural) use,[22] whereas their parts (at least some of them) can be useful to their possessors (e.g., eagles are not useful to anyone, but their wings are useful to them for flying).

Children may realize the differences between organisms and artifacts, but nevertheless think in teleological terms when they are asked to explain the use/function of their parts. It may be the case that children tend to explain the existence of specific parts of both organisms and artifacts in teleological terms, even if their overall perception of artifacts and organisms is different. While children may not provide teleological

---

[22] Humans use other organisms for various purposes but this is not, of course, why these organisms exist.

explanations for the existence of both airplanes and birds, they may do so for the wings of airplanes and the wings for birds. Certainly, further research is required, especially on understanding in more detail how children conceive of wholes and parts and the differences between these two conceptions. It is also necessary to study the content of children's teleological explanations in more detail (see Lennox and Kampourakis, 2013 for suggestions).

Based on all the above, I suggest that teleological, design-based, and artifact-like thinking about organisms *can* be an important conceptual obstacle to understanding evolution. If children explain the existence of the wings of birds in the same way they explain the existence of the wings of airplanes, it is important for them to realize from as young an age as possible that this is not the case. I suggest that education and public communication about the theory of evolution must clearly make this distinction in order to ensure that people do not think of organisms as artifacts.[23] Even if it is eventually shown that teleological thinking about organisms does not stem from an understanding of artifacts, it is useful to make the distinction between organisms and artifacts explicit and clear. I will return to suggestions about how this might be accomplished in the final section of this chapter. We now turn to another important obstacle that is also relevant to artifacts: psychological essentialism.

Psychological essentialism as a conceptual obstacle to understanding evolution

Both birds and airplanes have wings. Thus, we might plausibly subsume these under the same category: *objects with wings*. In a similar manner, we might also think of an elephant and a car sharing an important similarity that brings them under the same category: *objects without wings*. But we do not do this, because despite such similarities (having/not having wings) several other important differences exist between birds and airplanes, as well as between elephants and cars. Birds and elephants reproduce and develop, whereas airplanes and cars do not. Furthermore, as explained in the previous section, airplanes and cars are intentionally designed for a purpose whereas birds and elephants are not. Consequently, we would rather classify birds and elephants as organisms and airplanes and cars as artifacts. And we might distinguish between them due to some characteristic properties they have, which we might consider to constitute their *essence*. An essence can be defined as a set of properties that all members of the

---

[23] One might wonder at this point whether domesticated or genetically modified organisms should be considered as artifacts because they are intentionally modified by humans for some purpose. Human intervention in the case of artificial selection of domesticated animals differs enormously from genetic modification in the laboratory (resulting, for example, in the production of transgenic organisms). Humans may also use rocks in their original form to create a path in a river, or modify them extensively to produce objects of art. However, most natural entities (both organisms and non-living natural objects) are not modified by humans and most importantly come to existence in nature without any human intervention. Therefore, those cases in which humans modify organisms or non-living natural objects will be treated here as exceptions.

kind must have, and the combination of which only members of the kind do, in fact, have (Wilkins, 2013). I must note that in this definition these underlying properties are not necessarily fixed. Before I proceed I must also note that there exists some confusion around the proper meaning of the term essentialism and that different meanings have been given to the term (see Gelman and Hirschfeld, 1999; Gelman and Rhodes, 2012; see also Wilkins, 2013 for a detailed discussion).[24]

What could the essence of artifacts be? Could it be their appearance? Think of a kitchen fork. A kitchen fork has a prong. Can we claim that whatever has a prong is a kitchen fork? No, because pitchforks also have prongs. Does it make a difference whether the prong is made up of three or four tines or more? No, because although kitchen forks usually have four tines, those for babies may have fewer. And although a pitchfork may look like a kitchen fork, we would not call it one. The reason is that we know that a kitchen fork is an object we use in order to eat, whereas a pitchfork is an object we use in order to clean our yard. In a similar manner we can certainly distinguish between a knife and a sword, and we would never ask for a sword to cut the bread for dinner. How about chairs? We cannot define a chair as an object that has four legs, because tables also have four legs. And if we also think that there exist kitchen chairs, office chairs, wheel chairs, arm chairs, etc., we can realize that there is no single way we can provide a general description of chairs based on their appearance. What, then, is the essence of artifacts? It seems reasonable to suggest that the essence of artifacts is determined by their intended use (Bloom, 2004, p. 55). It is their intended use, or in other words what they were made for, that makes them distinct from each other.

As already explained in the previous section, artifacts are by definition objects created for an intended use. Thus, we distinguish between a kitchen fork and a pitchfork, or between a knife and a sword on the basis of their intended use, and not of their shape – it may be the case that a knife and a sword differ only in size, while they may be quite similar otherwise. We are also able to identify a chair from a table, although they both have four legs because of their intended use. We would not normally sit on a table and we would not put our dishes on a chair in order to have lunch. And if we used two chairs for dinner, one to sit on and another to put our dishes on instead of a table, the latter would still be a chair even though we used it as a table. Similarly, we can distinguish between a football, a basketball, and a volleyball; they not only have different colors, but different sizes and weights that are appropriate for the respective games. We might use a volleyball to play football, but it would still be a volleyball. How about using a chair and a ball and other materials to create a scarecrow? We might never use that chair again to sit on or the ball to play. Nevertheless, the chair would still be a chair and the ball would still be a ball. In short, the identity of artifacts is largely determined by their intended use because this in turn determines their particular

[24] There is a large philosophical literature about essentialism and its effect on taxonomy. I will not get into this discussion here, nor the discussion about the proper definition of species, as excellent book-length analyses of these topics are available (Wilkins, 2009; Richards, 2010). It should be noted, though, that the claim that biological taxa must have necessary shared characters no other kind or taxon does is described as taxic or biological essentialism and this is distinct from the psychological essentialism which is the focus of this chapter.

characteristics. Artifacts are designed for an intended use which must be served by their features, and so the latter reflect this use.

One might argue at this point that artifacts do change because they may rust, rot, decay, or more generally undergo changes in their appearance or internal structure. However, even then they still retain some properties relevant to their intended use. A rusted chair is still a chair even if we never sit on it again. Now, if we broke the chair into pieces and made a new artifact out of it, e.g., connect the legs to each other to make a billiard stick, most people might recognize that it is not a genuine billiard stick but one made of chair legs. If, on the other hand, we processed the chair legs entirely, then we would have made a new artifact because of our action that gave to the material a new intended use. In other words, the essence of artifacts would change only when the change in appearance was due to some human intervention and processing with a new intended use in mind. Otherwise, any change that artifacts undergo would be superficial; the initial parts would be there, even if they were somehow different from their initial state or condition. It is in this sense that artifact essences are considered to be fixed: The initial intended use is evident in their structure until they are consciously transformed with a new intended use in the mind of the human who makes the transformation.

What is the essence of organisms? We usually perceive particular activities/properties as essential for organisms, i.e., characteristic of them.[25] Organisms reproduce, develop, respire, digest food, excrete waste products of metabolism, react to stimuli, etc. We are also able to distinguish between particular types of organisms. It is often easy to distinguish between tigers and lions because the former have stripes whereas the latter do not. It is also easy to distinguish between rhinoceroses and hippopotami because the former have a horn whereas the latter do not. But are these "essential" characters? The answer is no. Otherwise, we might claim that all animals with stripes are tigers (but of course they are not; for instance, zebras also have stripes), as well as that all animals without stripes are lions (but many animals such as horses do not have stripes). Or, we might claim that all animals with horns are rhinoceroses (but what about bulls, goats, reindeer, etc.?) as well as that all animals without horns are hippopotami (they are not of course, as many animals such as horses, donkeys, sheep, etc. do not have horns). How about using more specific characteristics, such as those used in **taxonomy**? Could we distinguish between birds and mammals on the basis that birds have feathers whereas mammals do not, and that mammals have hair (or at least hair follicles) whereas birds do not? This is a more reasonable approach and one that actually helps us categorize individual organisms in classes. But this approach may also face problems: how should we classify *Archaeopteryx*? As a bird (because it has feathers) or as a reptile (because it has a dinosaurian skeleton)?[26]

[25] Essential could also mean necessary for them to live, but this is not the sense I am using here, although all the characteristic properties of organisms are indeed necessary for them to live.
[26] *Archaeopteryx* is today classified among Mesozoic birds, from which modern birds evolved, all belonging to the class Aves (see Prothero, 2007, pp. 257–268).

I will not argue that organisms definitely have essences, i.e., a distinctive set of properties that all members of some kind must have. Rather, I will argue what these essences could be, were they to exist at all. It should be emphasized at this point that organisms' essences have nothing to do with souls or any other transcendental notions. Apparently, if organisms have essences they must be at some deep level (that is by definition what an essence is about) and of course cannot coincide with external features. An interesting suggestion is that these essences are related to the developmental capacities of organisms.[27] The characters of organisms are the outcome of development on the basis of a particular **genetic material** expressed under particular environmental conditions: "Organismal natures – the goal directed capacities of organisms to develop and maintain viability, *given* the material resources at their disposal – play an ineliminable role in the explanation of adaptive evolution. Organismal natures are implicated in the explanation of crucial features of ontogeny – its robustness, plasticity and adaptiveness" (Walsh, 2006, pp. 444–445, emphasis in the original).[28]

So, assuming that both artifacts and organisms have essences, is there any difference between them? There are many differences: organisms are not designed and do not have intended uses; artifacts are not alive and do not develop, and so they do not have developmental capacities. A main consequence is that there are important differences in how their essences may change, both in the short and the long term. Artifact essences are more fixed compared to organisms' essences. A chair is designed to be used for sitting, and the intention of the artificer cannot change once it is created. We may adopt a chair for alternative uses, e.g., as a table if we have two chairs but no table. We might also modify a chair to make it look like a small table (by taking away its upper part). More generally, we might modify any chair and co-opt it for several desired uses. But it would still be a chair, albeit a modified one. Artifacts can also evolve, but this is a case of cultural evolution as artifacts are the products of human culture. For instance, think of the primitive tools humans once used for cutting. These initially were stone tools that were hewed simply by being hit on larger stones. Later, these stone tools were replaced by copper or iron ones which were more efficient cutters. Nowadays an enormous variety of knives and other cutting tools exists. As I understand it, today's artifacts are the outcome of artificial selection, which is a conscious, intentional process. Over the years, people have been modifying artifacts or creating new ones, and eventually selected to keep creating the ones which were more useful and stopped creating the less useful ones. None of the cars produced in the beginning of the

---

[27] Developmental capacities are usually considered as properties of multicellular organisms. This description of "organismal natures" usually overlooks the majority of organisms that are unicellular (see Duncan *et al.*, 2013 for a philosophical analysis). However, a wider perception of development might also include unicellular organisms. For example, changes from one phase to a morphologically different one occur regularly and predictably in many protozoans such as the *Trypanosoma*, so this could be regarded as development, too (see Minelli, 2011). I address this issue in Chapter 5.

[28] One important aspect of development is that it is characterized both by robustness (individuals exhibit the general characteristics of a species irrespective of the environment they live in) and plasticity (individuals of the same species with the same genotype may exhibit phenotypic variation depending on local conditions). Robustness and plasticity are complementary aspects of development (see Bateson and Gluckman, 2011, pp. 4–5). I will return to this topic in Chapter 5.

twentieth century are produced by car industries today, as no one would buy them except some romantic car lovers. The cars produced today are much faster, safer, and friendlier to the environment, and these are the outcomes of artificial selection in the car industry over the past 100 or so years. Finally, artifacts do not share a common ancestry. They were developed at different times, under different conditions for different purposes. Particular designs were implemented in each case, and the several types of artifacts have had independent origins from the others. And even if some primitive cutting tools evolved to more modern ones, their lineage is entirely independent from that of primitive means of transportation that evolved to contemporary cars. Common ancestry among artifacts is possible, but not obligatory.

Organisms differ from artifacts in all these respects. Organisms can undergo changes in their essences more easily and more drastically than artifacts. We cannot take away the upper part of an animal or plant to make it a different one, nor can we turn a pig into a lion, no matter how much we try. However, we are able to modify organisms genetically, either by selective breeding or by genetic engineering. These processes involve significant changes or alterations in the genetic material of the organisms, which is part of their essence (assuming there is one). If we insert human DNA into *Escherichia coli* in order to produce insulin, these bacteria will not be *E. coli* anymore. Rather, these will thereafter be genetically modified bacteria that upon further modification might end up differing significantly from the original strains. We are also able to make modifications to larger organisms such as animals, thereby producing chimeras, animals that consist of cells originating from different species. Consequently, organisms have fundamental (developmental) properties (essences) which are less fixed than those artifacts. It is exactly this that makes the evolution of organisms possible. With various phenomena that cause genome alterations (mutations, **horizontal gene transfer**, genome acquisitions, etc.) and through several processes (natural selection, drift, etc.)[29] individual organisms can change significantly and thus populations can evolve.[30] Evolution is a purposeless, unintentional process that depends on the genetic material and the developmental potential of organisms, as well as on their particular environment. What is also important is that all organisms, contrary to artifacts, share a common ancestry because they have evolved from common ancestors (see Chapter 5 for details). This is why some fundamental characters are common to all organisms. Perhaps the most important ones are that all organisms have DNA as their genetic material, that all organisms have cellular structure, and that many of them can undergo developmental changes (Figure 3.6).[31] Common ancestry among organisms is the outcome of biological evolution.

---

[29] These phenomena and processes are discussed in detail in Chapters 5 and 6.

[30] It should be noted that essences can be attributed to individual organisms, not populations. There are differences between individual essences and there is no point trying to describe an average essence. When individual essences change, however, population structure (in terms of genetic material and developmental potential) also changes and consequently evolution can occur.

[31] These fundamental features, common (but not identical) to all organisms, are those that can be considered to make up their essence.

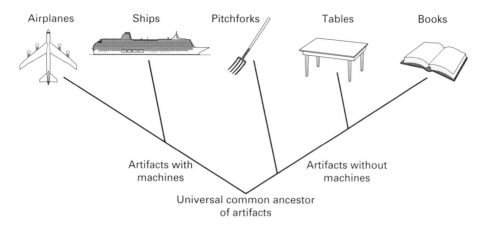

Figure 3.6 An imaginary depiction of several artifacts sharing common ancestry. This depiction is, of course, incorrect as these artifacts have independent origins. They were designed and made independently for different uses. However, do you realize that organisms as diverse in appearance as these artifacts all share a common ancestry and various fundamental common characters? Image © Simon Tegg.

There exists an enormous body of research which suggests that people tend to intuitively provide essentialist explanations for the characters of organisms from very early in childhood (see Gelman, 2003 for a detailed discussion, and Gelman, 2004 for a review). Gelman suggests that children think that internal causes are more appropriate for organisms and external causes are more appropriate for artifacts (Gelman, 2003, pp. 121–123). Interestingly enough, Gelman also suggests that children are not essentialist about artifacts, at least in the way they are for organisms. She concludes that **essentialism** is first and foremost characteristic of organisms, whereas it is more complicated to reach a conclusion about artifacts. But she also notes that essentialism is characteristic of both complex and simple artifacts (p. 138). This tendency is described as *psychological essentialism*, according to which certain categories are real rather than human constructions and they possess an underlying causal force, the essence, which is responsible for why category members are the way they are and share so many properties (Gelman and Rhodes, 2012, p. 5).

Several studies suggest that children generally think of organisms in essentialist terms. For example, in one study, four-year-old children were shown pictures of a colorful tropical fish, a gray dolphin, and a gray shark. Children were shown the picture of the fish and were told that it breathed underwater; they were shown the picture of the dolphin and were told that it should pop out of the water in order to breathe. They were then shown the picture of the shark (who looked similar to the dolphin) and were asked if it breathed like the fish or like the dolphin. Many children answered that the shark breathed like the fish, and more generally they based their answers on category membership (the tropical fish and the shark were both fishes) and not appearance (the shark and the dolphin were gray and had similar shapes and sizes). This is an important finding that has been replicated with younger children (two-year-olds). It seems that young children with no scientific or other relevant training make inferences about an animal based on another animal which they consider to belong in the same category (Gelman and Markman, 1986; Gelman, 2003, pp. 28–33). In a follow-up study, three- and four-year-olds overall made inferences based on category membership, not external appearance. Among the items shown were an insect (beetle), a leaf, and a leaf-like insect. The last two were both large and green, with striped markings. However, despite the similarities between them, children drew inferences from the leaf-like insect to the beetle because they could determine that both were bugs. In addition, they did not draw inferences from the leaf-like insect to the leaf because they noticed the antennae on the former. So, again, children's inferences were based on category membership and not apparent similarity (Gelman and Markman, 1987; Gelman, 2003, pp. 37–39). To sum up, these studies suggest that children seem to infer that members of a category share some underlying, non-obvious properties, and they use these to draw inferences.

In a very interesting study it was found that children thought that animals, contrary to artifacts, retain their essential properties despite transformations they may undergo. In particular, the following main types of transformations were shown to children (kindergarten, second grade, and fourth grade): animals into animals (e.g., horse into zebra), plants into plants (e.g., rose into daisy), non-living natural objects into non-living natural objects (e.g., salt to sand), artifacts into artifacts (e.g., bridge into table), animals into plants (e.g., squirrel into moss), machines into animals (e.g., toy mouse into real mouse), and animals into non-living natural objects (e.g., fish into stone). The main aim of this study was to assess how much children were basing their categorization of objects on apparent features. Both objects of each pair were portrayed in good-quality photographs against similar backgrounds, having similar orientations and size. In all cases, children were told that a scientist took the first object of a pair and performed appropriate operations to turn it into the second object of the pair. The results were extremely suggestive: children of all grade levels answered that kind was preserved in animal into non-living natural objects (e.g., although a hippopotamus looked like a big rock, it had not turned into one) and animal into plant transformations (e.g., although a squirrel looked like a moss plant, it had not turned into one), despite apparent similarities. They also thought that kind was preserved in machine into animal transformations (if an entity had a machine inside, it could not be a real animal, e.g., a toy bird could not turn into a real bird). In contrast, they thought that kind was not preserved in artifact into artifact transformations. Fourth-grade children also clearly thought that kind was preserved in animal into animal, plant into plant, and non-living natural object to non-living natural object transformations. However, this was not the case for younger children, especially kindergarten children, who overall thought that kind had changed. It should be noted, though, that children were not always consistent in their answers (Keil, 1989, pp. 197–215). Overall, it seems that second-grade and especially fourth-grade children think that kind identity persists despite changes in appearance. This is strong evidence that children think about organisms in terms of underlying, unchangeable essences.

An important question was whether kindergartners in the previous study did not think that identity was maintained because they did not understand the nature of the transformations performed. Thus, a new study was conducted with kindergartners, second-grade, and fourth-grade students. Although the same photographs were used, so that the beginning and end-state characteristics were the same for each pair as in the previous study, children were told that a different kind of transformation had taken place. Instead of being told that a scientist made changes in an animal (e.g., that he/she put black and white stripes on a horse, or that he/she taught it to run away from people and to live in the wild part of Africa rather than in a stable), children were told that the animal was dressed in a costume or that a superficial transformation had taken place so that it eventually resembled another animal (e.g., that a man was painting black stripes on his white horse every week). One important difference from the previous study was not only that the transformation was more superficial, but also the children were told that it had to be repeated regularly to ensure the animal would not revert to its initial state. These two new conditions (superficiality of transformation and the need for regular repetition) were expected to make even kindergartners suggest that kind had not changed. As was expected, it was found that even the youngest children denied that any change in kind had taken place. The results for artifacts were similar to those in the previous study. These findings seem to indicate that even pre-school children do not rely on external appearance but on some deeper properties in order to decide about whether a change in kind has occurred (Keil, 1989, pp. 217–236).

These results suggest that children do not consider external features as characteristic of organisms' identities. Does this mean that they consider internal parts of organisms as more privileged? To investigate this Gelman and Wellman (1991) asked four- and five-year-olds to consider particular transformations during which either internal or external parts were removed. For example, they asked whether the identity of a dog would change and whether it would still bark and eat dog food (a) if inside parts such as blood and bones were removed but the outside parts were left intact, or (b) if outside parts such as fur were removed but the inside parts were left intact. The questions also included containers, like refrigerators and jars, the insides of which are not their integral parts. Children answered that the identity of containers would not change if their inside parts were removed. However, in the case of entities for which inside parts are important, such as a dog or a car, the engine of which is more important for its function compared to its paint, children thought that inside parts were significantly more important compared to outside parts. These results again suggest that children consider internal features as more important than external ones. Again, it is shown that children think that some deeper features/properties are more important than external ones (Gelman, 2003, pp. 79–81).

An interesting question that comes next is: What do children think about biological transformations that occur in nature? Both evolution and development are processes of biological transformation, but it is the latter that is easily observed, even by young children. An interesting case is that of some animals whose development includes extensive transformations, known as metamorphoses (e.g., butterflies). Do children think that natural biological transformations lead to identity change? In a study, three- and five-year-old children were initially shown the picture of a caterpillar. Then they were shown another picture identical to the first one, and one that was the same but larger in size, and were asked which one represented the adult form. All of the children chose the larger figure, which shows that they thought that growth does not affect identity. In another task, children were shown a set of pictures: first the picture of a caterpillar and then the picture of the same caterpillar which was smaller in size, together with the picture of a moth which was larger but very different from the caterpillar. A significant number of five-year-olds chose the moth as the adult form of the caterpillar. This result suggests that by the age of five years children realize that organisms can undergo radical changes without a change in their identity (Rosengren *et al.*, 1991; Gelman, 2003, p. 64).

These research results primarily suggest that children intuitively think about organisms and artifacts in exactly the opposite way to the way they should. They think it is organisms and not artifacts that have fixed essences. They perceive organisms as capable of undergoing changes in their external features without undergoing a change in their identity. In contrast, children seem to think of artifacts as undergoing changes in identity when they simply change shape or form. Overall, children perceive that particular properties are characteristic ("essential") to organisms (something inside them) and to artifacts (their intended use), but think about them in exactly the opposite terms: They consider organisms as having fixed properties that they cannot change, and think exactly the opposite about artifacts.

Based on the research results presented so far, one might reasonably conclude that psychological essentialism, the intuition that organisms have fixed essences and that they do not undergo a change in kind even when their external features change, is a major obstacle to understanding evolution. Indeed, if children think that organisms have fixed essences and that they cannot change kind, then it is difficult for them to understand the idea of evolutionary change. However, this is part of the problem. Transformations such as those presented to children (horse to zebra – see Figure 3.7 – or caterpillar to moth) are very different from the changes that take place during

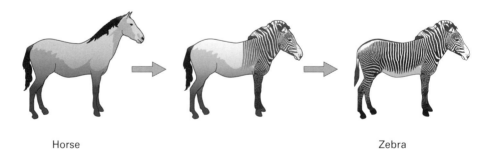

Horse   Zebra

Figure 3.7 A representation of stimuli used in Keil's transformation studies. Children thought that a horse would be a horse despite changes in external appearance that made it look like a zebra. It is important to note that this picture does not portray evolutionary change. It just presents changes in an individual, but this is not how evolution proceeds. Even if one imagined that this figure portrays changes across a lineage, it is still deficient as it neglects the other lineages evolving at the same time. So children's essentialist bias is actually a denial that individual essences undergo change. This is consistent with evolutionary theory, which suggests that evolutionary changes take place across generations and within lineages, not during individual lives. Image © Simon Tegg.

evolution. Evolution takes place through changes in populations and not in individuals. These studies reveal not only that children think that organisms cannot change significantly, but also that they do not realize that organisms of the same kind/category may exhibit an enormous amount of variation. Therefore, the major obstacle that essentialism poses to understanding evolution is not only that some "essential" properties are fixed, but also that these are perceived to be identical in the members of the same kind. However, this is not the case. Developmental processes produce both common (not identical!) characters and variable ones.[32]

Conceptual change in evolution

In the two previous sections, results from conceptual development research have been presented which suggest that young children exhibit particular teleological and essentialist intuitions, referred to as design teleology and psychological essentialism. These may form important conceptual obstacles to understanding evolution, and the researchers themselves acknowledge this. For example, Gelman and Rhodes (2012) have suggested that there are at least four ways in which essentialism may pose obstacles to understanding evolutionary theory: (1) the assumption that categories are stable and immutable is in conflict with the idea that species can evolve and change over time; (2) the tendency to intensify category boundaries makes it difficult to understand that two species may have a common ancestor; (3) essentialism may make people underestimate variation within a category and so make it difficult for them to understand how natural selection, which requires variation, operates; and (4) essentialism reinforces a focus on inherent causes within individuals rather than on the characteristics of a population, and this leads to a misunderstanding of evolution. Kelemen (2012) has also made particular suggestions about why teleological intuitions may form a conceptual obstacle to understanding evolution. One is that children may generally view natural phenomena as existing for a purpose due to underlying intuitions that make them believe that such phenomena derive from intentional design. Alternatively, children's generalized tendency to ascribe functions to natural entities may result from a basic, low-level cognitive mechanism which makes children come to view entities as made "for" a purpose based on simple cues about functional utility.

One problem with the conclusions made by researchers of conceptual development is that they usually identify evolutionary theory with Darwin's theory of natural selection. This is problematic, but I will overlook it for now. Darwin's theory and its development are presented in Chapter 4, and in Chapters 5 and 6 I explain in detail that there is much more in evolutionary theory than Darwin and natural selection. However, both Gelman and Rhodes (2012) and Kelemen (2012) make useful suggestions about how the intuitions revealed by their research may form serious obstacles to understanding evolution. There is concrete evidence that teleological and essentialist intuitions serve

---

[32] This is relevant to the concepts of robustness and plasticity that will be discussed in detail in Chapter 5.

as obstacles to an accurate understanding of evolutionary concepts (see Evans, 2008 for an overview and a developmental perspective; see Kampourakis and Zogza, 2007, 2008, 2009 for teleology; see Shtulman, 2006; Shtulman and Schulz, 2008 for essentialism). Overall, it seems that evolution instruction has been moderately effective (see Smith, 2010 for an overview). There are two distinct reasons for this: (1) strongly held intuitions produce misconceptions which are resistant to change and (2) the instructional approaches used so far to promote conceptual change may have not been appropriate to address these misconceptions. In this section I will refrain from discussing science education issues in detail and will rather focus on the conceptual obstacles and describe some requirements for conceptual change.[33]

In the two previous sections I described two major differences between organisms and artifacts, relevant to their origins and to their fundamental properties. Artifacts are designed and created for an intended use. Consequently, artifacts have parts that serve their intended use, and it is this which may be perceived as their "essence." In contrast, organisms are neither designed nor created for any intended use. If they have some parts that seemingly serve some use (which is entirely unintended), these have emerged through a long evolutionary history and may have been maintained through selection processes. The outcome of this history is a specific genetic material and developmental potential, which could be considered as their "essence." These differences are crucial for understanding in what sense organisms differ from artifacts. The studies from conceptual development research reviewed in the previous sections suggest that from a very young age children (1) provide teleological explanations for both organisms and artifacts, and (2) think that organisms have more fixed essences than artifacts.

What is the problem with design teleology? As already outlined in the respective section, teleological intuitions in combination with an early awareness of intentionality may intuitively make us think about the parts of organisms in the same way we think about the parts of artifacts. This does not necessarily mean that we consciously consider organisms as (divine) artifacts. Rather, it may mean that we unconsciously think of both organism and artifact parts in terms of intended uses, because this is what we perceive them to perform. For example, seeing an eagle flying does not make us think of it as an artifact, as we would do for an airplane. However, it may be the case that seeing an eagle flying by using its wings makes us think of them (the wings) as parts which exist for this particular use, in a similar way that we think of the wings of an airplane as existing for the same use (flying): as the product of some intention. Whether there is an intentional agent in the case of organisms as there is in the case for artifacts is a distinct question. It could be the case that the thought that an intentional agent may be involved in the emergence of organisms comes after the thought that their parts serve an intended use. Remember that in Chapter 2 I explained

---

[33] Overall, studies of conceptual change in evolution have largely focused on secondary and tertiary education. However, if conceptual obstacles have their roots in childhood, the focus of evolution education research should be on how these early biases can be addressed during instruction at elementary school (see Kampourakis *et al.*, 2012b for a proposed research program).

that Paley first acknowledged the analogy between a watch and organisms or between an eye and a telescope and then made the inference to the existence of a Creator. It is the unconscious idea of an intended use, *intended for* some purpose (and perhaps *intended by* someone) that makes design teleology an important conceptual obstacle to understanding evolution – in my view the most important one. The main issue is not whether a part of an organism exists for some purpose,[34] but whether it was intentionally made to fulfill this purpose.

What is the problem with psychological essentialism? As I have already outlined in the previous section, the problem is not only that we may think that change is impossible because organisms are perceived to have fixed essences, but also that the notion of a group of characteristic properties that also determine kind/category membership/ identity does not let us realize how enormous is the variation that exists within each kind/category. There also exists an important linguistic issue here that I would like to address, which has to do with how we refer to kinds/categories and their members. Consider the following sentences: (1) *The* eagle is flying by using its wings; (2) *An* eagle is flying by using its wings. Although these sentences seem similar, they nevertheless have different referents. "*The* eagle" implicitly refers to an exemplar, which possesses all those ("essential") properties required in order to classify a bird as an eagle. It also imposes the notion of a prototype to which all individuals of the kind/ category must be identical.[35] In contrast, the phrase "*an* eagle" implicitly refers to a particular individual only, which allows the possibility that it may differ from some other individuals. It seems that we intuitively tend to refer to prototypes or exemplars, and overlook the enormous within-group variation that actually exists in nature. We intuitively think that all individuals belonging to the same kind/category must be "essentially" the same. But if this was the case, how could evolution occur? The particular mechanism notwithstanding, both the direction and the outcome of evolutionary change depends on the existence of within-group natural variation. If all members of a kind/category all had the same fixed "essential" characters, small changes in "non-essential" ones might be possible, but more substantive changes that include "essential" ones would not. But this is what evolution is about: substantive changes that take place over long periods of time.

To sum up, particular teleological and essentialist intuitions produce the following (mis)conception: organisms of a particular kind/category have particular fixed characters which exist for some intended use. But this is true for artifacts, not for organisms, as I have already explained. So, here is one very important issue that causes misunderstanding of evolution: *artifact thinking, or (unconsciously) thinking about the parts of organisms as if they had specific intended uses.* Let us consciously oversimplify and describe this teleological–essentialist bias in a single sentence: *The eagle has wings in order to* fly. Note that I have put emphasis on the words that

---

[34] Natural selection can be invoked to develop a purely naturalistic explanation for this (see Lennox and Kampourakis, 2013). In this case there is a use for a purpose but *no intention* behind it.

[35] Gelman and Rhodes (2012, p. 14) argue that such a (Platonic) notion of an ideal essence may also cause misconceptions about evolution.

represent the essentialist bias (*the*) and the teleological bias (*in order to*). But my analysis above suggests that this sentence is wrong; it should be re-written as: Eagles fly *because* they have wings. The biases have now been eliminated and I have also added a causal connection between the parts (wings) and the use (flying). In a nutshell, this is what conceptual change in evolution is about: the explanation for the existence of wings has changed. There is no exemplar/representative/prototype eagle, but numerous eagles that differ from each other in various characters, including the ones we consider characteristic of their kind. Not all eagles have exactly the same wing length or equally efficient stereoscopic vision. In addition, eagles do not have wings in order to fly because wings are not always used *for* flying; penguins use them for swimming whereas ostriches do not use them at all. Eagles fly because they have the wings they do which are of the appropriate size (and they also have the appropriate skeletons, muscles, etc.).

I should note at this point that the teleological component of this explanation is rather tricky, and for this reason I think that teleology is the most important obstacle to understanding evolution. One can claim that eagles have *the particular wings they have* for flying because these wings may be maintained by natural selection as their effect (flying) has a positive effect for their possessors, which as a consequence survive and reproduce better. But this is different from stating that eagles have wings in order to fly, because given the human intuitions discussed above this may be taken to imply that all birds have wings in order to fly. Children tend to explain new instances based on what they already know. If they see a new airplane that they have never seen before, they might explain that it has wings in order to fly because this is what they know about the other airplanes they have seen so far. Such generalizations may not be sound all the time, but children could make them. Now, if children learn that eagles have wings in order to fly, they might think the same for every new bird they come across. But this is not the case: penguins use their wings for swimming, whereas those of ostriches do not seem to have any use at all. Consequently, such cases of birds which do not use their wings to fly make the proposition "*the* eagle has wings *in order to* fly" rather problematic.[36] In contrast, the proposition "eagles fly *because of* the wings they have" explains the contribution of wings and is also compatible with the conflicting observations (penguins, ostriches possessing wings but not flying). In this case, we explain function (flying) in terms of structure (wings): Eagles fly because they have appropriate wings. I suggest that it is better to avoid the implications of the statement that a structure (wings) exists in order to perform a function (flying) which, I repeat, is not wrong but can be problematic given human intuitions.

Now, how is conceptual change possible? Considering our explanations against the available evidence is very important. It is conceptual conflict – conflict between competing explanations for the available evidence – that may lead to conceptual change. In the case of evolution, as in the case of Copernicus, conceptual change does not

---

[36] But not necessarily wrong, as long as reference to natural selection is made (see Lennox and Kampourakis, 2013).

Dolphin

Shark

Figure 3.8 Dolphins and sharks have similar hydrodynamic shapes not because they were designed in order to be able to swim fast underwater, but because they have independently evolved similar shapes which have advantageous consequences for underwater life. Image © Simon Tegg.

involve introduction of new concepts but restructuring of an old explanatory scheme to produce a new one that accounts better for the observed phenomena. I should note that conceptual change in evolution can be understood in different ways. One way is to describe it as a change in the concept of adaptation. The concept of adaptation that children and adults intuitively form is one that assumes consciousness and intentional design. Thus, conceptual change in evolution could be the change in the concept of adaptation from being the outcome of conscious, intentional design to being the outcome of evolutionary processes, most importantly natural selection.[37]

In order to give a more philosophically sophisticated description of conceptual change, let us consider another example, comparing two characters of dolphins and sharks: their shapes and their way of breathing. By referring to dolphins and sharks, and not to "the dolphin" or to "the shark," the essentialist obstacle is addressed: reference is made to a population (i.e., dolphins – plural) and not to individuals (i.e., dolphin – singular). Having done this, the focus will then be on teleology and I will describe how conceptual conflict may occur that might lead to conceptual change. If one asked why dolphins and sharks have hydrodynamic shapes (Figure 3.8), an intuitive answer would be: *In order to swim fast underwater*. Dolphins and sharks are relatively large marine organisms which manage to swim fast thanks to their hydrodynamic shapes. So, one could intuitively think that it is no coincidence that dolphins and sharks have similar hydrodynamic shapes: These are useful in order for them to swim fast underwater. Dolphins and sharks have similar sizes, are predators, and face similar difficulties and challenges in the underwater environment in which they live. They can survive if they can catch their prey and overcome competition, so swimming fast is a way to achieve this. Therefore, both dolphins and sharks have similar shapes because they face the same problems and so they ended up with the same solution. We might summarize this

---

[37] I do not mean to suggest that the type of conceptual change I am describing here is the only one or the most important one. This is just the one which I consider more crucial.

account as follows: The explanation for the question "Why do organisms O have character A?" is:

[D] Organisms O have character A in order to perform function B.[38]

Is there any conflicting evidence for this proposition? The answer is yes. First, one might wonder why dolphins do not have gills as sharks and most other organisms living underwater do. Why is it that sharks, but not dolphins, have this additional useful feature? I remember once watching a documentary film about gray whales. At some point a big gray whale was swimming in the ocean, close to the surface, with its newborn that was barely the size of a big dolphin. The newborn was swimming very close to its mother's body. Then, suddenly, two orcas – or killer-whales – approached them and started trying to separate the newborn from the mother. The orcas did not get very close to the mother whale as it could hit them hard, and so tried for a long time to separate her and the newborn. Eventually they succeed, and then they repeatedly pushed the newborn into the sea until it drowned. But this would not have happened if gray whales had gills. So why don't they? Organisms may possess characters that are required for their survival; however, they have neither optimal characters nor ones that fulfill every possible need. More generally, the proposition that "Organisms O have structure A (or A_2 or A_3, etc.) in order to perform function B (or B_2 or B_3)" is not correct (unless natural selection is explicitly invoked – note that this is a psychological, not a philosophical assumption). Organisms are not intelligently designed, and not only do they not have what they need, but also exhibit numerous imperfections (Williams, 1996). In addition, dolphins and sharks differ significantly in many characters. Dolphins have forelimbs, whereas sharks have fins; dolphins have mammary glands whereas sharks do not; dolphins have lungs whereas sharks have gills; dolphins have blowholes whereas sharks do not; and many more. Why would two kinds of organisms that live in the same environment be so different from each other? Not to mention numerous other marine species which do not have hydrodynamic shapes and are thus very different from dolphins.

So, the initial explanation, which was based on proposition D, cannot sufficiently account for the differences observed. What one may then do is to look for alternative explanations and see how they account for the available data. One option is that dolphins and sharks happened to have shapes that provide an advantage in the environment in which they live. This advantageous feature has been maintained in their lineage *because* of the advantage it provides to its possessors. And here we are. This novel explanation can be summarized in the form of the following proposition:

[E] Organisms O have character A because it performs function B, which confers an advantage and consequently this character has been maintained in their lineage.[39]

In this case the explanation is not based on any intention (*in order to*) but on the consequences that the particular structure has (*because*). This new explanation is more

---

[38] I call this proposition D because it is based on the assumption of intentional design (*in order to*).
[39] I call this proposition E because it is based on the assumption of natural, evolutionary processes.

legitimate because it is free of the assumption of intentional design and only depends on the contribution of a feature to its bearers. Thus, it is compatible with imperfections, too. Apparently, this proposition relies on evolutionary history and this reliance will be discussed in detail in Chapter 6. Let us compare the two schemes with the following questions:

(1) Why do dolphins have hydrodynamic shapes?

[D1] Dolphins have hydrodynamic shapes in order to swim fast underwater.

[E1] Dolphins have hydrodynamic shapes because they help them swim fast underwater, which confers an advantage and consequently this character has been maintained in their lineage.

(2) Why do sharks have hydrodynamic shapes?

[D2] Sharks have hydrodynamic shapes in order to swim fast underwater.

[E2] Sharks have hydrodynamic shapes because they help them swim fast underwater, which confers an advantage and consequently this character has been maintained in their lineage.

(3) Why don't dolphins have gills?

[D3] Dolphins do not have gills, but have lungs in order to get more oxygen directly from the atmosphere.

[E3] Dolphins do not have gills because this character was not maintained in their lineage and because lungs evolved.

(4) Why do sharks have gills?

[D4] Sharks have gills in order to breathe underwater.

[E4] Sharks have gills because they help them breathe underwater, which confers an advantage and consequently this character has been maintained in their lineage.

Apparently, proposition D1 is compatible with D2 and E1 is compatible with E2. However, propositions D3 and D4 are incompatible. Why would two organisms, which both live underwater, have different organs for breathing had they been designed (or, more generally, were formed in a way that satisfies their needs)? On the other hand, propositions E3 and E4 are compatible with each other. So, when the explanatory scheme E is used, it produces propositions E1 to E4 which are all compatible with each other. In contrast, when the explanatory scheme D is used, some of the propositions produced (in particular propositions D3 and D4) are incompatible.[40]

It is exactly at this point that conceptual conflict arises. The old explanatory scheme (D) does not account sufficiently for the phenomena observed. If this happens and if also a new, more sufficient explanatory scheme (E) is proposed, then conceptual change (in the form of the restructuring of explanations) may occur. The realization that an old explanatory scheme is insufficient and that a new one explains better may lead to the

---

[40] This is what Paul Thagard (1992, pp. 64–65) has described as explanatory coherence. Two propositions P and Q cohere if they are analogous in the explanations they respectively give of some R and S. On the other hand, two propositions are incoherent if they contradict each other or if they offer competing explanations.

replacement of the former by the latter. Therefore, conceptual change in evolution consists of a restructuring of intuitive explanations. The initial design teleological-psychological essentialist explanations that had the form "The dolphin has a hydro-dynamic shape in order to swim fast underwater" changed to a new one that is compatible with the available evidence: "Dolphins swim fast underwater because they have evolved to have hydrodynamic shapes." In short, conceptual change in evolution as described here consists of the restructuring of explanations based on proposition D toward explanations based on proposition E.[41]

Conclusions

To summarize, I have argued that conceptual change in evolution has two important prerequisites: (1) understanding that organisms do not have useful characters *in order to* perform some function, but that organisms perform functions *because* they have particular characters which are useful, (2) realizing that organisms grouped together under the same category (kind, group but not necessarily species) may differ signifi-cantly from each other, and it is because of such differences, i.e., the existence of (inherited) variation among members of the same category, that evolutionary change can occur. In short, reference to populations and reliance on evolutionary processes produces better and more efficient explanations than reference to types or kinds and reliance on intentional design. In my view, a shift from the latter to the former is what conceptual change in evolution is about. Let me note again that teleology in the wider sense is characteristic of evolutionary explanations based on natural selection. Thus, generally speaking it is not wrong to state that we have legs in order to walk or that we

---

[41] Propositions E and D can, of course, be applied to the initial example as well:

(1) Why do eagles have wings?

[D1] Eagles have wings in order to fly.

[E1] Eagles have wings because they help them fly and as a result this feature was maintained in their lineage.

(2) Why do penguins have wings?

[D2] Penguins have wings in order to swim.

[E2] Penguins have wings because they help them swim and as a result this feature was maintained in their lineage.

(3) Why do ostriches have wings?

[D3] Ostriches have wings in order to ... (?)

[E3] Ostriches have wings because this feature was maintained in their lineage, although it does not help them do anything.

Again, propositions E1, E2, and E3 are perfectly compatible, whereas propositions D1, D2, and D3 are not. First, D3 makes no sense, as no intended use is described. In addition, D1 and D2 are incompatible because it is difficult to explain why the same feature was given to these organisms for different uses. Why don't penguins have hydrodynamic shapes similar to that of whales? No matter to which cases you apply these propositions, Es will be compatible with each other, whereas Ds will be incompatible in several ways.

have hearts in order to pump blood. However, research in conceptual development available so far suggests that children and adults relate teleology to intentionality. Consequently, it might be better to refrain from talking about natural design and purposes in evolution, unless there is explicit and detailed reference to natural selection and evolutionary history. In the next chapter I will explain how Charles Darwin underwent a conceptual change from his initial natural theological assumptions to eventually develop his theory of descent with modification. This is important in order to understand the effect of conceptual conflict on conceptual change, as well as that the process of conceptual change can be a long one.

Further reading

Two recent books address conceptual and epistemological issues relevant to the teaching and learning of evolution, so one had better start with them. The book *Evolution Challenges: Integrating Research and Practice in Teaching and Learning About Evolution*, edited by Karl Rosengren, Sarah Brem, Margaret Evans, and Gale Sinatra, includes chapters that discuss many of the issues addressed in this chapter. A similar book, which focuses on intelligent design but which is nevertheless useful, is *Epistemology and Science Education: Understanding the Evolution vs. Intelligent Design Controversy*, edited by Roger Taylor and Michel Ferrari. Beyond evolution, perhaps the best recent book on conceptual change is the *International Handbook of Research on Conceptual Change*, edited by Stella Vosniadou. Another interesting book, which is actually a collection of previously published papers, is Paul Thagard's *The Cognitive Science of Science: Explanation, Discovery, and Conceptual Change*. A detailed account of conceptual change, based on her cognitive-historical method, is given in *Creating Scientific Concepts* by Nancy Nersessian. Susan Carey's classic book *Conceptual Change in Childhood* is still a nice read, although it is now a bit dated. Similarly, another rather old but interesting book is Frank Keil's *Concepts, Kinds and Cognitive Development*. The book *Young Children's Naive Thinking About the Biological World* by Kayoko Inagaki and Giyoo Hatano presents their research and their own perspective, which is different but perhaps complementary to the previous ones. Since not only concepts but also explanations are important, and one should pay attention to the content and structure of students' explanations, *Explanation and Cognition*, edited by Frank Keil and Robert Wilson, might prove a useful resource. Finally, two books which discuss philosophical issues addressed in this chapter like functions, design, or the differences between organisms and artifacts are *Organisms and Artifacts: Design in Nature and Elsewhere* by Tim Lewens and *What Functions Explain: Functional Explanation and Self-Reproducing Systems* by Peter McLaughlin. My account of biological teleology is, despite some differences, strongly influenced by the writings of James Lennox, whose *Aristotle's Philosophy of Biology* is a highly recommended book.

4 Charles Darwin and the *Origin of Species*

A historical case study of conceptual change

There is no question that many people have heard of Charles Darwin (Figure 4.1) and the *Origin*. Especially since the 2009 anniversary, when several Darwin events took place all over the world, one can reasonably assume that many people have heard something about him and his famous book. But how many people have actually read what Darwin wrote in the *Origin?* I am not very confident that all those people who have something to say about Darwin and his book (both proponents and opponents) have actually read it. But is it necessary to have read the *Origin* given the amount of information about evolution currently available? I think yes. The *Origin* was written by Darwin as an abstract of his species theory for a general audience. It is written with clarity, and it is full of insight and evidence for evolution that make it a fascinating read. Moreover, so many different and conflicting claims have been made about what he wrote in the *Origin* that the only way for anyone involved in any discussion about Darwin and evolution to be sure is to read what he actually wrote. What is also interesting, and sometimes difficult to realize, is the particular political, cultural, social, religious, and scientific contexts in which Darwin's theory was developed and published. Darwin was not a prophet, nor did he ever intend to become the founder of a secular religion. He was a man of science; he developed his theory based on solid evidence, taking into account the best philosophical and scientific scholarship of his time. He was also anxious to convince his readers about natural selection as the mechanism of **transmutation**.[1] And he did this based on scientific work and not on speculation; he even explicitly discussed the questions which he was unable to answer, the "difficulties" of his theory.[2]

It may be a surprise for many to read that Darwin was initially trained as a clergyman at Cambridge,[3] after dropping out of his medical studies at Edinburgh.

---

[1] This is how the emergence of a species from a pre-existing one was called at the time. The word "evolution" at the time referred to progress and development rather than the process we nowadays mean. This is probably why no one used the word in the current sense: Lamarck used the term "tranformisme" and Darwin described the process as "descent with modification."

[2] Lustig (2009) provides an overview and a discussion of the relevant chapters in the *Origin*.

[3] Being enrolled in the divinity program was more or less standard for someone thinking of any kind of "naturalist" life. People were not allowed to teach at Oxford or Cambridge unless they took orders (which is why they all had the word "Reverend" in front of their names) (see Desmond and Moore, 1994; Browne, 2003a for details).

Figure 4.1 Charles Darwin. Photograph by Maull & Fox, c. 1857. (DAR 225:175). With permission from the syndics of Cambridge University Library.

Darwin initially accepted Paley's views,[4] already discussed in Chapter 2, but he eventually underwent a process of conceptual change to become an evolutionist in current terms. This was accomplished through his extensive, careful, and insightful study of nature, as well as his wide reading and reflection on that reading. Darwin underwent a shift from special creation to a theory of descent with modification, which he presented in the *Origin*. However, even his supporters could not accept some of its central arguments. Although many people were ready to accept the idea of evolution right from the start, Darwin's suggestion of natural selection as the major mechanism of evolutionary change was accepted by very few serious evolutionary thinkers until the 1930s. One reason for this was that natural selection had never been actually observed to result in the production of new species. It also seemed to conflict with almost everything that was known about inheritance. Consequently, there were many objections to Darwin's theory on scientific grounds – not only due to religious motivation as is commonly thought. The dismissal of the idea of natural selection which characterized the latter half of the nineteenth century

---

[4] But we should keep in mind that his father, Robert Darwin, was apparently a non-believer, and that his two grandparents, Erasmus Darwin and Josiah Wedgwood, were radical non-conformists and founders of the Lunar Society (see Desmond and Moore, 1994; Browne, 2003a for details).

(Bowler, 1983, 2005) was due to the problems identified in his theory presented in the *Origin* and not (only) to religious instinct.

This chapter serves as a historical case study of conceptual change, which was the topic of the previous chapter. Conceptual change has often been characterized as a paradigm shift. According to Thomas Kuhn, during a paradigm shift the new paradigm emerges at once in the mind of someone who is not committed to the old paradigm. Such a person is able to see that the old paradigm no longer works and so is able to conceive of a new one to replace it. The resulting transition to a new paradigm is a scientific revolution. Kuhn thought that a paradigm shift was a transition between incommensurable and competing paradigms, which could not be made gradually, but rather occurred at once as gestalt switch (Kuhn, 1996, p. 122). However, a careful study of the cognitive processes of past scientists points to a different conclusion: novel concepts do not emerge all at once and fully developed in the minds of scientists, but they rather are the products of lengthy cognitive processes under the influence of a combination of conditions (Nersessian, 2008, p. 5). This seems to be what happened in Darwin's case. Furthermore, the influences that Darwin's theory had were so many, so deep, so prolonged, and so various that there is no single transition that can be identified as a shift that replaced a pre-Darwinian with a Darwinian paradigm (Hodge and Radick, 2009b).

So, which are the characteristics of Darwin's own conceptual change? I argue that the study of history shows that Darwin underwent shifts due to conceptual conflicts. When he realized that the conceptions he held could not sufficiently account for the observed phenomena, he replaced them and accommodated new ones. Darwin developed his theory for 20 years and went through two major shifts from his initial views. The first shift was that from special creation to transmutation, which was completed around March 1837. The second and more prolonged shift was that from perfect adaptation to relative adaptation, which was completed around March 1857. In this chapter I first present the context in which Darwin's theory was developed. I also describe the development of his theory from 1839 to 1859 and the conceptual foundations of the *Origin*. Then I describe the conceptual shifts Darwin underwent during the time he was developing his theory from his initial acceptance of special creation to the theory presented in the *Origin*. Finally, I present the important scientific criticisms that his theory received from both supporters and opponents.

The development of Darwin's theory

Why did Charles Darwin develop his theory at a particular place and time? It seems that social context was important. Darwin had many intellectual and practical resources, characteristic of Victorian society, which were not available in earlier times and which were crucial for the development of his theory. Darwin's analogies and influences were distinct of the Victorian era: Thomas Malthus and Adam Smith were political economists who developed their theories in the English context; animal and plant breeding was a form of Victorian technology; John Herschel and William Whewell were among the first philosophers of science in a tradition based on Newton's science; and the

Anglican Church had a tradition, natural theology, that put emphasis on the idea of adaptation. Furthermore, industrialization and imperialism of the British Empire created many of the conditions that made Darwin's research possible. There was a public interest for natural history, which became very popular in Victorian England because it was seen both as amusement and as science. This complemented the widespread English enthusiasm for natural theology, and so it may have made doing natural history even more popular. The expansion of the railway network facilitated access to various sites where specimens could be collected. Moreover, the Penny Post made the exchange of specimens possible between naturalists from various parts of the country. As the practice of natural history required not only considerable skill but also specialist equipment, technological advancement also took place. Finally, cheaper books were soon produced, many of which were about natural history. Darwin made good use of all these resources. He also took advantage of the recent rapid growth of the empire. He famously travelled around the world for five years aboard the HMS *Beagle*. He also developed a vast network of correspondents, most of whom he never met in person, thanks to the British ships traveling around the world (see the Introduction in Endersby, 2009; Radick, 2009; but see also Hull, 2005).

Given all this, it may be no coincidence that Alfred Russel Wallace, who independently came up with a similar theory (but with important differences and 20 years later than Darwin), lived at the same time and in the same culture as Darwin. This, however, does not necessarily imply that had Darwin never lived or lived enough to write the *Origin*, Wallace or someone else would have developed a similar theory. It seems that Darwin's background, knowledge, experience, and skills brought him to a unique position to come up with his theory of descent with modification, by bringing together different pieces of evidence and different concepts. Not only the idea of evolution, but even the idea of natural selection was in the air before Darwin. For example, Patrick Matthew had referred to such a process as early as 1831, but never developed it further (Bowler, 2013, pp. 56–58). However, it was Darwin who carefully developed the theory of descent with modification that we read in the *Origin*.

Darwin was already wondering about transmutation as soon as he returned to England from his *Beagle* voyage in October 1836. Before this voyage he rather accepted the special creation of species. However, it seems that by March 1837 Darwin had become a convinced and confident transmutationist (Hodge, 2010). Darwin started his notebooks on transmutation in July 1837. What was very important and crucial for the development of his theory was his reading of and experience with breeding and artificial selection. Eventually, in September 1838, he read Malthus' *Essay on the Principle of Population*[5] and came up with the idea of natural selection, which he had

---

[5] Although Darwin wrote in his autobiography that he "happened to read for amusement Malthus on *Population*" (Barlow, 2005, p. 98), it seems that he had earlier been quite familiar with his views. While at Cambridge Darwin had read William Paley's *Natural Theology*, in which considerable attention was devoted to Malthus' essay – but Paley had read a different edition than the one Darwin read (Schweber, 1980; Ospovat, 1981, p. 63). In addition, through his brother Erasmus he had become well acquainted with Harriet Martineau, who had been promoting Malthusian doctrines and had built her literary fame on these (Desmond and Moore, 1994, pp. 201, 264; Browne, 2003a, pp. 385–386).

developed by March 1839 (Hodge, 2009). Malthus argued that while the natural tendency of human populations was to increase in numbers at a geometric rate, agricultural production increased at an arithmetic rate. Consequently, there would be a struggle for resources that slowed population growth and hence limited the increase of human population (Desmond and Moore, 1994, pp. 264–265; Browne, 2003a, pp. 386–387). Darwin thought that a similar process could be taking place in nature and eventually wrote in the *Origin* that:

Owing to this struggle for life, any variation, however slight and from whatever cause proceeding, if it be in any degree profitable to an individual of any species, in its infinitely complex relations to other organic beings and to external nature, will tend to the preservation of that individual, and will generally be inherited by its offspring. The offspring, also, will thus have a better chance of surviving, for, of the many individuals of any species which are periodically born, but a small number can survive. (Darwin, 1859, p. 61)

Darwin provided the explanation of this process, while crediting Malthus:

Hence, as more individuals are produced than can possibly survive, there must in every case be a struggle for existence, either one individual with another of the same species, or with the individuals of distinct species, or with the physical conditions of life. It is the doctrine of Malthus applied with manifold force to the whole animal and vegetable kingdoms; for in this case there can be no artificial increase of food, and no prudential restraint from marriage. Although some species may be now increasing, more or less rapidly, in numbers, all cannot do so, for the world would not hold them. (Darwin, 1859, pp. 63–64)

Darwin actually did more than just find "a theory by which to work," as he described in his autobiography (Barlow, 2005, p. 99). After reading Malthus, Darwin recognized the idea of struggle for existence as a driving force for natural selection. However, there seems to be a significant difference between the ways in which Malthus and Darwin conceived of this. Malthus' view of struggle was that of a species against its environment. However, Darwin conceived of two distinct concepts of struggle: the struggle of a species as a whole against its environment, but also the struggle that resulted from the competition between individuals of the same species. Darwin's theory was based on a combination of these two types of struggle. The important insight that Darwin added was that the struggle between individuals of the same species was a consequence of the struggle of species against their environments (Vorzimmer, 1969; Bowler, 1976).

For Darwin, the struggle between individuals of the same species had the effect of selection of variations that contributed to the survival and reproduction of their possessors. Darwin actually transformed Malthus' ideas in two ways. First, he expanded the concept of population checks from just the limitation of resources to include any factor of the environment that might limit population increase. Second, he made the idea of struggle, which for Malthus was just a force limiting population growth, the driving force behind adaptive change (Lennox and Wilson, 1994). It seems that Darwin initially held a natural theological, harmonious view of nature. His views changed after reading Malthus, but he did not altogether reject the idea of harmony. Darwin just abandoned the idea of perfect adaptation (that there is only one best possible form for any given set of conditions) for a quite similar view that nevertheless allowed the possibility of alternative forms and rudimentary organs (described as the idea of limited perfection) (Ospovat, 1981, pp. 33–37).

Darwin's reading of Malthus[6] was important for coming up with the idea of natural selection, but what was perhaps most crucial was his knowledge of artificial selection. Darwin explicitly referred to artificial selection in his definition of natural selection:

I have called this principle, by which each slight variation, if useful, is preserved, by the term of Natural Selection, in order to mark its relation to man's power of selection. We have seen that man by selection can certainly produce great results, and can adapt organic beings to his own uses, through the accumulation of slight but useful variations, given to him by the hand of Nature. But Natural Selection, as we shall hereafter see, is a power incessantly ready for action, and is as immeasurably superior to man's feeble efforts, as the works of Nature are to those of Art. (Darwin, 1859, p. 61)

Darwin was not the first to consider animal and plant breeding as a source for understanding natural history. Linnaeus, Buffon, and Lamarck had also considered the breeding of domesticated organisms for their conclusions, but contrary to Darwin they all assumed that it could only provide limited information (Cornell, 1984, pp. 305–306). As in the case of the struggle for existence, Darwin did not just borrow a concept, but insightfully transformed it. The analogy between artificial and natural selection was not at all obvious. Darwin was one of few among his contemporary men of science who was quite familiar with the work of breeders, although information on plant and animal breeding was widely disseminated in England at that time. Domestic breeding and the study of nature were pursued by separate individuals, and in separate organizations and publications. However, Darwin was able to bridge this gap. Many members of his family, such as his uncle Josiah Wedgwood, and several of those who influenced him, like Charles Lyell and John Henslow,[7] were involved in breeding. Darwin managed to establish an extensive network of contacts that involved breeders, most of whom he never met in person. These people provided him with valuable information for his studies, which he used after establishing its reliability. These people, on the other hand, also benefited from Darwin who, as a leading man of science, lent them status by referring to their work in his writings (Secord, 1985).

Darwin first made use of the concept of selection in his Notebook C that covers the period from February to July 1838, a few months before he read Malthus. After reading the pamphlets written by animal breeders John Sebright and John Wilkinson, who were explicit about the nature and power of artificial selection, Darwin realized that sustained selection for small changes could be taking place in nature. It was especially Sebright who mentioned natural selection, although by another name, and discussed the analogy

---

[6] A fact that highlights Darwin's originality is that it was only him (and ultimately Wallace) who realized that the idea of struggle could also be applied in nature, although Malthus was widely read at the time.

[7] Lyell and Henslow had an important influence on Darwin. Henslow was Darwin's mentor at Cambridge, where they talked a lot about natural history, and actually the one who suggested Darwin travel aboard the HMS *Beagle*. During this trip Darwin read Lyell's *Principles of Geology*, a book describing the view of uniformitarianism (that the surface of the Earth had undergone slow and gradual changes), which also had an enormous influence on Darwin (for these influences, see Walters and Stow, 2002; Herbert, 2005).

between that and artificial selection. Darwin wrote in the *Origin*, the first chapter of which (entitled "Variation under domestication") was devoted to artificial selection:

> That most skilful breeder, Sir John Sebright, used to say, with respect to pigeons, that "he would produce any given feather in three years, but it would take him six years to obtain head and beak." In Saxony the importance of the principle of selection in regard to merino sheep is so fully recognized, that men follow it as a trade: the sheep are placed on a table and are studied, like a picture by a connoisseur; this is done three times at intervals of months, and the sheep are each time marked and classed, so that the very best may ultimately be selected for breeding. (Darwin, 1859, p. 31)

Darwin's notes written on the Sebright pamphlet indicate that he paid special attention to them. It seems that Darwin read these several times and that they stimulated him to search for a mechanism in nature equivalent to the sustained gradual picking employed by breeders. Darwin also joined several pigeon-breeding clubs to see for himself how far selective breeding could go in producing new varieties. Hence, he realized that artificial selection could provide important insights about transmutation. And as soon as he read Malthus' essay, Darwin conceived of the mechanism by which a similar process – natural selection – might be operating in nature (Ruse, 1975a; Cornell, 1984; Evans, 1984). Darwin's analogy between artificial and natural selection was based on the assertion that breeders' selection resulted in modifications in the domesticated organisms that were permanent and that had not existed in their wild ancestors. He devoted the first chapter of the *Origin*, entitled "Variation under domestication," to convincing his readers that selection could produce new varieties which did not exist in the past (Largent, 2009, pp. 17–24).

Darwin's involvement in pigeon breeding supported his use of what was a popular hobby in England in order to establish the crucial analogy between artificial and natural selection. In doing so, Darwin had to refute the doubt of many naturalists that wild organisms did not vary as much as the domestic ones. To achieve this, Darwin referred to the difficulty of classifying organisms into distinct, appropriately defined species. He also followed Lyell in finding a familiar and observable example that could help readers understand the effects of unobservable processes. Lyell advanced uniformitarianism, the view that Earth's history was much longer than commonly thought at the time and that natural phenomena like earthquakes and volcanic eruptions were enough to bring about all the effects observed in the geological record. Lyell, like Darwin, had already faced the problem of how one could give a scientifically acceptable account of slow and, most importantly, unobservable processes. Lyell argued that for a historical science to be empirically acceptable, the causal processes it appealed to should be observable in principle. Thus, he insisted that geology should begin with a careful and detailed study of the forces actually shaping the Earth at the present, and argued that the characteristics observed in the Earth's crust should be explained by appeal to the same forces acting with the same intensities as currently observed. With these arguments he was countering the claims of catastrophists that there were periodic worldwide floods that explained various patterns in the fossil record (see Laudan, 1987). Darwin tried to do something like this, but he developed an analogical argument and tried to present a natural process (natural selection) by appeal to a similar process (artificial selection) with which most of his readers would be familiar.

It should be noted at this point that artificial and natural selection are different in one very important aspect. Artificial selection requires an intelligent external selector who picks variants according to particular aims or goals. No such selector exists in the process of natural selection, which is the outcome of an unmediated, unintentional natural process of struggle among variants. Hence, in this sense the analogy between artificial and natural selection is weak. But according to Darwin, divergence arises from competition between individuals of the same species that takes place simultaneously with competition between individuals of different species. In this view, it is the external environment, and the different types of competition that it entails, that takes the place of the intelligent selector of artificial selection and that makes natural selection strongly analogous to artificial selection. Individuals of the same species interact with each other but also with others from different species. In the long run, those individuals of a species that can compete more effectively in their environment with individuals of their own or of other species will be those that will live and reproduce and that will eventually be "naturally selected" (Kohn, 2009, pp. 93–94).

It has also been suggested that the Herschel–Whewell philosophies of science are important in order to understand Malthus' contribution to the theory of natural selection, based on the analogy with artificial selection. John Herschel and William Whewell were the two major representatives of the philosophy of science of that time, who both insisted that genuine science required an extensive evidential basis and the identification of a mechanism or cause that could explain phenomena in different areas. They believed that the aim of science was to find the laws of nature and then to identify the true causes (*verae causae*) that guided the workings of these laws.[8] Malthus provided Darwin with quantitative laws, the best kind of laws according to Herschel and Whewell, leading deductively to the idea of the struggle for existence on which Darwin could then base the process of natural selection. Consequently, Darwin was able to present natural selection as a possible mechanism of evolution in the light of the Herschel–Whewell philosophies (Ruse, 1975b; Radick, 2009). Then, Darwin tried to show that natural selection was indeed a true cause. He established the existence and competence of natural selection mostly based on the analogy from artificial selection, whereas its responsibility was based on the fact that it seemed more probable than any other theory

---

[8] It should be noted that there were important differences between the philosophies of Herschel and Whewell which should not be ignored. John Herschel was an astronomer and his main work on philosophy was his book *Preliminary Discourse on the Study of Natural Philosophy*, published in 1831. Herschel believed that the aim of science was to find the laws of nature and to identify the true causes (*verae causae*) that guided the workings of these laws. William Whewell's major work on the philosophy of science was *The Philosophy of the Inductive Sciences*, published in 1840. He adopted and emphasized most points made by Herschel. However, there was a difference with Herschel in at least one important aspect. Whereas Herschel thought that hypothetical reasoning could not identify the true causes, Whewell downplayed the role of direct experience in the identification of true causes and in his "consilience of inductions" theoretical causes could become true causes (Ruse, 1975b, 2000, pp. 5–6; Grene and Depew, 2004, pp. 169–171). Darwin knew Whewell from his Cambridge years and met Herschel in South Africa during his *Beagle* voyage. He later met them both again at the London Geological Society (Ruse, 1975b). Darwin took into account the philosophy of science of his day and wanted to be in compliance with the scientific standards of his time. This is probably the reason why in the first edition of the *Origin* he credited two scholars, famous for their studies on the "scientific method": William Whewell and Francis Bacon (Lewens, 2007, pp. 95–97).

in explaining several kinds of facts about species, like their adaptations or their geographical distribution (Hodge, 1977, 1992; but see also Waters, 2009). Darwin aimed to show that natural selection was competent to produce new species through a series of thought experiments which were imaginative narratives that tested the explanatory potential of natural selection. He tried to diminish the criticism his theory would receive by showing that it was based on the same principles as the established scientific theories of his time, e.g., Newtonian physics. However, Darwin never managed to show that natural selection was actually competent to produce new species through controlled observation and experiment (Lennox, 2005; Hull, 2009).

Darwin initially assumed that natural selection made every species perfectly adapted for the place it occupied, but he later took into account the developmental concepts and the generalizations of Karl Ernst von Baer and Henri Milne-Edwards. Von Baer suggested that animals developed by a progression from a common pattern to a more specialized one (Ospovat, 1981, pp. 117–124; Richards, 1992, pp. 55–61). Milne-Edwards argued that the diverging paths of development suggested by von Baer corresponded to the branching series of organisms in natural classification. Hence, the process of development revealed natural affinities which would best be represented with a branching arrangement (Ospovat, 1981, pp. 124–129; Richards, 1992, pp. 134–136). These ideas posed problems to Darwin because at least during the time he wrote the transmutation notebooks (1837–1839) (Ospovat, 1981, p. 151; Richards, 1992, pp. 152–166), he considered embryonic development as a process of recapitulation of evolutionary change (it should be noted, though, that there is some disagreement on this, see Bowler, 1975; Gould, 1977, p. 70; Mayr, 1982, p. 475). Then, Darwin came up with the idea of the principle of divergence that incorporated the views of von Baer and Milne-Edwards and at the same time was compatible with the idea of natural selection.

The core concept of this principle was the ecological division of labor, formulated around November 1854 to January 1855. By that time, Darwin had started organizing data from various sources and he then drew conclusions from and established hypotheses to be tested against these data. It seems that it was from these processes that the idea of the principle of divergence originated (Kohn, 1985b, pp. 249–250, 2009, p. 105). The principle of divergence was an important innovation as Darwin needed to explain how natural selection could give rise to the various branches of the tree of life. He wrote:

Here, then, we see in man's productions the action of what may be called the principle of divergence, causing differences, at first barely appreciable, steadily to increase, and the breeds to diverge in character both from each other and from their common parent. But how, it may be asked, can any analogous principle apply in nature? I believe it can and does apply most efficiently, from the simple circumstance that the more diversified the descendants from any one species become in structure, constitution, and habits, by so much will they be better enabled to seize on many and widely diversified places in the polity of nature, and so be enabled to increase in numbers. (Darwin, 1859, p. 112)

In short, according to the principle of divergence, natural selection could indefinitely produce better adapted forms by increasing the ecological specialization within groups, the members of which would eventually diverge from the initial form. Thus, natural

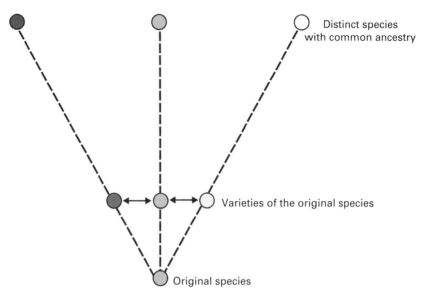

Distinct species
with common ancestry

Varieties of the original species

Original species

Figure 4.2 How varieties gradually diverge and become distinct species.

selection would automatically increase the ecological division of labor among animals found in competitive situations, by favoring those individuals which were most able to exploit new **niches**. In this sense, relative adaptation became a necessary implication of the principle of divergence (Ospovat, 1981, pp. 205–207).[9] This idea is depicted in Figure 4.2; this idea is actually depicted in the branching diagram, which is the only figure included in the *Origin*.

Darwin developed the two main principles of his theory, natural selection and divergence, based on three analogies: (1) between the struggle for existence in human societies and the struggle for existence in nature, (2) between artificial selection and natural selection, and (3) between the (physiological) division of labor and ecological specialization. The main arguments in the *Origin* are based on these two principles (see Figure 4.4, later). This required a significant amount of time. The development of Darwin's theory was a long process and nothing like a "Eureka!" moment. From July 1837 he started taking notes on transmutation. Although he came up with the idea of natural selection within two years, it took him much longer – until November 1854 – to come up with the wider (macroevolutionary) pattern in which natural selection would fit. This is a case of conceptual change, as well as innovation, and it is its details that I focus on in the next section.

---

[9] Interestingly enough, although the idea of the division of labor was used by the political economists of that time, Darwin cited the zoologist Milne-Edwards instead, perhaps in an attempt to provide scientific foundations for his theory (Schweber, 1980; Desmond and Moore, 1994, pp. 420–421). However, it has been suggested that Darwin's use of this term is closer to that of Adam Smith than to that of Milne-Edwards (Kohn, 2009, p. 88).

Darwin's conceptual change

Natural theology was quite popular in England at the beginning of the nineteenth century. The best known proponent of this idea was William Paley. As already discussed in detail in Chapter 2, he argued that the complexity and perfection of the natural world, as revealed through its empirical study, was the most powerful argument for the existence of God (Paley, 2006/1802). Darwin initially admired Paley's views. In his autobiography he wrote that:

In order to pass the B.A. examination, it was, also, necessary to get up Paley's *Evidences of Christianity*, and his *Moral Philosophy*. This was done in a thorough manner, and I am convinced that I could have written out the whole of the *Evidences* with perfect correctness, but not of course in the clear language of Paley. The logic of this book and as I may add of his *Natural Theology* gave me as much delight as did Euclid. The careful study of these works, without attempting to learn any part by rote, was the only part of the Academical Course which, as I then felt and as I still believe, was of the least use to me in the education of my mind. I did not at that time trouble myself about Paley's premises; and taking these on trust I was charmed and convinced by the long line of argumentation. (Barlow, 2005, p. 59)

Darwin initially held some natural theological assumptions which influenced his theory, such as that adaptation was perfect, that nature was a harmonious system, and that change served to maintain this harmony (Ospovat, 1981, pp. 2–3). It seems that Darwin embraced the idea of transmutation after his return to England, probably until March 1837 (Hodge, 2010). As described in the previous section, Darwin eventually developed the core of the theory of natural selection by March 1839 (Hodge, 2009). However, this theory was quite different from the one published in 1859, and there were several differences between Darwin's initial assumptions, the theory presented in the 1842 *Sketch* and the 1844 *Essay* (Darwin, 1909), and the theory presented in the *Origin* (Ospovat, 1981, pp. 208–209; Hodge and Kohn, 1985; Lennox, pers. comm.).

Overall, it seems that there are two important shifts in Darwin's views: (1) the shift from special creation to transmutation and, a while later, to natural selection; and (2) the shift from perfect adaptation to relative adaptation. I consider this as an exemplar case of conceptual change in science. Through a lengthy process and in the light of empirical data, a scientist eventually rejected his old conceptions to accommodate new ones, some of which he invented through this process. What is interesting, then, is to see what kind of conceptual change took place in each case and, most importantly for the aims of this book, what caused this change. As Nancy Nersessian (2008) has shown through her cognitive-historical approach, conceptual change and its causes can be understood through the study of the actual practices of scientists. I do not claim to implement a cognitive-historical approach like hers in this section; however, my account of Darwin's conceptual change is in this direction. Thus, I will describe the major shifts in Darwin's thinking that led to the theory we read in the *Origin*. It may be assumed that the first shift, from special creation to natural selection, was the most important one. After all, it is then that Darwin became an evolutionist, developing his theory soon after. However, it is one thing to conceive of natural selection, which indeed had happened in 1839, and another to understand it in detail and be able to provide a coherent and well-grounded account of how it takes place, which was accomplished several years later.

A significant aspect of the originality of Darwin's contribution was that he approached the problem of the origin of species through biogeography and the succession of species in the geographical space, and not only in the fossil record (Bowler, 2013, p. 37). Perhaps the most well-known case, and one that actually had a major influence on Darwin's changing views about the origin of species, was that of the Galápagos Islands. This change in his views did not happen when Darwin visited the islands in September and October 1835, but the next year, after his return to England. Hodge (2010) provides a detailed account of this change. During 1836 Darwin was wondering whether the birds he had collected from these islands were distinct varieties or distinct species. Concluding that they were distinct varieties would have important implications for the possibility of transmutation, because varieties have the potential to give rise to new species (see Figure 4.2). If this was not the case and the birds were distinct species, Darwin assumed that they could have simply migrated there from the American continent. However, in March 1837 ornithologist John Gould told Darwin not only that the birds he had collected at the Galápagos Islands were distinct species, but also that they were species which are not found on the American continent. Consequently, these species must have originated on the Galápagos Islands (Hodge, 2010, p. 94). For Darwin, this was crucial biogeographical evidence that supported an important conclusion: the bird species which had originated on those volcanic islands were very similar to species that had already originated in very different conditions on the nearest older continental land, rather than to species that had originated on other volcanic oceanic islands elsewhere in the world. Such similarities could only be explained as the outcome of common ancestry and not as adaptations to common conditions (Hodge, 2010, p. 99).

This was crucial evidence against the special creation of species. The Galápagos Islands provided several examples against such an idea. Why did those islands, which are far from a mainland, lack certain kinds of organisms, such as amphibians, although these could certainly live under the conditions there? It is probably not a coincidence that amphibian adults and eggs are killed in salt water and so they would not be accidentally transported across the ocean. Similarly, why are these remote islands characterized by a high proportion of **endemic species**, i.e., species which are found only there and not in other places in the world, compared to those close to a mainland? Most interestingly, why are these species always closely related to those found on the nearest mainland, in the case of the Galápagos similar to those found in the American continent, although the conditions on the islands are different from those on the mainland? According to Darwin, island populations emerged as a result of migration from the nearest mainland, and this explained the similarities with the organisms living there. But since migration did not take place all the time, island populations also diverged from those living on the mainland and became distinct species. In addition, despite the similar physical conditions on the islands, different organisms might migrate to different islands at different times and find different competitors or food already available there. Thus, the initial populations could evolve in different directions, and because organisms only rarely moved between islands, populations could eventually diverge and become distinct species. All these cannot be sufficiently explained if species were specially created and were designed to be perfectly adapted to their environments (Bowler, 2009b). Here is how Darwin summarized this central argument in favor of evolution and against the special creation of organisms, based on the study of biogeography:

> We are thus brought to the question which has been largely discussed by naturalists, namely, whether species have been created at one or more points of the earth's surface. Undoubtedly there are very many cases of extreme difficulty, in understanding how the same species could possibly have migrated from some one point to the several distant and isolated points, where now found. Nevertheless the simplicity of the view that each species was first produced within a single region captivates the mind. He who rejects it, rejects the *vera causa* of ordinary generation with subsequent migration, and calls in the agency of a miracle. [. . .] if the same species can be produced at two separate points, why do we not find a single mammal common to Europe and Australia or South America? The conditions of life are nearly the same, so that a multitude of European animals and plants have become naturalised in America and Australia; and some of the aboriginal plants are identically the same at these distant points of the northern and southern hemispheres? The answer, as I believe, is, that mammals have not been able to migrate, whereas some plants, from their varied means of dispersal, have migrated across the vast and broken interspace. The great and striking influence which barriers of every kind have had on distribution, is intelligible only on the view that the great majority of species have been produced on one side alone, and have not been able to migrate to the other side. Some few families, many sub-families, very many genera, and a still greater number of sections of genera are confined to a single region; and it has been observed by several naturalists, that the most natural genera, or those genera in which the species are most closely related to each other, are generally local, or confined to one area. What a strange anomaly it would be, if, when coming one step lower in the series, to the individuals of the same species, a directly opposite rule prevailed; and species were not local, but had been produced in two or more distinct areas! (Darwin, 1859, pp. 352–353)

This is compatible with the idea of relative adaptation: organisms are adapted to the particular environment they inhabit and to the local conditions there. The concept of relative adaptation that is characteristic of the *Origin* is not found in the first three transmutation notebooks that Darwin wrote from July 1837 until October 1838 (Ospovat, 1981, p. 37).[10] Rather, Darwin seems to have initially accepted the idea of perfect adaptation, according to which there was one best possible form for any given set of conditions. In other words, the Creator had adapted organisms to their environments in a strict sense; adaptations, or the fit between organisms and their environment according to Cuvier's conditions for existence,[11] explained their characters. Later, Darwin accommodated a quite similar view in its place, the idea of limited perfection. This idea retained the notion of harmony but allowed the possibility of alternative forms and rudimentary organs. In this case, adaptation was determined by laws set by the Creator. There was adaptation but not in a strict sense as several organisms could exploit the same niche. Eventually, Darwin came to accept the idea of relative adaptation,

---

[10] In this section I rely heavily on Ospovat's (1981) analysis of Darwin's writings, but I will add excerpts to show some signs of Darwin's shifts and of his process of conceptual change.

[11] See Reiss (2009).

which was a necessary implication of the principle of divergence: New individuals which could exploit new niches were favored by selection, and so there was divergence of form. But in this case, the direction of divergence of form was relative to the environment as only those who could adapt to it would survive (Ospovat, 1981, pp. 33–37, pp. 205–207).

According to Ospovat, the first shift in Darwin's views occurred after his reading of Malthus near the end of September 1838 (Ospovat, 1981, pp. 60–61). Darwin came up with the idea of natural selection, but in an entirely natural theological context. He assumed that natural selection made species perfect for the place they occupied, as well as that there was no variation among perfectly adapted forms. Natural selection was compatible with the idea of a benevolent Creator and a harmonious nature as long as it was perceived as a law of nature that was aimed to produce perfect adaptation. But as soon as Darwin came up with the principle of divergence, he realized that adaptation must be relative. Not only was the adaptedness of a species relative to the adaptedness of other species living in the same area, but also a well-adapted species could eventually become even better adapted through natural selection. This second shift seems to be evident for the first time in parts of his *Natural Selection* around March 1857 (Ospovat, 1981, pp. 205–207). Ospovat noted that "Its late emergence in Darwin's thought indicates that relative adaptation is not, as might be supposed, a necessary implication of the theory of natural selection" (Ospovat, 1981, p. 208).

There are other aspects in which the *Origin* differs significantly from Darwin's earlier writings. In particular, Ospovat identified several major differences between 1844 (when Darwin wrote the *Essay*) and 1859 (when the *Origin* was published). One is the difference between perfect and relative adaptation, already described above. A second difference has to do with the amount of variation available in nature. In 1844 Darwin thought that there was little available variation as perfectly adapted forms did not vary. Consequently, a change in external conditions was required for new variation to occur. In 1859 he assumed that no change was required and that variation was common. A consequence of this, and a third difference between 1844 and 1859, was that Darwin initially thought that transmutation could take place only in response to changes in external conditions (either because the environment itself changed or because organisms migrated to a new environment). In contrast, in 1859 Darwin believed that since variation was common, natural selection could take place anytime, leading to transmutation and to the production of better-adapted forms (Ospovat, 1981, pp. 83–86). A major reason for this second shift may have been Darwin's extensive study of barnacles from October of 1846 to September of 1854. This study made Darwin realize that there was more variation available than he initially thought. Given this, not only did the idea of perfect adaptation seem to be less plausible, but also there was adequate variation for natural selection to occur (Ospovat, 1981, p. 208; Kohn, 2009, pp. 102–103; see also Love, 2002).

In the first two of his transmutation notebooks (Notebooks B and C), which Darwin started writing in July 1837, there is no explicit reference to selection. Selection is mentioned for the first time in Notebook D. There is a pencil note, inserted later by

Darwin, that "Towards close [of this notebook] I first thought of selection owing to struggle." On September 28, 1838, Darwin wrote in his Notebook D on transmutation:

Population is increase at geometrical ratio in far shorter time than 25 years – yet until the one sentence[12] of Malthus no one clearly perceived the great check amongst men. – there is spring, like food used for other purposes as wheat for making brandy. – Even a few years plenty, makes population in Men increase & an ordinary crop causes a dearth. take Europe on an average every species must have same number killed year with year by hawks, by cold &c. – even one species of hawk decreasing in number must affect instantaneously all the rest. – The final cause of all this wedging, must be to sort out proper structure, & adapt it to changes. – to do that for form, which Malthus shows is the final effect (by means however of volition) of this populousness on the energy of man. One may say there is a force like a hundred thousand wedges trying force ~~into~~ every kind of adapted structure into the gaps ~~of~~ in the oeconomy of nature, or rather forming gaps by thrusting out weaker ones. (Darwin, Notebook D, 135e)

After reading Malthus, Darwin wrote that the final cause of the process described "must be to sort out proper structure, & adapt it to changes." And these proper structures are forced into the gaps in the economy of nature while new gaps are formed when other structures die out. This is a clear teleological argument, one based on design: Adaptation is the final cause of the struggle for existence. And this adaptation is perfect as it is not determined by the environment. It is already formed, it is sorted out, and it is "forced" into the environment. Thus, as Ospovat argued, at that time Darwin thought of natural selection as guiding form to perfection. This is quite different from the concept of adaptation we find in the *Origin*, where adaptation is relative. For instance:

As the individuals of the same species come in all respects into the closest competition with each other, the struggle will generally be most severe between them; it will be almost equally severe between the varieties of the same species, and next in severity between the species of the same genus. But the struggle will often be very severe between beings most remote in the scale of nature. The slightest advantage in one being, at any age or during any season, over those with which it comes into competition, or *better adaptation in however slight a degree to the surrounding physical conditions*, will turn the balance. (Darwin, 1859, pp. 467–468, emphasis added)

Finally, it has also been suggested that another reason for the shift from his initial natural theological assumptions, especially from the idea of perfect adaptation toward the idea of relative adaptation, may have been the death of his beloved daughter Annie in 1851. But this is not clear and certainly does not have anything to do with scientific understanding. After having probably suffered from tuberculosis, she died at the age of ten. Although this tragic event caused Darwin terrible pain, he did not experience any instant loss of faith. Instead, it seems that he rather went through fluctuations of belief that were also influenced by his social circle, which included people who could lead a moral life without embracing Christianity, and of course his own understanding of the world with natural selection producing suffering that was inconsistent with a benevolent

---

[12] The editors of Darwin's writings note that: "This note, written on 28 September 1838, makes it possible to identify the sentence in T. R. Malthus's *Essay on the Principle of Population* which enabled Darwin to see how the pressure of natural selection is inevitably brought to bear. It was in the 6th edition, London 1826, vol. 1, p. 6: 'It may safely be pronounced, therefore, that the population, when unchecked, goes on doubling itself every twenty five years, or increases in a geometrical ratio'."

- 1835 Darwin visits Galápagos
- 1836 Return from *Beagle* voyage
- 1837 Darwin convinced about transmutation   Perfect adaptation
- 1838 Selection in notebooks and reading Malthus   (strict sense)
- 1839 First outline of the theory, including natural selection

- 1842 *Sketch* of the theory

- 1844 Publication of *Vestiges* and *Essay* of the theory

- 1846 Study of barnacles begins
                                                              Perfect adaptation
                                                              (limited perfection)

- 1851 Annie dies

- 1854 Study of barnacles completed

- 1857 Abstract of theory sent to Gray
- 1858 Wallace's paper received   Relative adaptation
- 1859 Publication of the *Origin*

Figure 4.3 The major phases and shifting points in Darwin's thinking.

God. Annie's death caused terrible pain to Darwin and is said to have driven him to atheism; however, he preferred to describe himself as an agnostic (Keynes, 2001; Spencer, 2009; Brooke, 2009a, 2009b; Moore, 2009). The important question here is whether the pain from this terrible loss also influenced Darwin's understanding of nature. Could it be the case that this tragic loss made Darwin abandon his belief in a harmonious world? There is room for speculation here; there is no question that such a terrible loss might make one reconsider his/her views about harmony and perfection in the world. However, it may be the case that this incident was not as influential as commonly thought (van Wyhe and Pallen, 2012).

The major phases and points of shift in Darwin's thinking, based on the above, are presented in Figure 4.3. The important question is: What caused these shifts? Darwin realized that his initial conceptions of special creation and perfect adaptation were explanatorily insufficient. He considered this in his *Autobiography*:

The old argument of design in nature, as given by Paley, which formerly seemed to me so conclusive, fails, now that the law of natural selection has been discovered. *We can no longer argue that, for instance, the beautiful hinge of a bivalve shell must have been made by an intelligent being, like the hinge of a door by man.* There seems to be no more design in the variability of organic beings and in the action of natural selection, than in the course which the wind blows. Everything in nature is the result of fixed laws. (Barlow, 2005, p. 87, emphasis added)

Thus, whereas Darwin initially believed that organisms and their characters are intelligently designed, and so specially created, he later concluded that this is not possible. He became fully convinced when he realized that adaptation is relative:

Such suffering, is quite compatible with the belief in *Natural Selection, which is not perfect in its action, but tends only to render each species as successful as possible in the battle for life with other species*, in wonderfully complex and changing circumstances. That there is much suffering in the world no one disputes. Some have attempted to explain this in reference to man by imagining that it serves for his moral improvement. But the number of men in the world is as nothing compared with that of all other sentient beings, and these often suffer greatly without any moral improvement. A being so powerful and so full of knowledge as a God who could create the universe, is to our finite minds omnipotent and omniscient, and it revolts our understanding to suppose that his benevolence is not unbounded, for what advantage can there be in the sufferings of millions of the lower animals throughout almost endless time? This very old argument from the existence of suffering against the existence of an intelligent first cause seems to me a strong one; whereas, as just remarked, the presence of much suffering agrees well with the view that all organic beings have been developed through variation and natural selection. (Barlow, 2005, p. 90, emphasis added)

Why do organisms suffer? They may suffer because they do not always possess the characters necessary for their wellbeing, survival, or reproduction; because they are prey for other organisms; because their environment may undergo violent changes. Darwin's extensive study of nature of more than 30 years (from his Edinburgh natural history studies with Robert Grant until the time the *Origin* was published) showed him that species cannot be specially created, as well as that the world is not as harmonious as he initially believed. Following Paley, Darwin initially believed that organisms are intelligently designed and perfectly adapted, but then he realized that many organisms could be maladapted or not adapted at all. This made him change his initial views. He accepted transmutation in the place of special creation and he developed a theory of descent with modification based on natural selection. He also incorporated imperfections in the idea of perfect adaptation which was eventually replaced by the idea of relative adaptation. His conclusion was that populations were not adapted in any absolute sense, but only relatively to the environment they inhabited. And these populations became well adapted through natural selection. His explanation for the origin of characters based on their evolutionary history was more efficient than design (remember the discussion in the previous chapter about the hydrodynamic shape of dolphins and sharks and the comparison between propositions D and E).

What are the implications of these conclusions for the existence of God? In my view, there are no direct implications (see my Concluding remarks). Darwin's conclusion was that organisms are neither perfectly nor intelligently designed; he made no conclusion that God does not exist. And indeed, he wrote: "I cannot pretend to throw the least light on such abstruse problems. The mystery of the beginning of all things is insoluble by us; and I for one must be content to remain an Agnostic" (Barlow, 2005, p. 94). Darwin's conceptual change was from a view of organisms as artifacts, intelligently designed for a purpose, to their view as natural entities, not from theism to atheism. Too much has been written about the implications of evolution for religion (see Alexander, 2013; Ayala, 2013 for nice overviews). In the final section I show that religious belief notwithstanding, the *Origin* received sound and valid criticisms by both proponents and opponents. But before that, let us consider in some detail the arguments in the *Origin*.

The publication of the *Origin of Species*

As described earlier in this chapter, Darwin had developed some core concepts of his theory by 1839, but he did not publish it until 1859. It seems that this was not just a matter of fear to publish (van Wyhe, 2007). In the previous section it was explained that the theory eventually published in the *Origin* was different from the one Darwin had initially conceived of 20 years earlier. This suggests that Darwin probably had questions to answer and problems to solve. This apparently took him some time and also resulted in a theory quite different from the one he initially came up with.

Darwin was aware of the reaction to previously published theories of evolution. For instance, Lamarck's theory had been severely criticized by Lyell in his *Principles of Geology*. This book contains a long and careful exposition of Lamarck's theory, pointing to an "important chasm in the chain of the evidence" (Lyell, 1832, p. 8). There was a fiercer reaction against the *Vestiges of the Natural History of Creation*, anonymously published by Chambers in 1844. Huxley was very critical in his 1854 review of a later edition of the *Vestiges*: "we find reason to doubt if the author ever performed an experiment or made an observation in any one branch of science" (quoted in Secord, 2000, p. 500). The public reaction to the *Vestiges* made Darwin anxious and uncomfortable about the prospect of publishing his own evolutionary views. Darwin was also aware that his book would be judged in comparison to the *Vestiges*. For many people, the latter was an attack on Christianity, not only due to its content but also due to the political instability of the time. It seems that Darwin's "delay" in publishing also allowed this instability to subside.

However, and most importantly, Darwin was aware that these older theories had received fierce criticism, such as that quoted above, because they were mostly speculative and were not based on solid scientific research. As a consequence, Darwin tried to gather more data in support of his theory; his work on the classification of barnacles, which lasted eight years, was in part a response to the reviews of the *Vestiges*. Thus, he behaved in a truly scientific manner as there were important scientific questions he had to answer before proceeding to publication. His study of the barnacles was also crucial for a major scientific problem he had to resolve: that of limited variability and of consequent weak natural selection (Kohn, 2009, p. 102). Darwin learned a variety of lessons from his barnacle research: that there was a high degree of variation in all external characters; that there were several homologies in modified animal structures that had changed function; that there was evidence of transformism in the bizarre sexual characteristics of barnacle anatomy; and that a hermaphroditic origin of life was possible. Darwin also developed practical skills in dissection (Love, 2002). Eventually, the study of the barnacles also gave Darwin scientific credibility as he won the Royal

Medal for Natural Science of the Royal Society. Classification was seen as the foundation of natural history, and so by establishing his expertise in this field, Darwin
established his expertise and competence as a naturalist (Endersby, 2009, p. xxxii).

This, of course, does not mean that Darwin was not concerned about possible
reactions to the publication of his theory. He was concerned about the reaction of
the leading scientific figures of his day. This is why he tried to comply with the scientific
standards of his time, as described in the previous section, crediting Whewell and Bacon
in the opening pages of the *Origin*. Darwin also feared the reaction of religious people,
who might consider his theory as an attack on the established beliefs of the time.
His theory might seem to challenge everything that had previously been thought about
humanity's place in nature. According to that, humans were just one species among the
others, closely related to primates, and not one specially created by God in His image.
One of these people was his wife, Emma Wedgwood, whom he had married in 1839.
Darwin was afraid that he might hurt her feelings as she was deeply religious, and
firmly believed in resurrection and salvation. Because Darwin's scientific findings on
the origin of humanity and Emma's own devout Christian beliefs were in conflict,
she was afraid that his ideas would keep them apart in life after death[13] (Desmond and
Moore, 1994; Browne, 2003a, 2006). In an undated letter, written shortly after their
marriage, Emma wrote (Barlow, 2005, p. 199):

Everything that concerns you concerns me and I should be most unhappy if I thought we did not
belong to each other for ever.

And Darwin wrote in his annotation at the end of the letter:

When I am dead, know that many times, I have kissed and cryed over this. C.D.

In 1844 Darwin gave Emma an essay containing an outline of his theory. It was an
enlarged version of a sketch he had written in 1842. It contained some arguments also
included in the *Origin* (see Glick and Kohn, 1996, pp. 87–117), but as explained in the
previous section, it was also different from that. Darwin also gave Emma a letter in
which he asked her to publish this essay in the event of his sudden death. Initially, he
seemed to prefer giving up credit for his ideas during his lifetime rather than hurting her
feelings, or even worse to be the cause of her and their children's social ostracism
(Browne, 2003a, pp. 446–447). Darwin wrote in that letter:

I have just finished my sketch of my species theory. If, as I believe, my theory in time be accepted
even by one competent judge, it will be a considerable step in science. I therefore write this, in
case of my sudden death, as my most solemn and last request, which I am sure you will consider
the same as if legally entered in my will, that you will devote £400 to its publication and further
will yourself, or through Hensleigh, take trouble in promoting it. I wish that my sketch be given to
some competent person, with this sum to induce him to take trouble in its improvement, and
enlargement. (Darwin, 1995/1902, p. 171)

---

[13] However, it is important to note that existential questions and questions on the authority of scripture were
already in the air before the *Origin* was published. Tennyson's *In Memoriam* is indicative of that.
Geological findings such as fossils, revealed during quarrying, mining, railway construction, and canal
cutting, gave rise to questions and conclusions that challenged religious authority.

Darwin also shared his views with Joseph Dalton Hooker, a botanist and director of Kew Gardens (see Endersby, 2008), in a letter that he wrote in the same year (Desmond and Moore, 1994, pp. 313–316):

I have been now ever since my return engaged in a very presumptuous work, and I know no one individual who would not say a very foolish one. I was so struck with distribution of the Galápagos organisms &c. &c., and with the character of the American fossil mammifers, &c. &c., that I determined to collect blindly every sort of fact, which could bear any way on what are species. I have read heaps of agricultural and horticultural books, and have never ceased collecting facts. At last gleams of light have come, and I am almost convinced (quite contrary to the opinion I started with) that species are not (it is like confessing a murder) immutable. (Darwin, 1995/1902, pp. 173–174)

In 1856 Darwin started working on a big book that would be called *Natural Selection* (Stauffer, 1975). However, this was interrupted by the receipt in 1858 of a letter from Alfred Russel Wallace. This was the incident that eventually forced Darwin to proceed to the publication of the *Origin*. Wallace was one of Darwin's numerous correspondents from around the world who knew that Darwin was interested in the question of how species originate, and trusted his opinion on the matter. Thus, he sent him his essay in which he presented his own answer to this problem and asked him to review it. While Wallace's essay did not employ Darwin's term "natural selection," it did outline a process of evolutionary divergence of species from pre-existing ones due to environmental pressures. In this sense, it seemed the same as Darwin's theory, although it was quite different in some crucial aspects (Hull, 2005; Bowler, 2013, pp. 58–66).[14] Darwin considered it as being the same as the theory he had worked on for 20 years but had yet to publish; he wrote in a letter to Charles Lyell: "if Wallace had my MS. sketch written out in 1842, he could not have made a better short abstract! Even his terms now stand as heads of my chapters!" (Darwin, 1995/1902, p. 185). Darwin's priority was eventually saved as Lyell and Hooker arranged for a joint presentation at the Linnean Society of both Darwin's and Wallace's papers (Darwin and Wallace, 1858). Wallace found out about this several months later, and sent Darwin a letter of approval that arrived early in 1859. Wallace was not at the presentation, but neither was Darwin because one of his children was seriously ill (Desmond and Moore, 1994, pp. 467–472; Browne, 2003b, pp. 14–23, 33–53). It should be noted that Darwin's priority in conceiving natural selection was certified by an abstract of his theory that he had sent to the American botanist Asa Gray as early as September 1857, which was part of what was presented to the Linnean Society (Ospovat, 1981, p. 188; Glick and Kohn, 1996, pp. 152–155; Kohn, 2009, p. 106).[15] Eventually, the *Origin* came out on November 24, 1859.

---

[14] One interesting question is whether the development of the theory of natural selection by Darwin and Wallace was entirely independent. It seems that particular characteristics of Victorian society were crucial for the development of this theory. Both versions of this theory were developed in a rather similar context as, for instance, both Darwin and Wallace had read Lyell and Malthus. Hence, their theories could be seen, despite their differences, as products of Victorian culture. However, it seems that only Darwin was in a position to develop his theory and publish it, the way and when he did (see Radick, 2009; Bowler, 2013).

[15] In addition, while it was believed that Wallace's essay was sent in March and arrived at Down House on June 18, 1858, a letter by Wallace to Bates leaving on the same steamer arrived in Leicester on June 3,

As already described in the previous sections, the concepts of struggle for existence, artificial selection, and divergence form the conceptual foundations of the *Origin*. The argumentation in the *Origin* involved two central concepts: the tree of life and natural selection. According to the first concept, species changed over time; some went extinct while others continued to exist or gave rise to multiple descendent species. This concept involved two other distinct concepts: transmutation (one species changing into another) and common descent (one species splitting into two or more species). The second central concept, natural selection, explained how species changed through a process of selection similar to that applied by breeders on domesticated varieties of plants or animals. The idea of common descent is logically distinct from the idea of transmutation because individual species might change significantly over time without splitting into two new ones. The idea of common descent is also distinct from the idea of natural selection because the latter might occur without the splitting of one species into two. In addition, such a splitting might take place due to a process other than natural selection. Finally, although Darwin assumed that natural selection was the dominant mechanism of transmutation, he presented examples where the latter might take place without appealing to natural selection (Waters, 2009).

The idea of transmutation was based on the existence of intra-species variation. Over the course of time, varieties of a single species could gradually become distinct species (see also Figure 4.2):

We have, also, seen that it is the most flourishing and dominant species of the larger genera which on an average vary most; and varieties, as we shall hereafter see, tend to become converted into new and distinct species. The larger genera thus tend to become larger; and throughout nature the forms of life which are now dominant tend to become still more dominant by leaving many modified and dominant descendants. But by steps hereafter to be explained, the larger genera also tend to break up into smaller genera. And thus, the forms of life throughout the universe become divided into groups subordinate to groups. (Darwin, 1859, p. 59)

The gradual transformation of the varieties of a species would lead to the emergence of new species and to greater diversity, something that was in accordance with the principle of divergence. An immediate consequence of this process would be that all species would have a common ancestry. If a species could give rise to new ones, then for each group of organisms a common ancestor should exist. Many arguments in the *Origin* are based on the assumption that some phenomena are better explained with common descent. Thus, Darwin assumed that all animals and plants had descended from a small number of species:

Therefore I cannot doubt that the theory of descent with modification embraces all the members of the same class. I believe that animals have descended from at most only four or five progenitors, and plants from an equal or lesser number. (Darwin, 1859, pp. 483–484)

1858. Darwin was thus accused of keeping the essay secret for some time in order to revise his theory. However, new evidence shows that Wallace in fact sent his essay in April 1858, for which the postal connections actually indicate the letter to have arrived precisely on June 18 (van Wyhe and Rookmaaker, 2012).

He also extended this assumption to suggest that there might be a single, universal common ancestor:

Analogy would lead me one step further, namely, to the belief that all animals and plants have descended from some one prototype. But analogy may be a deceitful guide. Nevertheless all living things have much in common, in their chemical composition, their germinal vesicles, their cellular structure, and their laws of growth and reproduction. [. . .] Therefore I should infer from analogy that probably all the organic beings which have ever lived on this earth have descended from some one primordial form, into which life was first breathed. (Darwin, 1859, p. 484)

The idea of a common ancestor was not entirely new in Darwin's time. However, a specific mechanism for the divergence of life forms from pre-existing ones was more difficult to conceive than common ancestry. Darwin focused on explaining how this might have taken place. He suggested that the divergence of life forms could be explained with an entirely natural process of intra-population change:

Owing to this struggle for life, any variation, however slight and from whatever cause proceeding, if it be in any degree profitable to an individual of any species, in its infinitely complex relations to other organic beings and to external nature, will tend to the preservation of that individual, and will generally be inherited by its offspring. The offspring, also, will thus have a better chance of surviving, for, of the many individuals of any species which are periodically born, but a small number can survive. I have called this principle, by which each slight variation, if useful, is preserved, by the term of Natural Selection. (Darwin, 1859, p. 61)

If such [variations] do occur, can we doubt [. . .] that individuals having any advantage, however slight, over others, would have the best chance of surviving and of procreating their kind? On the other hand, we may feel sure that any variation in the least degree injurious would be rigidly destroyed. This preservation of favourable variations and the rejection of injurious variations, I call Natural Selection. (Darwin, 1859, pp. 80–81)

In Darwin's view, his theory could thus serve as a unifying theory that would bring together two major principles: Geoffroy's idea of the Unity of Type and Cuvier's idea of the Conditions for Existence – in current terms, morphology and adaptation.[16]

On the theory of natural selection we can clearly understand the full meaning of that old canon in natural history, "Natura non facit saltum." This canon, if we look only to the present inhabitants of the world, is not strictly correct, but if we include all those of past times, it must by my theory be strictly true. It is generally acknowledged that all organic beings have been formed on two great laws – Unity of Type, and the Conditions of Existence. By unity of type is meant that fundamental agreement in structure, which we see in organic beings of the same class, and which is quite independent of their habits of life. On my theory, unity of type is explained by unity of descent. The expression of conditions of existence, so often insisted on by the illustrious Cuvier, is fully embraced by the principle of natural selection. For natural selection acts by either now adapting the varying parts of each being to its organic and inorganic conditions of life; or by having

---

[16] Cuvier and Geoffroy famously disagreed enormously on the importance of form and function for the explanation of organisms' characters. The central question of the Cuvier–Geoffroy debate was whether animal structure was better explained as a consequence of function or whether animal structure was independent of function, although it eventually became modified because of functional requirements. Cuvier believed that function determined structure while Geoffroy believed that structure was based on a common plan of organization from which function was derived (for the Cuvier–Geoffroy debate, see Appel, 1987).

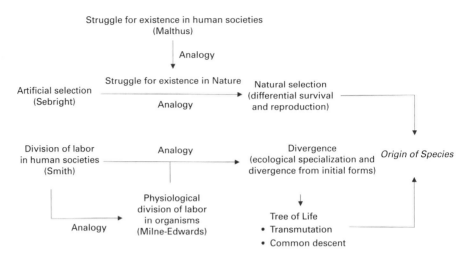

Figure 4.4 The conceptual foundations and the arguments in the *Origin*. Note that Darwin's theory was based on particular analogies.

adapted them during long-past periods of time: the adaptations being aided in some cases by use and disuse, being slightly affected by the direct action of the external conditions of life, and being in all cases subjected to the several laws of growth. Hence, in fact, the law of the Conditions of Existence is the higher law; as it includes, through the inheritance of former adaptations, that of Unity of Type. (Darwin, 1859, p. 206)

Thus, Darwin's theory of "descent with modification" was an attempt to provide a coherent account of both shared characters and species-specific ones. The existence of shared characters (which Geoffroy explained through the idea of Unity of Type) could be explained by common descent since distinct species, no matter how different they were, could still have common characters derived from a common ancestor. In addition, the existence of special adaptations which facilitated organisms' survival under particular conditions could be explained through gradual modification in particular environments by means of natural selection. Thus, Darwin's theory was a synthesis that would coherently explain both common and distinctive characters. The conceptual foundations and the central arguments of the *Origin* are presented in Figure 4.4.[17]

---

[17] There are interesting views about the structure of the *Origin*. Darwin presented natural selection first, although common descent was more evident and independent from whether or not natural selection took place. As Elliott Sober put it: "So, did Darwin write the *Origin* backwards? The book is in the right causal order; but evidentially, it is backwards" (Sober, 2009, p. 10055, 2011, p. 44). It seems that Darwin wanted to establish the existence of natural selection and to do this he started his book with a discussion of artificial selection, with which readers would be familiar. James Lennox (pers. comm.) also argues that between his presentation of the case for natural selection and the presentation of what he took to be the evidence that supported it, Darwin carefully considered the difficulties that the theory faced. The "Recapitulation and conclusion" also begin with a summary of the objections to the theory, and not with the theory itself or the evidence that supports it. Lennox argues that Darwin thought that his readers would not find his positive evidence convincing unless his responses to objections were really effective and he decided to address the objections first.

Science and religion in the reviews of the *Origin of Species*

Not only had Darwin managed to establish himself as an important man of science of his time, but he had also developed a reputation of modesty and generosity, which eventually led to a rather courteous reception of the *Origin* (Endersby, 2009, p. xxxi). In addition, through his extensive reading and correspondence he had managed to accumulate numerous detailed examples to support his arguments. This fact made the *Origin* an interesting and valuable book, even for those who entirely disagreed with its arguments. There was an enormous public reaction to the *Origin*, but not (only) because of religiously motivated instincts as is commonly thought. In this final section, I present criticisms, both by proponents and opponents of Darwin, which were founded on scientific grounds. As I described in Chapter 2, one should try to distinguish between knowledge and belief, or between scientific arguments and religious sentiment. It is therefore important to note that some critics of the *Origin* raised important scientific objections, independently of their religious beliefs. The aim of this section is thus twofold: (1) to show that even ardent Darwinians like Huxley did not blindly accept all of Darwin's propositions, as well as that clergymen like Wilberforce criticized the *Origin* solely on scientific grounds, and (2) that it is indeed possible for scholars to debate on scientific–epistemic grounds despite their (sometimes entirely contrasting) religious beliefs (see also Lennox, 2010 for such an account).

Thomas Henry Huxley was one of the prominent supporters of Darwin, who is sometimes described as "Darwin's bulldog." He wrote an interesting review that was anonymously published in the *Westminster Review* (Huxley, 1860). Huxley, despite his support for Darwin's theory, which he considered the best up to that time, raised some serious criticisms:

There is no fault to be found with Mr. Darwin's method, then; *but it is another question whether he has fulfilled all the conditions imposed by that method. Is it satisfactorily proved, in fact, that species may be originated by selection? that there is such a thing as natural selection?* that none of the phænomena exhibited by species are inconsistent with the origin of species in this way? If these questions can be answered in the affirmative, Mr. Darwin's view steps out of the rank of hypotheses into those of proved theories; but, so long as the evidence at present adduced falls short of enforcing that affirmation, so long, to our minds, must the new doctrine be content to remain among the former an extremely valuable, and in the highest degree probable, doctrine, indeed the only extant hypothesis which is worth anything in a scientific point of view; but still a hypothesis, and not yet the theory of species.

After much consideration, and with assuredly no bias against Mr. Darwin's views, it is our clear conviction that, as the evidence stands, *it is not absolutely proven that a group of animals, having all the characters exhibited by species in Nature, has ever been originated by selection, whether artificial or natural.* Groups having the morphological character of species distinct and permanent races in fact have been so produced over and over again; but there is no positive evidence, at present, that any group of animals has, by variation and selective breeding, given rise to another group which was, even in the least degree, infertile with the first. Mr. Darwin is perfectly aware of this weak point, and brings forward a multitude of ingenious and important arguments to diminish the force of the objection. We admit the value of these arguments to their fullest extent; nay, we will go so far as to express our belief that experiments, conducted by a skilful physiologist, would very probably obtain the desired production of mutually more or less infertile breeds from a common stock, in a comparatively few years; but still, as the case stands at present, this "little rift within the lute" is not to be disguised nor overlooked. (Huxley, 1860, pp. 567–568, emphases added)

Huxley acknowledged the importance of Darwin's contribution, but yet pointed out that even if there was enough evidence for the competency of natural selection to produce new species, there was no actual evidence that it had indeed done so. As a result, more work was necessary to establish Darwin's "hypothesis" as a "theory of species."

This is an important point, and interestingly enough the same criticism was anonymously made by Samuel Wilberforce, Bishop of Oxford, who also wrote a review of the *Origin* in 1860. There exists a widespread myth about the encounter between Wilberforce and Huxley. According to this, Wilberforce attempted to ridicule Darwin and his theory at a meeting of the British Association in Oxford on June 30, 1860. There he faced Huxley, who is said to have succeeded in defeating Wilberforce and, through that, any attempt to dictate to scientists the conclusions they were allowed to reach. However, careful historical analysis has shown that the legend overlooks the fact that Wilberforce's speech, rather than reflecting prejudice and religious sentiment, included many of the scientific objections of Darwin's contemporaries. It may also be the case that Joseph Dalton Hooker's contribution in defending Darwin was more successful than Huxley's (see Lucas, 1979; Livingstone, 2009). Rather than being an instance of a wider conflict between science and religion, the Huxley–Wilberforce debate reflects trends and developments in Victorian society that had to do with the formation of science as a profession, particular divisions within the Church, reactionary voices from inside the Church such as Baden Powell, the emergence of new scientific methodologies, and the challenges that publications such as the *Vestiges* caused (Brooke, 2001; Livingstone, 2009).

Eventually, in his review of the *Origin*, Wilberforce actually expressed the same criticism as Huxley:

We come then to these conclusions. *All the facts presented to us in the natural world tend to show that none of the variations produced in the fixed forms of animal life, when seen in its most plastic condition under domestication, give any promise of a true transmutation of species*; first, from the difficulty of accumulating and fixing variations within the same species; secondly, from the fact that these variations, though most serviceable for man, have no tendency to improve the individual beyond the standard of his own specific type, and so to afford matter, even if they were infinitely produced, for the supposed power of natural selection on which to work; whilst all variations from the mixture of species are barred by the inexorable law of hybrid sterility. *Further, the embalmed records of 3000 years show that there has been no beginning of transmutation in the species of our most familiar domesticated animals*; and beyond this, that in the countless tribes of animal life around us, down to its lowest and most variable species, no one has ever discovered a single instance of such transmutation being now in prospect; no new organ has ever been known to be developed – no new natural instinct to be formed – whilst, finally, in the vast museum of departed animal life which the strata of the earth imbed for our examination, whilst they contain far too complete a representation of the past to be set aside as a mere imperfect record, yet afford no one instance of any such change as having ever been in progress, or give us anywhere the **missing links** of the assumed chain, or the remains which would enable now existing variations, by gradual approximations, to shade off into unity. (Wilberforce, 1860, pp. 247–248, emphases added)

In the same sense as Huxley, Wilberforce questioned the competence of natural selection to produce new species as there was no evidence that it has ever been actually responsible for doing so. That the same criticism was made both by Huxley, "Darwin's bulldog," and the Bishop of Oxford is important for two reasons. First, it shows that scientific judgments and criticisms can be made on objective grounds by people with vast differences in their worldviews. Their motivations notwithstanding, both Huxley and Wilberforce raised a question that Darwin already knew would be raised: Does natural selection actually produce new species? Second, this shows that there was more in the reaction to Darwin's theory than religious instinct and fundamentalism. Not all religious people are ignorant and fundamentalists. Some of them may confuse what they actually know with what they believe, but proponents of evolution must take their (scientific and philosophical) arguments seriously when they point to actual problems and difficulties of scientific theories. Wilberforce wrote in his review that:

Our readers will not have failed to notice that we have objected to the views with which we have been dealing solely on scientific grounds. We have done so from our fixed conviction that it is thus that the truth or falsehood of such arguments should be tried. We have no sympathy with those who object to any facts or alleged facts in nature, or to any inference logically deduced from them, because they believe them to contradict what it appears to them is taught by Revelation. We think that all such objections savour of a timidity which is really inconsistent with a firm and well-instructed faith. (Wilberforce, 1860, p. 256)

Darwin himself acknowledged the quality of the review just a few days after the debate, in a letter he wrote to Hooker:

P.S. I have just read Quarterly R. It is uncommonly clever; picks out with skill all the most conjectural parts, & brings forwards well all difficulties. – It quizzes me quite splendidly by quoting the Anti-Jacobin versus my grandfather. – You are not alluded to; nor, strange to say, Huxley, & I can plainly see here & there Owen's hand. (Darwin, C. R. to Hooker, J. D., July 20?, 1860)

Richard Owen, one of the leading scientific figures of the time, also wrote his own anonymous critical review of the *Origin*. Owen again questioned the competence of natural selection to produce new species:

Individuals, it is said, of every species, in a state of nature annually perish, and "the survivors will be, for the most part, those of the strongest constitutions and the best adapted to provide for themselves and offspring, under the circumstances in which they exist." [...] *The element of "natural selection" above illustrated, either is, or is not, a law of nature. If it be one, the results should be forthcoming*; more especially in those exceptional cases in which nature herself has superadded structures, as it were expressly to illustrate the consequences of such "general struggle of the life of the individual and the continuance of the race."[18] The antlers of deer are expressly given to the male, and permitted to him, in fighting trim, only at the combative sexual season; they fall and are renewed annually; they belong moreover to the most plastic and

---

[18] Darwin had written in the *Origin* that "Individual males have had, in successive generations, some slight advantage over other males in their weapons, and have transmitted these advantages to their male offspring" (Darwin, 1859, p. 89).

variable parts or appendages of the quadruped. *Is it then a fact that the fallow-deer propagated under these influences in Windsor Forest, since the reign of William Rufus, now manifest in the superior condition of the antlers, as weapons, that amount and kind of change which the succession of generations under the influence of "natural selection" ought to have produced?* Do the crowned antlers of the red deer of the nineteenth century surpass those of the turbaries and submerged forest-lands which date back long before the beginning of our English History? Does the variability of the artificially bred pigeon or of the cultivated cabbage outweigh, in a philosophical consideration of the origin of species, those obstinate evidences of persistence of specific types and of inherent limitation of change of character, however closely the seat of such characters may be connected with the "best chance of taking care of self and of begetting offspring?" If certain bounds to the variability of specific characters be a law in nature, we then can see why the successive progeny of the best antlered deer, proved to be best by wager of battle, should never have exceeded the specific limit assigned to such best possible antlers under that law of limitation. *If unlimited variability by "natural selection" be a law, we ought to see some degree of its operation in the peculiarly favourable test-instance just quoted.* (Owen, 1860, pp. 519–520, emphases added)

In the same spirit as Wilberforce, Owen questioned the competence of natural selection to produce new species as there was no evidence of significant change over the years in animals or plants under domestication, or at least living in a place where they could be observed by humans.

Critical reviews also appeared for subsequent editions of the *Origin*. There is no need to quote more than the above reviews in order to reach an important conclusion: that it is indeed possible to distinguish between what one knows and what one believes. Huxley described himself as an agnostic and he was trying to turn science to a profession so that people other than clergymen could get employment in universities. His main aim was to liberate the practice of natural history from the domination of the Church. Wilberforce was a bishop so there is no need to get into details about his religious beliefs. There were clergymen such as Charles Kingsley and Baden Powell who gladly accepted the idea of evolution; Wilberforce was for various reasons opposed to that. Owen was one of the greatest anatomists of his time, a prominent expert on fossils, the first superintendent of the natural history department of the British Museum, and also the scientist who convinced the government to found the London Natural History Museum. He made major contributions to science by revealing homologies which he ended up explaining as instances of the archetype in God's mind. And yet, as is shown by the above quotes, all these people came to the same conclusion regarding Darwin's theory and questioned the competency of natural selection in producing new species, as it was never shown to be responsible for doing so.

In my view, these quotes are good examples of how people should talk about science. Their (probably religious) motivation notwithstanding, Wilberforce and Owen raised important questions that Darwin did not manage to answer. His explanation for the origin of species, which he described as descent with modification, was questioned on rational grounds. Natural selection seemed to be competent of producing new species, and thus was a plausible explanation for some observations (e.g., biogeography), but it was never shown that it was responsible for doing so. Thus, natural selection could not be a true cause according to Herschel and Whewell. But even today, no scientist would accept an explanation as definitive if it was never shown that what is presented as a cause does indeed have particular effects. But since Darwin's time numerous examples of evolution in general, and natural selection in particular, in action have been described (see Chapter 1 for the HIV virus; see also Coyne, 2009; Dawkins, 2009; Rogers, 2011). This is not to deny that evolutionary theory may have implications for personal world-views. But here is what I think rationality demands: Let us first examine the evidence available, decide whether a theory has a solid evidential basis, and then wonder about its implications beyond the realm of science. Of course, no one can tell what Wilberforce and Owen would think today given the available evidence. But what they did, i.e., discuss scientific issues only, is what we should all do when it comes to science (see also my Concluding remarks on this point).

Conclusions

For 20 years Darwin accumulated evidence that contradicted the widely accepted explanation that species were created and which also formed the basis for a new theory that, if true, would more plausibly explain that evidence. In this two-step process (conceive a new theory that explains data better than older ones and confirm it) Darwin was fully successful in the first step only. Confirmation usually takes time and later generations of scientists showed that the basic arguments of the *Origin* were correct. But in order to do this, people found more evidence and evolutionary theory was thus further refined. Indeed, the evolutionary theory of the mid-twentieth century was in some respects very different from Darwin's theory. Darwin's was a force-based (*vera causa*) theory, whereas the theory of the mid-twentieth century was statistical (Depew and Weber, 1995; Depew, 2013). And there are even more recent advances in evolutionary theory that make it even more different (Pigliucci and Muller, 2010). In the next two chapters I will describe in detail core evolutionary concepts and processes, while I will also focus on the structure and nature of evolutionary explanations. The focus of these two chapters is on what we currently know and serve as an introduction to the main concepts of contemporary evolutionary theory.

Further reading

There are numerous books about Darwin and the history of evolutionary thinking. Perhaps the best book to start with is Janet Browne's *Darwin's "Origin of Species": A Biography*, which provides a nice overview of Darwin's life until the *Origin* was published. Browne has also co-authored with Adrian Desmond and James Moore, the other major Darwin biographers, a rather short biography of Darwin for the Oxford series VIP (very interesting people). Desmond and Moore are the authors of a major biography, entitled *Darwin: The Life of a Tormented Evolutionist*, published in the early 1990s. Janet Browne later published a massive two-volume biography of Darwin; the first volume is entitled *Charles Darwin: Voyaging* and the second one *Charles Darwin:*

The Power of Place. Desmond and Moore have more recently published an account of how Darwin's hatred for slavery influenced his view on human evolution in *Darwin's Sacred Cause*. If one wants to start with the *Origin*, then a highly recommended edition is the recent Cambridge University Press version, edited by Jim Endersby. This edition includes a lengthy introduction that nicely sets the context in which the book was written. I have also always liked the Harvard facsimile edition of the *Origin*, edited by Ernst Mayr. All of Darwin's books, manuscripts, and much more are freely available online at Darwin Online (http://darwin-online.org.uk). Similarly, most of his correspondence is freely available by the Darwin Correspondence Project online at www.darwinproject.ac.uk. As is probably evident in this chapter, Dov Ospovat's book, *The Development of Darwin's Theory: Natural History, Natural Theology and Natural Selection, 1838–1859*, provides a detailed account of Darwin's writings from his manuscripts until the publication of the *Origin*. For those wishing to go further and deeper into historical and philosophical issues, two must-read books are *The Cambridge Companion to Darwin* (2nd edition, 2009) edited by Jonathan Hodge and Gregory Radick, and *The Cambridge Companion to the Origin of Species*, edited by Michael Ruse and Robert J. Richards. Another valuable, but old and perhaps difficult to obtain, book is *The Darwinian Heritage*, edited by David Kohn. Perhaps the best history of evolutionary thought is *Evolution: The History of an Idea* by Peter J. Bowler. Quite interesting is also the recent book *Darwin Deleted: Imagining a World Without Darwin*, by the same author. Another interesting but shorter history is *Evolution: The Remark- able History of a Scientific Theory* by Edward Larson. Books about evolution and religion are suggested at the end of Chapter 2; however, Nick Spencer's *Darwin and God* provides a nice overview and is more relevant to the topics discussed in this chapter. Finally, *The Cambridge Encyclopedia of Darwin and Evolutionary Thought*, edited by Michael Ruse, is a useful resource on almost everything about Darwin.

5 Common ancestry

In the previous chapter I described the development of Darwin's theory, which established the foundations of modern evolutionary theory. It was explained that Darwin's theory was in particular ways incomplete because, for instance, there was no evidence that natural selection could indeed produce new species from pre-existing ones. As discussed in the previous chapter, this raised important criticisms, from both proponents and opponents. However, the major arguments in the *Origin* were sound and pointed to two important scientific facts: (1) that all organisms living on Earth are related through descent from common ancestors, thus forming a tree (or rather, as I describe later, a network) of life, and (2) that all organisms living on Earth evolve through natural processes (often, but not exclusively, through natural selection) and in doing so species may persist, evolve to new ones, or die out. Despite the advancements in evolutionary biology in the more than 150 years since the publication of the *Origin*, Darwin's description of evolution as *descent with modification* remains at the core of contemporary evolutionary theory. These two ideas, common descent and evolutionary change through modification of extant species, will be the focus of this and the next chapter, respectively. Let us start with the idea of common descent. In Chapter 3 I argued that organisms are not designed in the way artifacts are. So, contrary to artifacts, we have to look for the origin of species in pre-existing ones, and should not postulate any intentional plan behind that. All species have evolved through natural processes, and the outcome is that both extinct and extant species are more or less related. Actually, time matters and so two species usually are more similar the more recent their common ancestor is. Does this necessarily imply that there must be some universal common ancestor(s) from which all life evolved? Yes, it does.

Scientists are certain that there must have been one or a few universal common ancestor(s) because all organisms share some crucial characters: (1) they consist of (one or more) cells; (2) they exhibit the characteristic properties of life (metabolism, reproduction, homeostasis, etc.[1]), which are the outcome of intra- and/or intercellular processes; (3) proteins have central roles in these processes; and (4) these proteins are synthesized inside cells on the basis of specific DNA sequences and their interaction with their cellular contexts. These common characteristics are fundamental to life on

---

[1] Viruses and prions may exhibit some but never all of these properties and therefore they are usually not considered as alive. Thus, we tend to exclude these from living forms, although this exclusion may be quite subjective (see Cleland and Zerella, 2013; see also Moreira and López-García, 2009).

Earth, but how could they have emerged? One intuitive explanation would be that they are products of design; these were characteristics of an archetypal plan, on the basis of which living organisms were formed. However, in Chapter 3 I explained that some peculiar characters of organisms cannot be the product of design because no competent and rational designer would design something with fundamental problems or imperfections (remember the comparison between dolphins and sharks). But couldn't it be the case that the aforementioned characteristics (1–4) which are common in all organisms and fundamental to life on Earth are the products of design? Couldn't there be a designer who designed the molecular and cellular foundations of life and then let it evolve?

This could be the case, but again there is such diversity in the molecular and cellular levels that is explained more sufficiently and coherently as the outcome of evolution rather than the product of intentional design. For example, at least 5% of the human genome consists of families of DNA sequences which are more than 90% identical to each other. At first thought, this could be perceived to be the product of design, since the presence of multiple copies of a sequence makes sense when these are useful. The DNA sequences which are implicated in the synthesis of rRNA are present in more than 400 copies in the human genome, and so an adequate amount of rRNA and eventually of ribosomes, where such a fundamental process as protein synthesis takes place, is produced. Also, several DNA sequences are involved in the production of the subunits of oxygen-carrying proteins (hemoglobin A, hemoglobin A_2, hemoglobin F, myoglobin), which are specialized for particular tissues or particular stages of development. So, it seems that having multiple copies of similar DNA sequences is an advantage. However, these multiple copies of similar DNA sequences are responsible for a number of disorders. Because these DNA sequences are very similar to each other and because they are also arranged close to each other, they cause the abnormal pairing of the respective chromosomes during meiosis. As a result, various deletions, additions, inversions, or translocations[2] of DNA sequences occur and they are associated with a number of disorders such as α-thalassemia, hemophilia A, neurofibromatosis type 1, red–green color blindness, Prader–Willi syndrome, and others (Avise, 2010, pp. 108–112). Thus, it seems that having multiple copies of DNA sequences in our genomes is not always good. Their presence, which results in genetic disorders, is sufficiently explained as the outcome of evolution (e.g., unequal crossing over has produced extra copies of existing sequences which later accumulated changes and might have acquired new roles or not). There is no need to invoke non-natural processes or any kind of rational design.

All similarities between organisms at the molecular level are best explained through evolution as a consequence of common descent. A nice description of why common descent is a sound inference of what we observe has been given by Francisco Ayala (2009, p. 136). The genomes of all organisms are like books which contain text written with the same letters (DNA nucleotides), which are combined to produce more or less

---

[2] These are changes in the structure of chromosomes; see Griffiths *et al.* (2012) for the details of these and other similar phenomena.

the same words (almost universal genetic code). What is more important is that the different texts are quite similar, not only because they contain the same words, but also because these are used to form quite similar sentences (DNA and protein sequences). Let me elaborate on this argument. Consider this: the English and French alphabets are identical in terms of the letters they include, although they are pronounced differently. We know of course that these two alphabets are derived from Latin and thus their similarity is not a coincidental one: it is due to common descent. Nevertheless, these two identical alphabets give rise to very different words for the same concepts. Imagine the numbers 1, 2, 3, and 4: The words that correspond to these numbers in English are "one," "two," "three," and "four," respectively, whereas the respective words in French are "un," "deux," "trois," and "quatre." Not identical for sure. Despite the fact that some words are spelled identically (although they may be pronounced differently, e.g., the word "table") most words that correspond to the same concept are different. Compare "walk" and "marcher," "sea" and "mer," or "sky" and "ciel." Not surprisingly, sentences are, as a consequence, very different, too. Thus, the sentence "I walk by the sea and I look at the sky" becomes "je marche prés de la mer et je regarde le ciel."[3] Now, if two closely related languages can be so different in terms of words and sentences, the only sound conclusion is that the very similar DNA language used in all forms of life which is framed using exactly the same alphabet (A,T,C,G) and almost the exact same words (DNA, RNA, and protein sequences) can only be evidence for their close relatedness and common ancestry.

In this chapter I first describe what we currently know about the common ancestry of all life on Earth and I also explain why this is the only plausible explanation for the common characters of organisms we currently observe. Then I focus on the concept of **homology** and I explain how similar characters are found in organisms because the latter have a common ancestor. I also distinguish between homologies and homoplasies, similar characters which are not due to common descent but which have evolved independently, and briefly comment on the fact that there are deeper homologies between distantly related organisms. Finally, I describe how evolutionary developmental biology, one of the most active fields of contemporary biological research, sheds new light on the evolution of life on Earth and actually explains much that was until recently left unexplained or was explained by extra assumptions.

The evolutionary network of life

Imagine two families, one coming from Europe and another coming from Africa. Imagine also that each family consists of four members: a father, a mother, a daughter, and a son. The European parents are white with brown, straight hair. Their children are

---

[3] As a fan of *The Beatles*, I can't help thinking of the lyrics of "Michelle, ma belle." Paul McCartney sings that "these are words that go together well," having previously sung in French that "[ce] sont des mots qui vont très bien ensemble." This is the same message, transmitted through very different words which are nonetheless based on very similar alphabets.

(a)

(b)

Figure 5.1 Two families of (a) African and (b) European origin. Children usually resemble their parents in several characters such as hair type or skin color, as is the case here. Image © Alashi.

neither identical to their parents, nor to each other, but they exhibit the aforementioned characters (white skin and brown, straight hair). In contrast, the African parents have brown skin and black, curly hair. Again, their children are neither identical to their parents, nor to each other, but they exhibit the aforementioned characters (brown skin and black, curly hair) (Figure 5.1). Why do children resemble their parents? Genetics provides the answer: Offspring develop from a fertilized ovum, a single cell that emerges from the fusion of the reproductive cells of their parents. Consequently, the genetic material of that first cell consists of the DNA molecules contained in the spermatozoon and the ovum of the parents, and each offspring possesses a unique combination of one half of the maternal DNA and one half of the paternal DNA. Eventually, during development a multicellular organism emerges and particular parts of this DNA interact with their cellular environment to drive the formation of tissues and organs. Offspring resemble their parents in some respect because they have inherited part of their genetic material, but they will also be different from either or both of them due to the specific interactions of paternal and maternal DNA molecules, as well as due to several important phenomena taking place during development.

Family members are usually depicted in family trees. This is a way to provide an overview of the members of a family and their characters, and such trees are used in

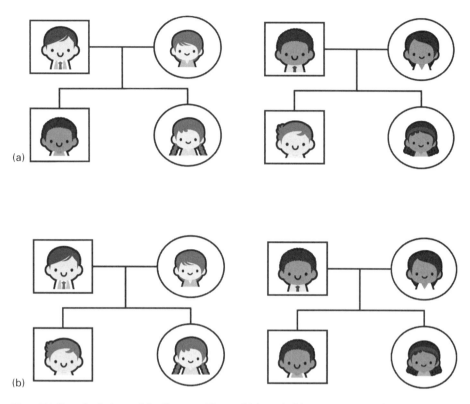

Figure 5.2 Two depictions of family trees. Figure (b) is probably more accurate than (a) because in the latter case all family members in each of the families have the same skin color. Image adapted from Figure 5.1, © Alashi.

genetic counseling. Parents are connected to each other with a horizontal line, whereas both of them are connected to their children with vertical lines. Males are depicted with rectangles and females with circles. Family trees are useful because one can infer shared characters based on relationships, as well as infer relationships based on shared characters. It is the latter that is of utmost importance for evolutionary biology. Scientists do not know the exact relationships between (extant or extinct) organisms and so they rely on common characters to make inferences[4] about relationships. Let us see how this works; the families of Figure 5.1 are depicted in family trees (Figure 5.2a and Figure 5.2b). Assuming you had never seen Figure 5.1, could you infer which of the family trees is more accurate? Yes, you probably could. Figure 5.2a provides a less probable case because in both families there is a family member with different skin color than his/her parents. This is not entirely impossible; however, the depiction in

---

[4] Inferences are central in science. An inference is made when a conclusion is drawn from a set of premises, and includes both the psychological process of drawing conclusions and the rules that entitle or justify drawing conclusions from certain premises (Psillos, 2007, p. 122). A major explanatory process in biology in general and evolutionary biology in particular is inference to the best explanation (see Lipton, 2004; see also Chapter 6).

Figure 5.2b is more probable because in both cases all offspring resemble their parents in terms of skin color, which is an inherited character.[5] Relatedness is in this way inferred from shared characters.

The point made here is that if we knew nothing about these families (e.g., if we had never seen Figure 5.1), we would nevertheless be able to group these individuals into families based on their skin color. Thus, here is a first conclusion: we can group individuals on the basis of shared characters. Those individuals that share a character can be grouped together and consequently will be found in a different group from those with whom they differ in that character (in this example their skin color). The inference then made is that the members of each family are related, because this gives a good explanation for their resemblance; kinship explains resemblance because the character under consideration is an inherited one. This brings up a new question: Their difference in skin color notwithstanding, could these families, and consequently their family members, be somehow related? To answer this question, we need to identify characters which are shared by all of them. If we find some, then the new inference will be that these individuals are related. Indeed, this is the case. All members of these families are humans; they have two eyes, two ears, a nose, hair on the top of their heads, and numerous other characters shared by humans. Consequently, each member of the European family is related to each member of the African family. And the difference in relatedness between any two individuals is a matter of degree, not of kind. For instance, the girl of the African family has the same skin color as her brother and differs from the European girl who has a lighter skin color. Thus, in terms of skin color only, the African girl is more closely related to her brother than to the European girl.

It should be noted at this point that the grouping performed here was based on a character chosen arbitrarily: skin color. If we decided to use another inherited character, e.g., human blood groups, the eventual grouping could be different or we might end up with different kinds of groupings. For instance, let's assume we decided to group these individuals on the basis of their blood groups. Let's assume that African individuals all had blood group A and that the European individuals all had blood group O, as well as that we knew nothing about the skin color of these individuals. Could we then conclude that all individuals with blood group A are members of the same family and that all individuals with blood group O are members of the same family as well? The answer is no. Although parents with blood group O can only give birth to children with blood group O, it may be the case that parents with blood group A give birth to a child with blood group O. Consequently, either of the European children could, in terms of grouping based on blood groups only, belong to the African family. The important conclusion here is that we cannot use any character for grouping, or that some characters may be more appropriate than others (see the next section).

---

[5] The inheritance of skin color is a case of polygenic inheritance: the color of the skin is affected by several DNA sequences which have an additive effect (the more the DNA sequences that contribute to dark color, the darker the color will be).

Returning to our earlier conclusion that all individuals are related because, despite the difference in skin color, they share a number of other common characters, the conclusion to be made is that all of these individuals belong to a wider group: humans. The European children belong to the European family, which is distinct from the African family to which the African children belong. The reason for this is that the parents of the European children (their most recent ancestors) are different from the parents of the African children (again, their most recent ancestors). This explains the difference in skin coloration. However, both the European and the African children and their parents belong to a wider group, humans, the members of which share a number of common characters (two ears, two eyes, one nose, more than 99% of their DNA sequences, and so much more). These can be explained by accepting that both the African and the European children have, except for their parents who are their most recent ancestors, some more remote common ancestors from whom these common characters are derived: the first humans. This makes the two families related as well. Each child is more related to the members of his/her own family and less related (but related nevertheless) to the members of the other family (Figure 5.3). Of course, in this way they are also related to every other human family in the world.

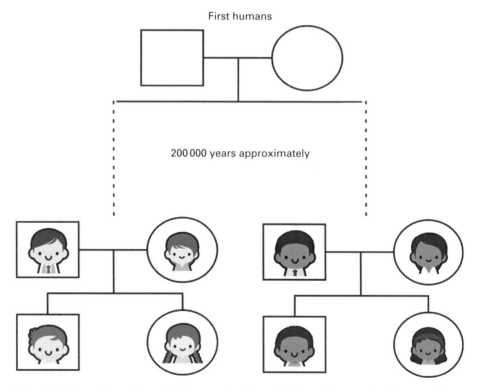

Figure 5.3 All members of these families are related to each other and to any other human, and as a consequence they all share some common "humane" characters. Scientists estimate that modern humans (*Homo sapiens*) emerged about 200 000 years ago. However, they still lack several details of human evolution (see Figure 1.4). Image adapted from Figure 5.1, © Alashi.

What has been described so far is the human branch, a single branch in the evolutionary network of life.[6] In order to depict the evolution of life on Earth and the relatedness of the various taxonomic groups (taxa[7]) scientists use a different means of representation which, however, has some apparent similarities with family trees: evolutionary trees (see Figure 1.3). Evolutionary trees are depictions of evolutionary relationships among taxa.[8] The branching points (nodes) correspond to common ancestors. What is important here is that evolutionary trees indicate historical relationships, and not just similarities. Closely related species generally tend to be similar to one another; however, this is not always the case. For instance, crocodiles look more similar to lizards but they are more closely related to birds when DNA sequences are used for the comparison. In this case, relatedness refers to common ancestry: two species are more closely related the more recent is their common ancestor (Baum et al., 2005). This could be illustrated in the family trees above (Figure 5.3): the European girl is more closely related to her brother than to the African girl because the (most recent) common ancestors she shares with her brother, their parents, are more recent than the (more remote) common ancestor she shares with the African girl. Thus, both family trees and evolutionary trees provide historical information so that one can infer relatedness from how old the common ancestor is.[9] The other common element is that both facilitate grouping. In family trees we can group individuals into families; a family could consist of a couple and their children, or of a couple and their children, grandchildren, and great-grandchildren, or of an individual and his/her parents, his/her grandparents, and his/her great-grandparents, etc. In evolutionary trees taxa can be grouped in clades, which are hierarchically nested groups that include a common ancestor and all its descendants. To illustrate this in terms of a family tree, a group that includes a couple, all their children, and all their grandchildren would be a clade; if the group included a couple, one of their children and his/her offspring but not the other one and its offspring, then this would not be a clade (for a concise and comprehensive discussion of how evolutionary trees are constructed and read, see Baum and Offner, 2008; Gregory, 2008; for a book-length treatise on the topic, see Baum and Smith, 2013).

---

[6] Stephen J. Gould advanced the idea that life is like a bush, by writing that "Life is a copiously branching bush, continually pruned by the grim reaper of extinction, not a ladder of predictable progress" (Gould, 2000/1989, p. 35). It is much preferable to refer to a bush rather than a tree for two reasons: (1) bushes do not grow upwards like trees, which gives the impression of some kind of progress, and (2) the branches of a bush are closer to each other and to the roots compared to those of a tree. However, as will soon be explained, the complexities of relatedness among organisms are so many that the metaphor of the bush is insufficient. I think the metaphor of an evolutionary network is more appropriate.

[7] The term *taxon* (plural *taxa*) will be used here to refer to a taxonomic group of organisms such as a phylum, a class, a species, etc. (phylum, class, species are categories; individual phyla, classes, and species are taxa).

[8] Scientists distinguish between phylogenetic trees and cladograms. Phylogenetic trees depict actual evolutionary histories and their branches have different lengths to show the relative age of extinction. The branches of cladograms constitute hypotheses of relative recency of common ancestry and ancestral or extinct taxa are located at the tips of terminal branches like extant taxa (Sereno, 2005). The nuances of these differences are not important for our purposes here. Thus, although most of the trees that will be used in this book are actually cladograms, in all cases reference will be made to evolutionary trees.

[9] Of course, we should not overlook the major differences between family trees and evolutionary trees, the most important one being that the common ancestor in an evolutionary tree is a single taxon, whereas the common ancestor in a family tree is a couple.

What is important within a clade is that all its members (the common ancestor and all its descendants) share at least one common (usually identifying) character that is derived from the common ancestor. What is more interesting, and occasionally more confusing, is that different evolutionary trees can be constructed depending on the characters used for classification. Remember that in Chapter 1 it was shown that humans are more closely related to chimpanzees when molecular data are considered, whereas humans are more closely related to orangutans when structural, behavioral, and physio-logical data are considered (see Figure 1.3; in the same sense it was explained above that if the members of the families in Figure 5.1 were classified in terms of their blood groups and not their skin color, a different classification from the one in Figure 5.2 would have emerged). Thus, clades and ancestry as depicted in evolutionary trees are always relative to the criterion used for classification. This, of course, does not mean that all organisms within a clade share exactly the same characters. Roughly put, there are two main types of characters: homologies, which point to common ancestors, and homoplasies, which have evolved independently in different clades. These will be the topics of the next two sections.

Based on these, let us now draw an analogy between family trees and evolutionary trees. Again, it should be noted that these trees are not equivalent. However, familiar family trees may help one understand the less familiar evolutionary trees. To make the analogy work, some assumptions are required: We can think of clades as families and of species as individual members of families; births will correspond to speciation events, i.e., to the production of new species, and both parents will correspond to the common ancestor. Thus, let's imagine a couple (I1 and I2) that gives birth to two children (II2 and II3), each of which also gives birth to two children (III1, III2, III3, III4 – to illustrate the differences, the offspring of daughter II2 will be two girls whereas the offspring of son II3 will be two boys) (Figure 5.4a). This family tree can, under the above assumptions, be represented as an evolutionary tree (Figure 5.4b – circles are used for both males and females here) that is equivalent to an actual one (Figure 5.4c), which can also be depicted in a branching form (Figure 5.4d). Grandparents (I1 and I2) correspond to the earlier common ancestor (G) of all species A–D; parents (II2 and II3) correspond to the most recent common ancestor P of species A and B, and to the most recent common ancestor Q of species C and D, respectively. Finally, children (III1, III2, III3, III4) correspond to species A–D. Species are genetically more similar to their more recent common ancestor than to the older one, exactly as children are genetically more similar to their parents than to their grandparents. Species A and B, as well as species C and D, are not identical to each other as two brothers or two sisters are not identical to each other (except in the case of identical twins, which is not the case here).

Let us now examine what an evolutionary tree of all life, corresponding to the family tree in Figure 5.3, would look like.[10] Do we know what the universal common ancestor was like? It seems that we have a good idea of what this common ancestor could be. Organisms are usually divided in two major groups, based on the morphological characters

---

[10] An overview and access to detailed references are available at www.tolweb.org (Tree of Life Web Project)

(a)

(b)

(c)

(d)

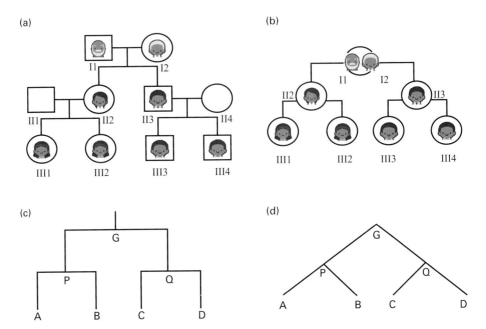

Figure 5.4 Analogy between family trees and evolutionary trees. A family tree (a) is gradually (b) transformed to an evolutionary tree (c) in order to show the similarity in the process of reconstructing human and species ancestry (under some abstract assumptions of course – see text for details). Family trees and evolutionary trees are not equivalent, but they are both constructed to represent genealogies. Image adapted from Figure 5.1, © Alashi.

of their cells: prokaryotic (cells without nuclei; this group practically includes bacteria only) and eukaryotic (cells with nuclei – this group practically includes everything else except for bacteria). Following a proposal first made by Woese and his colleagues (1990), organisms are nowadays classified in three domains: Archaea, Bacteria, and Eukarya (or archaeobacteria, bacteria, and **eukaryotes**, respectively). It seems that the first universal common ancestor was not a single cell, but rather "[i]t was communal [. . .], a loosely knit, diverse conglomeration of primitive cells that evolved as a unit, and it eventually developed to a stage where it broke into several distinct communities, which in their turn become the three primary lines of descent" (Woese, 1998, p. 6858).

The first two domains, Archaea and Bacteria, include unicellular, prokaryotic organisms only. It seems that the first nucleated cells, which later gave rise to eukaryotes, were cells without mitochondria that emerged from the merging of an archaeobacterium (similar to today's *Thermoplasma acidophila*) and a eubacterium (similar to today's *Spirochaeta*). It is from such a **symbiogenesis** that eukaryotic cells later emerged. Some of these first cells ingested and retained oxygen-breathing (aerobic) bacteria, which evolved to eukaryotic cells with mitochondria, such as those of animals. Finally, some of these aerobic cells ingested and retained cyanobacteria and evolved to cells with plastids, such as those of algae and plants. Archaeobacteria and bacteria are generally described as prokaryotic organisms, and morphologically they do not seem to be different. However, research in molecular and cellular biology has revealed that they are physiologically and genetically very different. In contrast, eukaryotes include both unicellular and multicellular organisms; organisms seemingly very diverse from each other such as protists, fungi, algae, plants, and animals belong to the same domain. This should not be surprising given that most organisms on Earth are unicellular. What is more interesting is that in some respects (e.g., in terms of the transcriptional and translational machinery of cells), archaeobacteria are more similar to eukaryotes than to bacteria.[11] This initially seems to suggest that eukaryotes evolved from archaeobacteria. If this were true then eukaryotic DNA sequences similar to bacterial ones should only exist within mitochondria and chloroplasts, and would be relevant to either respiration or photosynthesis. However, this is not the case as DNA sequences similar to bacterial ones have been found in the nuclei of eukaryotes. This suggests that the evolution of eukaryotes was more complicated than it was initially thought; it was not a linear process, but involved extensive exchange of genetic material between cells, described as horizontal gene transfer.[12] In short, the evolutionary relationships of eukaryotes to the other two domains are not entirely clear as they exhibit characters of either archaeal or bacterial origin, as shown in Tables 5.1 and 5.2 (Doolittle, 2000; Margulis and Sagan, 2009; Koonin, 2010a, 2010b; Cavicchioli, 2011).

Currently classification is based more on molecular sequences rather than on morphological features, mostly because there are similarities in molecules due to common ancestry for which there are no corresponding, apparent morphological similarities. It is **homologous DNA sequences**, i.e., sequences derived from some common ancestor, which are actually considered for classification. There are three different types of homologous DNA sequences: **orthologous** ones have evolved from a common ancestral DNA sequence through speciation events, i.e., events that lead to the emergence of new species; **paralogous** ones have evolved from a common ancestral DNA sequence through duplication events; and **xenologous** ones have emerged through horizontal transfer of DNA sequences between different species. It is the comparison of orthologous DNA sequences that yields information for phylogenetic relationships because these are by definition related to speciation events (see Fitch, 2000).What is then done is that sequences are carefully compared to each other, differences between all pairs of sequences are counted, and then relationships are inferred based on the number of

---

[11] It is indeed counter-intuitive to read that eukaryotic, complex, multicellular organisms like ourselves can be more related to some **prokaryotes** than these are to other prokaryotes. However, this seems to be the case (see Tables 5.1 and 5.2).

[12] When a bacterium reproduces, it divides by a process called binary fission, and gives rise to two bacteria that should be genetically identical to each other and to the maternal cell (unless some kind of mutation took place). However, during their cell cycle bacteria can exchange DNA molecules and consequently sequences which affect some process, e.g., confer resistance to antibiotics. Thus, if there are two strains of bacteria, e.g., the ABCD and the EFGH ones, where A–H are DNA sequences which are related to some cellular process, binary fission would only produce cells with these genetic structures. However, exchange of DNA molecules among bacteria (e.g., plasmids) allows the emergence of new genetic combinations (e.g., ABCDG or EFGHD, etc.). Thus, after several generations of extensive transfer of DNA molecules from cell to cell, numerous genetically different bacteria could emerge. Thus, genome evolution in prokaryotes is not entirely tree-like; it is best represented by a complex network that combines branches of a tree corresponding to evolution of multiple DNA sequences with numerous horizontal connections (Koonin, 2010a, 2011). This makes the construction of a universal phylogeny difficult (O'Malley, 2012).

Table 5.1 Comparison of some characters of Bacteria, Archaea, and Eukaryota (refers to the majority of cases; adapted from Cavicchioli, 2011)

Character	Bacteria	Archaea	Eukarya
Nucleus	No	No	Yes
Metabolism	Bacterial	Bacterial-like	Eukaryotic
Transcription apparatus	Bacterial	Eukaryotic-like	Eukaryotic
Translation elongation factors	Bacterial	Eukaryotic-like	Eukaryotic
Organelles	No	No	Yes
Methanogenesis	No	Yes	No
Pathogens	Yes	No	Yes

Table 5.2 Apparent origins of some key molecular systems of eukaryotes (adapted from Koonin, 2010b)

System/complex	Inferred origins
DNA replication and repair machinery	Mostly archaeal
Transcription machinery	Archaeal
Translation apparatus, including ribosomes	Mostly archaeal
Cell division and membrane remodeling	Primarily archaeal
Cytoskeleton	Primarily archaeal
Chromatin/nucleosomes	Complex mix of archaeal and bacterial
Endomembrane system/endoplasmic reticulum	Complex mix of archaeal and bacterial
Mitochondrion/electron transfer chain	Bacterial

sequence differences – the less the differences are, the more related are the sequences compared and consequently the corresponding organisms considered. Of course, this process is much more complicated than a pencil-and-paper activity and requires computer programs to perform the necessary calculations.

Moreover, conclusions may sometimes differ depending on the sequences compared. There are many sequences that can be compared but the most appropriate ones are those that exist in all organisms, are orthologous, and extremely conservative (i.e., have changed slightly or not at all during evolution). The DNA sequences that correspond to ribosomal RNA molecules seem to be the most appropriate for this purpose (Pace, 2009). However, a more recent study attempted to use as much genomic information as possible, by relying both on orthologous and xenologous sequences (the latter were usually excluded in older studies). Phylogenetic analysis of this data yielded a tree of life that distinguishes between Archaea, Bacteria, and Eukarya as three monophyletic groups, i.e., groups the members of which are derived from a common ancestor (Lienau et al., 2011). The findings of this study were in remarkable agreement with those of an older study based on genomic data only (Ciccarelli et al., 2006). However, the exact relationship between Archaea and Eukarya is not clear (Gribaldo et al., 2010). The resulting image in both of these articles does not look at all like a tree; it is more like a network. This is important to note because there is no special place in this network of

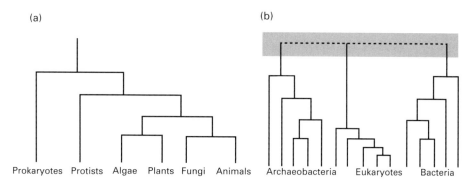

Figure 5.5 (a) The tree of life with six kingdoms, one including prokaryotes and five including eukaryotes. (b) The evolutionary network of life with the three domains. The gray shading indicates our uncertainty for what happened during the early stages of evolution and the exact relation between the three domains. Only vertical descent is depicted in the evolutionary trees. Instances of horizontal gene transfer both within and between domains are not depicted (adapted from Koonin, 2010a).

life not only for humans, mammals, or vertebrates, but not even for eukaryotes. This is very counter-intuitive indeed: most of life with which we are familiar is no more than a small part of the evolutionary network of life (Figure 5.5).

All life on Earth shares a common ancestry and as a result many shared characters between taxa are derived from their common ancestor. This is obvious for some multicellular organisms, but less obvious for unicellular organisms. Relatedness becomes more evident when DNA sequences are compared. There are DNA sequences (e.g., for the elongation factor α-1) which are similar among organisms in all domains. However, this does not mean that one can easily infer relationships from these (Roger *et al.*, 1999). Evolutionary relationships are inferred from shared characters, derived from a common ancestor. These characters are called homologies and they are the topic of the next section. In the rest of the chapter I will focus on the evolution of multicellular organisms. This is not to imply that we should ignore microbial life; quite the contrary, microbial life and its evolution are of enormous importance as I have shown so far. However, it is the evolution of complex, multicellular organisms that people find difficult to understand, and consequently tend to explain the origin of their characters under the assumption of intentional design (see Chapter 3). In this and the next chapter I explain how evolutionary biology provides efficient and detailed explanations for the evolution of all organisms, making the assumption of intentional design entirely unnecessary.

Homology and common descent

Let me clarify the meaning of the term homology right from the start: a homology is more than a similarity in structure. There must be some connection through common ancestry for characters in two different organisms to be considered as homologous.

Generally speaking, one can identify four different concepts of homology. Thus, two characters can be considered as homologous if: (1) they are variants of the same archetype (non-historical concept of homology), (2) they stem from the same common ancestor (historical concept of homology), (3) they are generated from the same underlying genetic network (proximal-cause concept of homology), and (4) they are developmentally homologous because they have independently co-opted the same developmental process present in their most recent common ancestor, although they may be non-homologous in terms of structure (factorial or combinatorial concept of homology). The latter concept indicates that homology may not be evident in structure but may exist at a deeper (molecular or developmental) level. What is important to keep in mind is that homology is not an all-or-nothing relation but, rather, a relative one (Minelli and Fusco, 2013).

It is important to note that apparently homologous structures can be formed by different developmental paths or that the development of apparently non-homologous structures may be under homologous genetic control. For example, all tetrapods (vertebrates with four limbs) have digits in their limbs. Digits are considered as homologous characters between tetrapods; however, the developmental processes that produce them may differ. In all tetrapods, except salamanders, digits separate from each other during embryonic development as the result of apoptosis, a process of cell death that creates inter-digital spaces. In salamanders, however, it is not apoptosis but the differential growth of the digits that produces them (Hall, 2003). Thus, whether or not the digits of tetrapods are considered as homologous depends on the definition of homology that is being used. In addition, different structures can be under homologous genetic control: butterflies, flies, and beetles belong to the clade of winged insects and have two pairs of dorsal appendages that are homologous, in the development of all of which the *Ubx* (Ultrabithorax, a *Hox*[13] DNA sequence) is implicated. These appendages are the forewings, which are flying organs in flies and butterflies but protective organs in beetles, and the hind wings, which form functional wing blades in butterflies and beetles, but are sensory organs (halteres) in flies (Wagner, 2007).

For our purposes here we can define homology as a relation of sameness[14] between two or more characters in two or more organisms, and homologous

---

[13] *Hox* are the DNA sequences, first identified in *Drosophila*, which affect arrangement of structures along the main body axis. They are implicated in the production of transcription factors, a group of proteins that influence the transcription of specific DNA sequences determining which ones are "turned on" or "turned off," and they are grouped together in two clusters, the *Antennapedia* complex that comprises five DNA sequences that affect the front half of the body, and the *Bithorax* complex comprising three DNA sequences that affect the back half of the body. The relative order of these DNA sequences corresponds to the relative order of the body parts they affect. Despite their differences they all contain the same 180 bp sequence which was called the homeobox, and so the DNA sequences were later called *Hox*. Such sequences are found in all animals with the same structure and organization (Carroll, 2006a, pp. 61–72). The body structure of vertebrates is characterized by an increased complexity compared to that of all chordates and it seems that duplications of *Hox* clusters were associated with the formation of novel (cis-) regulatory sequences, which in turn were crucial to the evolution of many vertebrate characters (Soshnikova *et al.*, 2013).

[14] It should be noted that *sameness* is different from *similarity*. Sameness is not simply similarity of structure or function, but implies a historical continuity through evolution (Wagner, 2007).

characters as those that derive from the same character in the most recent common ancestor of those organisms.[15] Having been explicit about the importance of micro-bial life in the previous section, for the purpose of clarity and comprehensibility I will use examples from the search for homologies in vertebrates. All people are familiar with vertebrates and so you will be able to observe on your own some of the characters discussed here if you go to a zoo or even while you are eating your fish or chicken. Furthermore, humans are vertebrates and it is important to realize the enormous sameness between ourselves and other animals. Hence, the question is: What inferences can we make from sameness? Recall the comparison of sharks and dolphins in Chapter 3. They both have hydrodynamic shapes which certainly facilitate swimming underwater, although they significantly differ in other characters, i.e., in how they breathe. What is the conclusion about the relatedness of organisms who share similar characters?

Vertebrates are traditionally divided into seven major groups: jawless fish, cartilagin-ous fish, bone fish, amphibians, reptiles, birds, and mammals.[16] There are various ways in which these groups can be compared to and distinguished from each other: limbs/no limbs; hair/no hair; mammary glands/no mammary glands; feathers/no feathers; lungs/ no lungs, etc. If we choose one character only, e.g., limbs, we only manage to distin-guish between two groups each time: those having and those not having limbs. However, depending on their conception of each vertebrate class, two people might come up with different topologies, i.e., different patterns of branching. For instance, if one had in mind snakes as the exemplar of reptiles, then these would be classified among vertebrates without limbs (Figure 5.6a). But if one had in mind crocodiles as the exemplar for reptiles, then the latter would be classified among vertebrates with limbs (Figure 5.6b). So it seems we face two kinds of problems here: (1) which character(s) should one choose to construct evolutionary trees, and (2) how many characters are adequate? The answer to the first question is that homologies, common characters derived from a common ancestor, are the appropriate ones to use for constructing evolutionary trees. These could be either morphological characters or DNA sequences. The answer to the second question should be obvious: the more, the better. The larger the amount of the available data is, the more accurately we are able to group the various taxa. However, it should be noted that resolving phylogenetic relationships is not just a matter of obtaining more data; the kind of data and the tools of analysis also matter (Rokas and Carroll, 2006; more details about this in the next section).

---

[15] There also exist specific concepts of homology such as serial homology (repetitive structures of the same individual, e.g., vertebrae), positional homology (different, non-homologous structures localized in hom-ologous positions in individuals of two species), and special homology (the same homologous structure is localized in non-homologous positions in individuals of two species) (Minelli and Fusco, 2013).

[16] There are different views on how these groups should be classified. A recent taxonomy is to divide the Subphylum *Craniata* of Phylum *Chordata* into the following classes: *Myxini, Cephalaspidomorphi, Chondrichthyes, Actinopterygii, Dipnoi, Crossopterygii, Amphibia, Reptilia,* and *Mammalia.* This classifi-cation comprises both extinct and extant species. For the purpose of comprehensibility I have opted to use a more popular classification so that the reader is not lost in the details, such as that birds are not considered as a distinct class but are included in *Reptilia.*

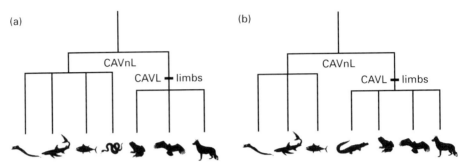

Figure 5.6 Evolutionary trees of vertebrates based on a single character (existence/absence of limbs). Reptiles are classified differently depending on whether one thinks of reptiles as not having (e.g., snakes in (a)) or having (e.g., crocodiles in (b)) limbs (CAVL: common ancestor of vertebrates with limbs; CAVnL: common ancestor of vertebrates without limbs). Such depictions are clearly inadequate, and more accurate criteria are required for constructing evolutionary trees. Image adapted from "300 vector silhouettes of animals (mammals, birds, fish, insects)" © Shutterstock.com/ntnt; "Big collection of vector" © Shutterstock.com/Pavel K; "66 pieces of detailed vectoral (fish silhouettes) © Shutterstock.com/pinare; "vector animals silhouettes" © Shutterstock.com/lilac.

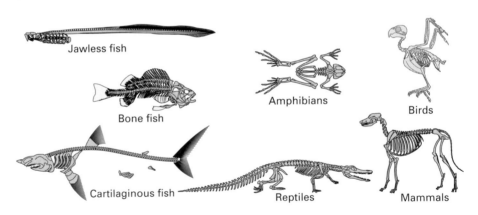

Figure 5.7 Skeletons of the seven major vertebrate groups (not to scale). Image © Simon Tegg.

One might also think that vertebrates vary significantly as, e.g., fish are very different from mammals because they lack limbs, lungs, and other major mammalian characters. In Chapter 1 I described the discovery of *Tiktaalik*, which supports the evolution of tetrapods (limbed vertebrates) from sarcopterygian (lobe-finned) fish. This transition may seem to be a large one. However, most of the major anatomical characters of all vertebrates, including humans, existed in fish like *Tiktaalik*. The skeletons of the various vertebrates exhibit significant similarities and are considered to be homologous (Figure 5.7). It seems that about 90% of the human anatomical structure was formed during the Devonian period, some 380 million years ago. Figure 5.8 provides a visual confirmation of this. In this figure a human skeleton (left) is compared to the skeleton of the "Gogonasus man" (right). The latter consists of the bones of a Devonian advanced lobe-finned fish, which are also present in the human skeleton. These bones have been drawn to the same scale as the human ones.

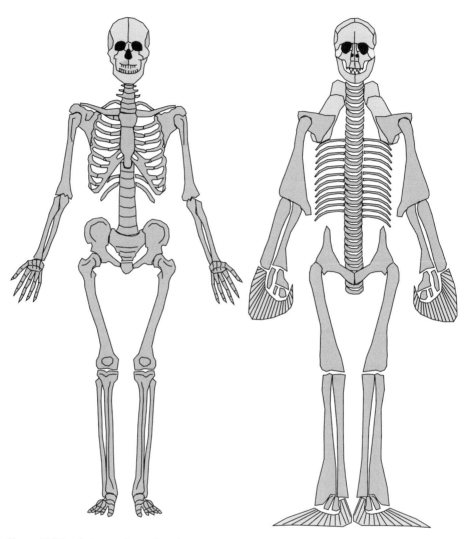

Figure 5.8 The skeleton of a modern human compared to the skeleton of the imaginary Gogonasus man, which consists of the bones present in both humans and Devonian advanced lobe-finned fish, drawn to the same scale as humans (reproduced from Long J. (2012) with permission from Cambridge University Press).

It seems that the lack of digits is the only major difference; the evolution of vertebrates otherwise includes rearrangements of the same basic skeleton (Long *et al.*, 2006; Long, 2012).

This example is used to show that organisms may actually be less different than what we usually think. It seems that vertebrates do not differ significantly in terms of structure, or that their differences are less significant than one might assume. Before getting into more detail, let us further clarify the concept of homology. To start with, we need to carefully specify how we distinguish between different forms of the same structure and between different structures. Homologies are similar characters, structures, properties, processes, modules,[17] or sequences which are derived from a common ancestor and which are common among the members of a **taxon**. If these currently are in the same primitive condition in which they are also found in the common ancestor they are called plesiomorphies; if they currently are in a different, derived condition then they are called apomorphies. In other words, homologies can be shared characters in ancestral (plesiomorphic) form (and they are called symplesiomorphies), or they can be shared characters in derived (apomorphic) form (and they are called synapomorphies). For example, the feather, considered as an epidermis derivative, is a bird apomorphy within the clade of amniotes, while it is a **plesiomorphy** within the clade of birds. Eventually, it is synapomorphies that matter for phylogenetic classification. Finally, if two characters are similar due to convergence, **parallelism**, or reversal to an ancestral condition, they are described as homoplasies. For example, the wings of birds and bats can be considered homologous as tetrapod forelimbs (and in this sense they are synapomorphies), but they are not homologous as tetrapod wings[18] (see Minelli and Fusco, 2013; see also the next section for a discussion of homoplasies).

Why is it that synapomorphies are useful for phylogenetic inference and the construction of the respective trees? This is illustrated in Figure 5.9. If a character is in a plesiomorphic form, i.e., in the same primitive condition both in the common ancestor and in its descendants, no inference can be made about which one evolved first or how closely related these descendents are to each other. The reason for this is that both the common ancestor and its descendants possess exactly the same character. For example, lampreys, sharks, and trouts are all fish which lack limbs, as the common ancestor of all vertebrates did. In this case, the lack of limbs (which is a plesiomorphy) provides no information about which of these taxa are more closely related. Other characters should also be studied in order to reach such a conclusion. Characters in apomorphic form can be informative because a shared **apomorphy** (**synapomorphy**) suggests common derivation from a common ancestor in which the apomorphic character first appeared. As is illustrated in Figure 5.9, lizards, eagles, and hens all have four limbs, which is a character also shared by a common ancestor that might look like a lobe-finned fish such as *Tiktaalik*. Possessing limbs is an apomorphic character given that the primitive state is not having limbs. Then, eagles and hens have forelimbs which have evolved to wings, a character also shared by their common ancestor, which was a feathered dinosaur like *Archaeopteryx*. But how do we know whether it was forelimbs that evolved to wings or wings that evolved to forelimbs? The answer is generally given by outgroup comparison; in the case of features of the vertebrate skeleton, this can be supported by fossil evidence. Lobe-finned fish like *Tiktaalik* are estimated to have lived approximately 380 million years ago, whereas feathered dinosaurs like *Archaeopteryx* are estimated to have lived approximately 150 million years ago (Prothero, 2007, p. 265).

---

[17] Modules are distinguishable, partially independent, interacting units at several hierarchical levels (e.g., segments). **Modularity** allows for evolutionary change to occur in one character without detrimentally affecting another character or the entire organism (Gass and Bolker, 2003).

[18] An interesting distinction that is useful here is that between character identities (e.g., wings and legs) and character states (size, shape, or color) (Wagner, 2007).

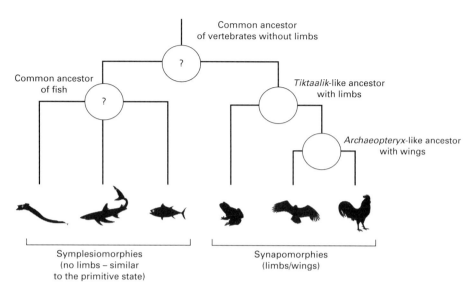

Common ancestor
of vertebrates without limbs

?

Common ancestor
of fish

?

Tiktaalik-like ancestor
with limbs

Archaeopteryx-like ancestor
with wings

Symplesiomorphies
(no limbs – similar
to the primitive state)

Synapomorphies
(limbs/wings)

Figure 5.9 Symplesiomorphies and synapomorphies, or how the study of form of extant taxa as well as that of extinct taxa in dated fossils can help clarify evolutionary relatedness. Note: *Tiktaalik* and *Archaeopteryx* are *not* the common ancestors; however, the common ancestor in each case could be similar to *Tiktaalik* or *Archaeopteryx* and so these are included in the evolutionary tree. Not possessing limbs is the plesiomorphic character and a **symplesiomorphy** for lampreys (jawless fish), sharks (cartilaginous fish), and trouts (bone fish); possessing limbs is the apomorphic character and a synapomorphy for frogs (amphibians) and birds; wings are a synapomorphy for eagles and hens. Image adapted from "300 vector silhouettes of animals (mammals, birds, fish, insects)" © Shutterstock.com/ntnt; "Big collection of vector" © Shutterstock.com/Pavel K; "66 pieces of detailed vectoral (fish silhouettes) © Shutterstock.com/pinare; "vector animals silhouettes" © Shutterstock.com/lilac.

However, and unfortunately, fossils are not available for all groups of organisms. Although they are precious when available, the use of fossils in reconstructing phylogeny is very limited outside vertebrates and a few smaller taxa. Nevertheless, when they are available they can be very important; not only do they provide information about extinct life forms, but they also serve as points of reference in order to establish a chronology of phylogenetic splittings. Despite the DNA sequence data that may be available, fossils are important as they allow for the dating of evolutionary events. Fossils can provide important information such as characters of extinct taxa, characters of basal taxa, the polarity of evolution, divergence times for taxa, times of appearances of homologous characters, times of extinction, life-history data for extinct taxa, sister clades to living single-taxon clades, and some extinct genomes. In contrast, fossils cannot provide a full array of characters of fossil taxa, equivalent sampling of all clades, unequivocal ancestor–descendant links, or very ancient genomes (Raff, 2007). Eventually, it seems that the best approach is to combine fossil and molecular data, which can be used to complement each other in calibrating evolutionary events. The fossil record is imperfect and **molecular clock** methods are uninformative on their own. However, together they can be used to construct reliably dated trees (Donoghue and

Benton, 2007; for details about the age of 30 divergences among key genome model organisms, see Benton and Donoghue, 2007).

In this sense, we can study living taxa, compare them to extinct ones, and construct evolutionary trees. In each case, we can indicate which new characters evolved and were distinctive of more recent taxa. These new, derived characters called apomorphies are the informative ones for constructing evolutionary trees. Plesiomorphies cannot be used for the "grouping together" of taxa because they do not offer a "signal" of common and exclusive derivation from a specific ancestor as apomorphies do. Of course, this is possible as long as we have some epistemic access to the past. This is often possible through fossils, but only for organisms that could be *and were* fossilized. If this is the case, predictions are possible like the one that Shubin and his colleagues made about where an organism like *Tiktaalik* could be found (see Chapter 1).

To make clearer how epistemic access to the past is possible, here is a thought experiment by way of analogy. Imagine you have several soft balls with a sticky surface on which small pieces of paper can very easily be attached while the balls are rolling over them. Imagine also that you have two intersected, curve-shaped slides on which you have put small pieces of paper with different colors (corresponding to each of the slides, black and gray, respectively) and numbers (corresponding to the vertical distance from the point they were released, let's say 1–5). When you release balls consecutively, each of them will follow different routes and produce different sequences. Suppose you release four balls and get the following pieces of paper on each (Figure 5.10): (a) black – 1, black – 2, black – 3, black – 4, black – 5; (b) black – 1, black – 2, gray – 3, gray – 4, gray – 5; (c) gray – 1, gray – 2, gray – 3, gray – 4, gray – 5; and (d) gray – 1, gray – 2, black – 3,

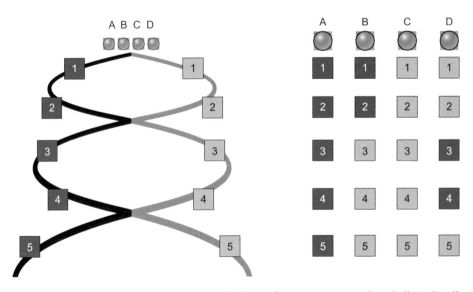

Figure 5.10 A thought experiment of how to infer history from contemporary data. Balls A–D roll over the black and gray planes and colored, numbered papers stick on the balls as they roll over them. At the end, depending on which papers are found on each ball one can infer the actual route of each of the balls. Image © Simon Tegg.

Table 5.3 Some derived characters among vertebrates (not all of these groups correspond to classes). Such tables can be constructed by using DNA, RNA, or protein sequences and used for inferring evolutionary relationships

Taxonomic groups	Apomorphic characters						
	Jaws	Bone skeleton	Four limbs	Astragalus bone	Diapsid skull	Wings	Synapsid skull
Jawless fish	—	—	—	—	—	—	—
Cartilage fish	✓	—	—	—	—	—	—
Bone fish	✓	✓	—	—	—	—	—
Amphibians	✓	✓	✓	—	—	—	—
Reptiles	✓	✓	✓	✓	✓	—	—
Birds	✓	✓	✓	✓	✓	✓	—
Mammals	✓	✓	✓	✓	—	—	✓

black – 4, gray – 5. From the color and the number of the pieces that will be found on each ball one can infer the exact route of the ball on the slides. In other words, from the "present" characters of the balls (which pieces of papers are found on them) and from their "history" (where on each slide each piece of paper was initially put, and in what order compared to the other papers) we can infer the route taken. In the same sense, from the present (apomorphic) characters of taxa and from their first appearance in the fossil record, one can infer the evolutionary history of taxa and consequently relationships.

In order to construct an accurate evolutionary tree of vertebrates, we should not rely on a single character only, e.g., the existence/absence of limbs, as in Figure 5.6, but on more than one apomorphic (derived) character. Then we could construct a table like Table 5.3. Based on this table it is possible to construct the evolutionary tree in Figure 5.11. It is important to note that the tree in Figure 5.11 can be drawn in different ways, as shown in Figure 5.12. All trees in this figure are equivalent. In human family trees it is usually the case that older offspring appear leftmost, whereas their youngest siblings appear rightmost. This is not the case for evolutionary trees, where it does not matter which group is on the left or the right, or whether two groups are horizontally close to each other or far apart. Consequently, the apparent progression from one group to another, such as of jawless fish to mammals in Figure 5.11 (from left to right), is an illusion.

Thus, it is synapomorphies which can be used to reveal evolutionary relationships among taxa. Evolutionary biologists study not only morphological characters, but also DNA, RNA, or protein sequences. For example, in a study that aimed to resolve the evolutionary tree of vertebrates by identifying the phylogenetic position of turtles within amniotes, and the relationships between the three major extant amphibian groups, an evolutionary tree for vertebrates was produced based on 75 protein-coding DNA sequences for 129 taxa. This evolutionary tree is practically the same as the one in Figure 5.11 (Fong *et al.*, 2012). The rationale for constructing evolutionary trees with DNA sequences is more or less the same as that which relies on morphological characters. Fossils are used again to confirm or reject hypotheses about evolutionary relationships. However, nowadays molecular data are being accumulated at a fast pace. Thus, one would expect that homologous DNA sequences will help us reconstruct all

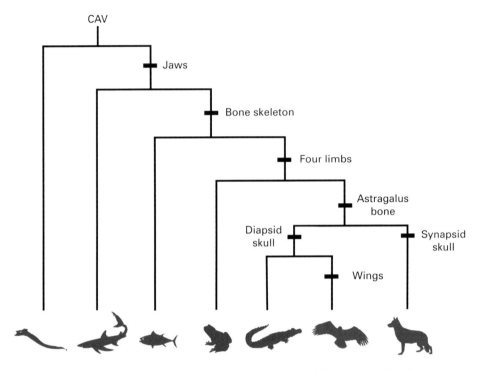

Figure 5.11 An evolutionary tree of vertebrates based on apomorphic characters (CAV: common ancestor of vertebrates). Note that the apparent progression from jawless fish to mammals (from left to right) is an illusion (see next figure). Image adapted from "300 vector silhouettes of animals (mammals, birds, fish, insects)" © Shutterstock.com/ntnt; "Big collection of vector" © Shutterstock.com/Pavel K; "66 pieces of detailed vectoral (fish silhouettes) © Shutterstock.com/pinare; "vector animals silhouettes" © Shutterstock.com/lilac.

kinds of phylogenies. However, this is not the case. One reason for this, as already explained in the previous section, is that although the study of orthologous sequences can be informative, paralogous and xenologous sequences may nevertheless blur the whole picture. Another reason for this is our difficulty distinguishing homologies from those characters described as homoplasies. This is the focus of the next section.

Homoplasy and convergence

Similar characters are not always due to common descent. It is often difficult to distinguish between similarity and sameness – in other words, between simply similar characters and similar characters derived from a common ancestor. Let us illustrate this. Figure 5.13 depicts an imaginary evolutionary tree of names and of their "evolution." "Jonathan" is the primitive state of the character. This can gradually "evolve" to "Nathan" but also to "Jon." However, "Nathan" can also "evolve" from Nathanael. Thus, different "Nathan" states can be derived from a common ancestor or may have "evolved" independently, and these are described as homologies or homoplasies, respectively. There are two important

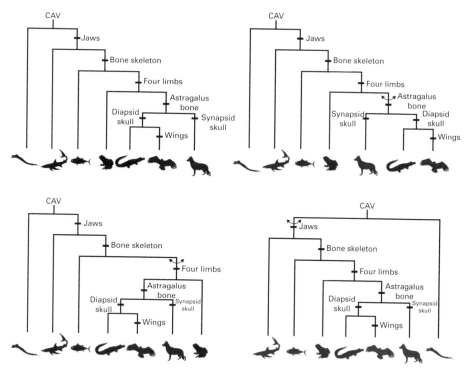

Figure 5.12 The phylogenetic relationships of vertebrates of Figure 5.11 can be represented in various ways, as shown here. All these evolutionary trees are equivalent and represent the same relationships (the double arrows indicate the nodes in which the taxa belonging in a clade have just changed relative places; CAV: common ancestor of vertebrates). Image adapted from "300 vector silhouettes of animals (mammals, birds, fish, insects)" © Shutterstock.com/ntnt; "Big collection of vector" © Shutterstock.com/Pavel K; "66 pieces of detailed vectoral (fish silhouettes) © Shutterstock.com/pinare; "vector animals silhouettes" © Shutterstock.com/lilac.

questions here. The first is how one can decide whether two "Nathan" states are homologies or homoplasies. The second is whether taxa "Jonathan" and "Nathanael" are closely related or not, and whether the "nathan" suffix of "Jonathan" and the "nathan" prefix of "Nathanael" could in fact be a homology at some deeper level.[19]

Keeping this example in mind, let us now turn to organisms and try to answer these questions. In the previous section, I mentioned that bird and bat wings are homologous as tetrapod forelimbs, but are not homologous as tetrapod wings. This means that both kinds of wings when considered as forelimbs, i.e., only in terms of structure and not in terms of function, are homologous since they were derived from their common ancestor. However, when their function is taken into account, one can no longer talk about homologous structure since the use of bird and bat forelimbs as wings for flight is not due to their being derived from their common ancestor. In contrast, the evolution of bird and bat forelimbs to wings took place independently from each other through a phenomenon described as

---

[19] My discussion here implies nothing at all about the names themselves. I made an abstract choice of names just to make my case.

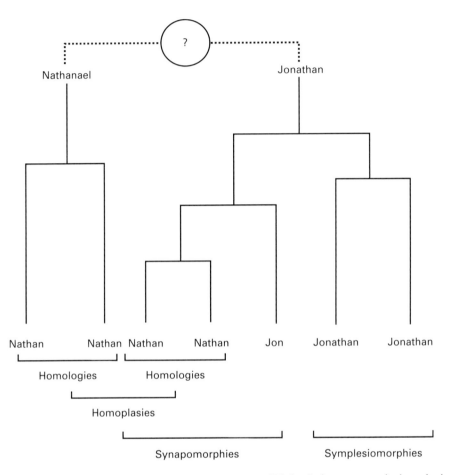

Figure 5.13 The "evolution" of "Nathan." Contemporary "Nathan" characters can be homologies if they have evolved from a common ancestor (either "Nathanael" or "Jonathan"), or homoplasies if they have evolved independently from different ancestors ("Nathanael" and "Jonathan"). "Nathan" and "Jon" are synapomorphies, i.e., derived characters from an ancestral one ("Jonathan"), whereas contemporary "Jonathan" are symplesiomorphies.

convergence. Characters like these, which evolved independently and which are not derived from a common ancestor, are called homoplasies (Figures 5.14–5.15).

Homologies and homoplasies have been considered as antithetical concepts, but they are rather quite complementary. It seems that there is a continuum of biological processes from one to the other, including parallelism and convergence. Hall (2003) notes that one may intuitively think of homologies as the product of conserved developmental processes and homoplasies as the product of independent ones. However, this is not actually the case. Similar developmental processes may produce different structures, but also different developmental mechanisms may produce similar structures, even in distantly related organisms (see, respectively, the examples of insect dorsal appendages and tetrapod digits in the previous section). Since homology has been defined here as a relation of sameness between two or more characters that derive

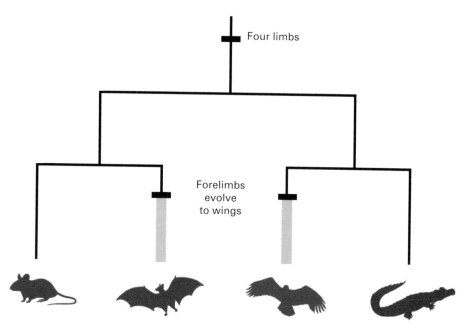

Four limbs

Forelimbs
evolve
to wings

Figure 5.14 Bird and bat wings are homologies as vertebrate forelimbs but they are homoplasies as vertebrate wings. Whereas their structure is due to the common ancestor of vertebrates, their functionality as wings evolved independently. Image adapted from "300 vector silhouettes of animals (mammals, birds, fish, insects)" © Shutterstock.com/ntnt; "Big collection of vector" © Shutterstock.com/Pavel K; "66 pieces of detailed vectoral (fish silhouettes) © Shutterstock.com/pinare; "vector animals silhouettes" © Shutterstock.com/lilac.

from the same character in their most recent common ancestor, **homoplasy** can be defined as a relation of similarity between two characters in two or more organisms that are not derived from the same character in their most recent common ancestor. In this section I focus on convergence because it is the clearest case of homoplasy and as such can be clearly distinguished from homology. Convergence refers to the emergence of the same character through independent evolution, i.e., from different ancestral characters (McGhee, 2011).[20] In the case of bat and bird wings the forelimbs evolved to different kinds of wings. The wings of bats consist of their elongated digits, which are connected via a webbed membrane of skin, whereas the wings of birds consist of their whole forelimb that is covered by feathers (see Figure 5.16 for the skeleton structure of birds and bats).[21]

---

[20] Whether parallelism is a case of homoplasy or homology has been debated (see Hall, 2012). The main difference between convergence and parallelism is that whereas convergence refers to the evolution of the same character, C, from two different ancestral characters, A and B, in two different lineages, L1 and L2, parallelism refers to the emergence of C from the same ancestral character, A, in two different lineages, L1 and L2. However, the latter can be considered as a special case of the former (see McGhee, 2011, pp. 2–5).

[21] The wings of insects are very different from those of tetrapods. Insect wings have not evolved from their legs but from modified gill branches. The proteins Apterous and Nubbin are required for building wings and they are also expressed in the respiratory lobe of the outer branch of crustacean limbs. Thus, these structures must be homologous; insect wings have evolved from the gills of crustaceans (Carroll, 2005a, pp. 175–179).

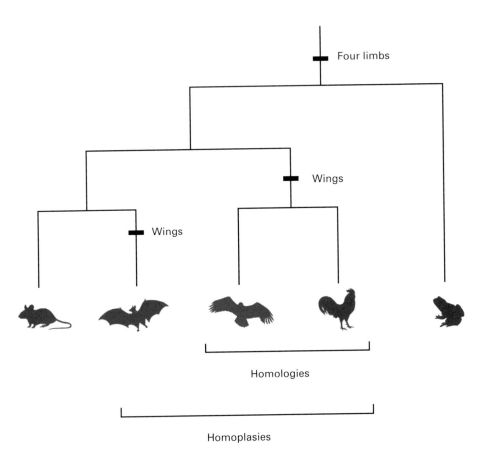

Figure 5.15 Eagle and hen wings are homologies as they are derived from a common ancestor; both of them and bat wings are homoplasies. Image adapted from "300 vector silhouettes of animals (mammals, birds, fish, insects)" © Shutterstock.com/ntnt; "Big collection of vector" © Shutterstock.com/Pavel K; "66 pieces of detailed vectoral (fish silhouettes) © Shutterstock.com/pinare; "vector animals silhouettes" © Shutterstock.com/lilac.

Figure 5.16 Bat (a) and bird (b) skeletons. Bat skeleton © istock.com/ilbusca; bird skeleton © Simon Tegg.

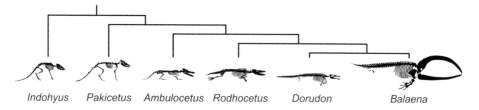

Indohyus   Pakicetus   Ambulocetus   Rodhocetus   Dorudon   Balaena

Figure 5.17 Transitional forms in the evolution of modern whales (*Balaena*) from tetrapod ancestors (adapted from Coyne, 2009, p. 54). Image © Simon Tegg.

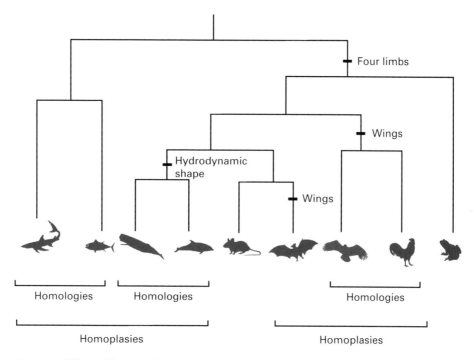

Figure 5.18 Wings of bats and birds and hydrodynamic shapes of fish and sea mammals as homoplasies and homologies. Image adapted from "300 vector silhouettes of animals (mammals, birds, fish, insects)" © Shutterstock.com/ntnt; "Big collection of vector" © Shutterstock.com/ Pavel K; "66 pieces of detailed vectoral (fish silhouettes) © Shutterstock.com/pinare; "vector animals silhouettes" © Shutterstock.com/lilac.

Another interesting case of convergence is the body shape of sea mammals, such as dolphins and whales. As already discussed in Chapter 3, sharks and dolphins have the same hydrodynamic shape. However, this is not due to common ancestry, because dolphins evolved from a tetrapod ancestor (see Coyne, 2009, pp. 51–55). Whereas the lack of limbs and the hydrodynamic shape was a character common among fish groups, most of the other vertebrate taxa have limbs. However, some mammals turned from life on land to life in the sea and evolved hydrodynamic shapes. The fossil record provides

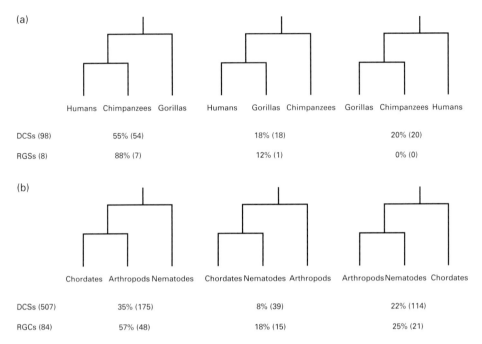

(a)

	Humans Chimpanzees Gorillas	Humans Gorillas Chimpanzees	Gorillas Chimpanzees Humans
DCSs (98)	55% (54)	18% (18)	20% (20)
RGSs (8)	88% (7)	12% (1)	0% (0)

(b)

	Chordates Arthropods Nematodes	Chordates Nematodes Arthropods	Arthropods Nematodes Chordates
DCSs (507)	35% (175)	8% (39)	22% (114)
RGCs (84)	57% (48)	18% (15)	25% (21)

Figure 5.19 The evolutionary trees show all possible topologies for each set of taxa ((a) humans/ chimpanzees/gorillas, 5–8 million years ago; (b) chordates/arthropods/nematodes, >550 million years ago). Below each topology, the percentage and number (in parentheses) of DNA coding sequences (DCSs), and rare genomic changes (RGCs) supporting that topology are shown (based on maximum likelihood analyses). A number of DNA coding sequences in each case are uninformative: (a) 6 of 98 DCSs; and (b) 179 of 507 DCSs (adapted from Rokas and Carroll, 2006). What the resulting trees indicate is that we cannot be 100% certain about which topology is the accurate one. However, RGCs increase the level of certainty compared to DCSs.

adequate evidence about how this transition could have been possible. Thus, the lack of limbs can be considered as a homology between sharks and trouts, or between whales and dolphins. However, the hydrodynamic shape of sharks and dolphins (see Figure 3.8) is a case of homoplasy, not homology. Although both of them are vertebrates and as such share common ancestors, their shape is not derived from them. In the case of whales and dolphins it has evolved independently from an ancestor with four limbs (Figure 5.17). The evolutionary tree of Figure 5.15 can now be enriched with more examples of homologies and homoplasies (Figure 5.18).

In the case of vertebrates, scientists can rely on the fossil record to infer phylogenies and reconstruct the evolution of each taxon. However, there are cases where this can be very difficult, especially when only data from the present (e.g., DNA sequences) and not from the past (e.g., fossils) are available. Homoplasy, in general, and convergence, in particular, can act as confounders of genuine homology in evolutionary studies. Molecular characters (e.g., DNA or protein sequences) typically have a few alternative states. Consequently, the probability that different species acquire the same nucleotide or amino acid independently is significant and can confound evolutionary history. Thus, one cannot simply claim that by comparing the DNA sequences of organisms we can infer evolutionary history: Are all similarities observed due to common descent or due to convergence – in other words, are they homologies or homoplasies? One strategy to overcome this problem has been to compare rare genomic changes. These are changes due to rare mutational events, such as insertions and deletions in coding sequences,[22] which are less likely to occur independently in the same way. However, phenomena such as horizontal transfer of DNA coding sequences and hybridization can still complicate the picture. Figure 5.19 shows alternative evolutionary trees based on DNA coding sequences and rare genomic changes, showing that in some cases we cannot be certain about the actual evolutionary history, mostly due to homoplastic events. This poses both a conceptual and methodological challenge (Rokas and Carroll, 2006). It should be noted, though, that this does not pose any concerns about the methods of evolutionary biology, but rather about the approaches needed to understand phenomena at a finer level of detail.

Another point that is important to keep in mind is that due to constraints some, but not all, possible morphologies of organisms are realized, and this in turn increases the likelihood of homoplasy. By comparing genetic regulatory networks and conducting experiments to alter them, it could be possible to understand how fundamental morphological changes evolve (Wake *et al.*, 2011). Thus, to answer the first question it is not always easy to distinguish between homologies and homoplasies, and certainly understanding the underlying developmental mechanisms and their evolution can prove crucial. The point here is that one should try to have some kind of epistemic access to the history of each lineage in order to realize how old a common ancestor is. To achieve this, multiple lines of evidence are required: fossils (when available), comparative genomics, and the comparative study of developmental processes. Conclusions can be difficult to draw, but this is what makes science fascinating. As shown in Figure 5.19, rare genomic changes are more informative than coding DNA sequences, so perhaps finding new targets of analysis would be important. A challenge that is more difficult to cope with is, as also shown in Figure 5.19, that the older the lineages are, the more homoplastic events may have taken place and this might make the actual history more difficult to discern. The exact relatedness between chordates, arthropods, and nematodes seems to be less certain than that between gorillas, humans, and chimpanzees because evolutionary events in the former case are much older than in the latter.

Thus, it is not always easy to distinguish homologies from homoplasies. This brings us to the second question, which is perhaps the most difficult one to answer. Can two homoplasies exist due to a homology at some deeper level? In the first section of this chapter I described how all living forms are related through common ancestry. If all organisms are more or less related, can we actually talk about independent evolution of characters? Even if we know, in the example of Figure 5.13, that two of the "Nathan" states have evolved from an ancestral "Nathanael" state and the other two have evolved from a "Jonathan," are we sure that "Nathanael" and "Jonathan" are not related through a shared common ancestor in the deep past? An example of this kind is the evolution of eyes, which were long thought to be the outcome of evolutionary convergence. More

---

[22] See Griffiths *et al.* (2012) for the details of these and other similar phenomena.

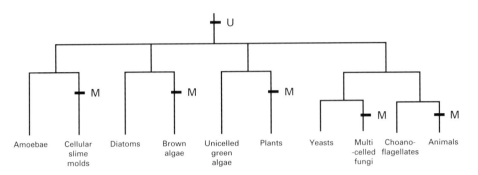

Figure 5.20 The possible independent evolution of multicellularity (adapted from Arthur, 2011, p. 12). The important conclusion here is that not all multicellular organisms are derived from a single, first multicellular common ancestor (U: unicellular state; M: multicellular state).

recent studies have revealed the existence of shared genetic networks for eyes in otherwise different animal taxa. It has been found that a particular set of transcription factors, such as those produced by members of the *eyeless*, *atonal*, and *eyes absent* families of DNA coding sequences in *D. melanogaster* and their homologues in vertebrates, are involved in the specification and formation of various types of animal eyes. In addition, the ability to detect light in all light-sensing organs in animals depends on a set of chemical reactions that involves opsin proteins, and so it has been assumed that all modern variations of light sensing in bilaterians[23] can be traced to the existence of photosensitive cells in a common ancestor with PAX6 and other transcription factors involved in a regulatory pathway leading to opsin production. This phenomenon, in which the development and evolution of morphologically disparate organs depends on homologous genetic regulatory circuits, has been described as *deep homology*. It seems that there are many, striking developmental similarities between diverse taxa. But there also exist interesting differences. For example, all animal eyes are composed of photoreceptors that have distinct phototransduction signaling reactions. The eyes of insects and other invertebrates have different photoreceptors and types of reactions from vertebrates. This seems to suggest that these have evolved independently. However, it could be the case that both cell types co-existed in the common bilaterian ancestor of vertebrates and invertebrates, and that different cell types for light detection were eventually used in the visual systems of each lineage (Shubin *et al.*, 2009).

Many characters have evolved independently many times. An interesting case is multicellularity (Figure 5.20), the property of an organism consisting of many cells (Arthur, 2011, p. 12). How can this be explained? Is this due to some deep homology, because unicellular organisms have the inherent tendency to evolve, forming multicellular ones? Or is it a case of homoplasy and multicellular states are selectively favored and evolve once they arise? There is no simple answer to this question. To answer it, we first need to distinguish between two kinds of multicellularity: simple multicellularity and

---

[23] Bilaterians are organisms with twofold symmetry that gives them definite front and rear, as well as left and right, body surfaces. This group includes most animal phyla except for organisms with radial symmetry, such as jellyfish and sponges, which belong to the group Radiata.

complex multicellularity. Simple multicellular organisms have the form of filaments, clusters, balls, or sheets of cells. Although they may have differentiated somatic and reproductive cells, they do not exhibit more complex patterns of differentiation. Simple multicellular eukaryotes consist of cells which are connected to each other by adhesive molecules, but there is not much communication or transfer of resources between cells. In contrast, complex multicellular organisms exhibit both cell to cell adhesion and intercellular communication, as well as tissue differentiation mediated by networks of regulatory DNA sequences. Simple multicellularity has evolved many times among the eukaryotes, but complex multicellular organisms belong to only six groups: animals, land plants, two groups of algae, and two groups of fungi. Although it has not been possible so far to resolve all phylogenies, it seems that at least in both animals and plants multicellularity evolved with adhesion, which gave rise to simple multicellular structures from a single progenitor cell. The next step seems to have been the evolution of bridges between cells, which facilitated the transport of nutrients and signaling molecules between cells, something that is found in all groups with complex multicellularity, but not in others. It seems that further evolution of multicellularity involved both co-optation and de novo evolution of signaling molecules, molecules involved in transmitting information between cells, and transcription factors which led to more complex body structures (Knoll, 2011).

Homoplasy confounds evolutionary histories and what has long been explained as the outcome of convergence may actually be the outcome of homology at a "deeper" level. Here are two conclusions that you should *not* make: (1) that we cannot know the evolutionary histories of particular taxa, and (2) we cannot be certain about our evolutionary explanations since we cannot definitely distinguish common descent from convergence or homology from homoplasy. One way to deal with this could be to distinguish between different levels. It could be the case that convergence occurs at the level of the organism due to the fact that particular individuals survive and reproduce in a particular environment. But, at the same time, some kind of homology can exist at a deeper level, as the organismal forms which are being selected differ from those which are not in their developmental mechanisms. Apparently, there is a lot that we still need to know. In the case of the evolution of vertebrates, which has been the focus of my discussion so far, light can be shed by the very active research field of evolutionary developmental biology, which is the focus of the next section.

Evolutionary developmental biology

Let me note again that I have so far focused on multicellular organisms, and particularly vertebrates – which include humans – because it is their evolution that many people find difficult to understand. In Chapter 3 I noted that artifacts have more fixed essences and if organisms have essences, these are more plastic. Here I will explain how this is possible due to the interrelation of evolution and development. The evolution of organisms is characterized both by robustness (the consistency of phenotype despite genetic or environmental perturbations) and plasticity (the potential of organisms with the same **genotype** to produce different phenotypes during development as a response to genetic or environmental perturbations) (Bateson and Gluckman, 2011, p. 8). Such phenomena are studied by evolutionary developmental biology (usually dubbed as evo-devo). In the two previous sections, the study of homology and homoplasy relied a lot on this approach as it was mentioned that what matters is not only the evolution of form, but also that of the underlying developmental processes. **Evo-devo** studies both the evolution of development (i.e., how developmental processes evolve) and the developmental basis of evolution (i.e., how development structures the evolution of organismal characters) (Love, 2013). In this section I will explain some significant features of the evolution of multicellular organisms in order to show both how complexity is possible as the outcome of natural processes without any assumption of design and how evolution of very different forms is possible due to changes in shared underlying developmental processes.[24]

The focus of most public discourse about evolution has been on whether natural selection, Darwin's main principle, is adequate to drive evolution: The divergence of populations of the same species which are modified and eventually become new species descending from a common ancestor (this is what Darwin described as *descent with modification*). According to Darwin, natural selection was the process that drives this modification, not simply by eliminating the individuals which cannot survive or reproduce in a given environment, but also by accumulating favorable variations (see Chapter 6 for natural selection and processes of evolutionary change). Although Darwin was not able to show that natural selection was indeed competent to produce new species (see Chapter 4), he drew the analogy from artificial selection to convince his readers that this could be the case. In a later book, *Variation of Animals and Plants Under Domestication* (1868), he used another analogy to illustrate the process of selection. According to this:

> Let an architect be compelled to build an edifice with uncut stones, fallen from a precipice. The shape of each fragment may be called accidental; yet the shape of each has been determined by the force of gravity, the nature of the rock, and the slope of the precipice, – events and circumstances, all of which depend on natural laws; but there is no relation between these laws and the purpose for which each fragment is used by the builder. In the same manner the variations of each creature are determined by fixed and immutable laws; but these bear no relation to the living structure which is slowly built up through the power of selection, whether this be natural or artificial selection. If our architect succeeded in rearing a noble edifice, using the rough wedge-shaped fragments for the arches, the longer stones for the lintels, and so forth, we should admire his skill even in a higher degree than if he had used stones shaped for the purpose. So it is with selection, whether applied by man or by nature; for though variability is indispensably necessary, yet, when we look at some highly complex and excellently adapted organism, variability sinks to a quite subordinate position in importance in comparison with selection, in the same manner as the shape of each fragment used by our supposed architect is unimportant in comparison with his skill. (Darwin, 1868, pp. 248–249)

Darwin suggests that it is selection that matters. Variation is less important as it is "accidental," and so it is selection that is responsible for shaping the characters of organisms. Variation is important for selection, because there can be no selection if there is no variation. The builder would not make any selection if all stones were the same; but beyond that, according to Darwin, variation does not affect the outcome as it is the builder who will select the stones which are appropriate for the building. In the same

---

[24] Several good introductions to evo-devo exist in papers (Raff, 2000, 2007; Arthur, 2002; Müller, 2007, 2008) and books for a wider audience (see the further reading section at the end of this chapter).

sense, selection in nature drives evolution. Variation is necessary for selection, but less important as it cannot shape the outcome, because it is subordinate to selection. The importance of the nature and extent of available "chance" variation was thus underestimated for a long time (see Beatty, 2010 for a detailed discussion).

Since then, evolution has been described as a two-step process: the "random" emergence of variation and the "non-random" process of selection (e.g., see Mayr, 2002, p. 119). This is where the intuitive question arises. Let's assume that selection is a natural process which can produce adaptations by accumulating favorable variations. How is it possible to randomly produce the appropriate variation for this process? To use Darwin's own example, let's assume that our architect is competent enough to build a house from stones. What guarantees that the stones fallen from a precipice will be appropriate for this? What if the small stones fall first and the big ones follow much later? Can one build a house with small stones below and big stones on the top? And what if only small stones are available? Will the architect be able to build a stable house with small stones only? Darwin's analogy is not a good one for two reasons. First, the conscious and skillful builder is not a good analogue of natural selection, which is an unintentional process. Second, and most importantly for the topic of this section, variation is neither random, nor as limited as the stone example might indicate. In the case of organisms, variation is in a sense biased. Organisms have an inherent potential in their genetic material which drives the development of some characters but not of others. This seems to limit the potential of evolution. However, this is circumvented by the fact that variation is not as limited as the stones example indicates. In Darwin's examples, stones are stones and they differ only in size. In contrast, in actual organisms particular DNA sequences involved in forming a particular structure or affecting a particular process can eventually be "co-opted" to form a different structure or to affect a different process.

Let me use another example to illustrate this, before turning to a description of these processes in scientific terms. Imagine someone starts a small shop selling pizzas. The first pizzas produced are only made with dough (flour and water), cheese, and tomato. Gradually, more kinds of pizzas are produced, with sausage, onions, bacon, garlic, and more types of cheese. But the shop is still selling pizza. However, the owner decides that she also wants to sell cookies. Then the dough used for pizza is also used for making cookies, with the addition of sugar and eggs. Pizzas and cookies do not look similar and taste very different, but they have the same basis. They are made of dough and several other materials. You might think of the processes of making cookies and pizzas as distinct developmental processes, where pizzas and cookies "develop" by molding the various materials together. The shop now sells both pizzas and cookies. However, in the particular neighborhood customers also buy pizzas elsewhere, as more pizzerias have opened in the area in the meantime. Nonetheless, they find the cookies very good and they mostly come to the shop to buy these, instead of pizzas. This makes the owner of the pizza shop gradually change it to a cookie shop because cookies were a more successful commodity than pizzas, as customers were mostly selecting them. What has been described in this example could be perceived as equivalent to an evolutionary process that took place due to a change in development and selection. This is what evo-devo is about. Changes in the initial developmental process (dough + cheese + tomato → pizza) gradually produced a new developmental process (dough + sugar + eggs → cookies). Then selection favored the latter

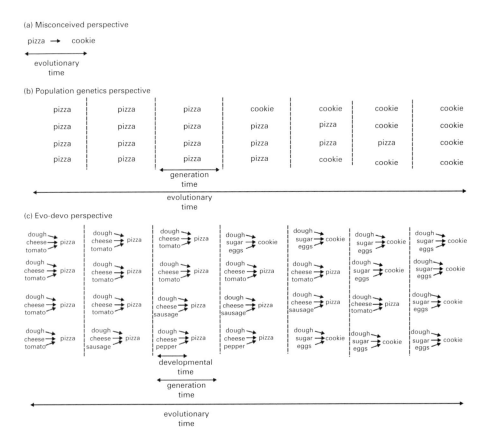

Figure 5.21 Three perspectives of evolution: (a) changes in individuals, which is wrong; (b) changes in the genetic structure of populations, which is inadequate; and (c) changes in the developmental trajectories of individuals, which is the evo-devo perspective.

developmental process over the former and eventually the first population (think of the pizza shop as a population of pizzas) evolved to another (the cookie shop can be considered as a population of cookies). This is a simple example which I hope makes clear the core of evo-devo. The question is not whether a pizza can change into a cookie (this is how evolutionary transitions are often described), but how a population of pizzas (represented by the pizza shop) can evolve to a population of cookies (represented by the cookie shop) due to a change in developmental processes (materials available are used to make cookies and not pizzas)[25] (Figure 5.21).

The main point intended by the above example is that evolution in multicellular organisms proceeds not by changes in adult forms (e.g., ancestral adult A is transformed to descendant adult D), but by changes in the developmental processes which produce these adult forms (e.g., development of ancestral adult A changes to development of descendant

[25] I must note that in this example, the "evolution" of cookies is driven by the intentions of the shop owner. There is no such intentionality in nature and this is important to keep in mind. Thus, this example of selection is not a good analogue and not equivalent to natural selection. It is only used in a metaphorical sense to illustrate how changes in development can affect evolution, as well as the differences between the different perspectives.

adult D). This is possible because DNA sequences that are involved in regulating development are conserved across very different groups (e.g., as in the case of *Hox* DNA sequences which influence the development of body parts in insects and vertebrates). The important issue here is that evolution occurs not only due to changes in DNA sequences that directly affect characters (e.g., limbs) but also to DNA sequences that are implicated in the development of these characters. This is important to understand. To use a simple example, when we are studying the inheritance of stem length in a plant species we may realize that there are roughly two alternative adult forms: tall plants and short plants.[26] We usually refer to *genes for* long stems and short stems, i.e., alternative DNA sequences (alleles) that somehow produce the respective phenotypes. The new perspective advanced by evo-devo is that in such a case only a single DNA sequence may be implicated and that changes in the phenotype (e.g., short/long) could be due to regulation of that sequence: when, where, and for how long it is activated and with what kind of outcome.

Let's consider two cases already discussed in the previous section: the loss of limbs in whales and dolphins and the evolution of wings in bats. Modern cetaceans, such as whales and dolphins, are characterized by the absence of hind limbs. However, hind limb bud development is initiated in their embryos but is not maintained due to the absence of Hand2, a regulator sequence that affects initial limb outgrowth in amniotes. Thus, it seems that the initial reduction in hind limb size[27] was driven by changes in regulatory control sequences which affect development (Thewissen *et al.*, 2006). In the case of bat wings, the digits in bats are initially similar in size to those of mice during embryonic development, but then bat digits lengthen enormously. It seems that bone morphogenetic protein 2 (Bmp2) can stimulate cartilage proliferation and differentiation and increase digit length in the bat embryonic forelimbs as its expression is increased in bat forelimb embryonic digits relative to mouse or bat hind limb digits. This affects developmental elongation of bat forelimb digits, and probably their evolution (Sears *et al.*, 2006). In both of these cases, it is the change in the expression of particular DNA sequences, and not in the DNA sequences themselves, that promote evolutionary change. Changes in development produce novel phenotypes which may be favored by selection.

It has been suggested that there are four cases of differences in development that may contribute to evolutionary change: (1) *heterochrony*, differences in the timing of developmental events, (2) *heterotopy*, differences in the spatial location of developmental events, (3) *heterotypy*, differences in the type of developmental events, and (4) *heterometry*, differences in the amount of activity in developmental events (Arthur, 2002, 2004, 2011). In other words, changes in the timing, positioning, amount, or type of expression of a DNA sequence, implicated in some developmental process, may produce novel characters. Arthur has described such changes as **developmental repatterning** (see Figure 5.22 for simple illustrations of these changes in

---

[26] This is not the case most of the time, as the range of genetic and phenotypic variation in populations is much larger than is commonly assumed by Mendelian genetics. However, high-school genetics mostly refers to characters with two alternative states, and this is what you have probably learned in high school (see Burian and Kampourakis, 2013; Jamieson and Radick, 2013 for details).

[27] Snakes are another example of absence of limbs. In this case, it is changes in the domains of expression of *Hox* sequences which correlate with the great expansion of thoracic identity in the axial skeleton of snakes and the consequent limblessness (Cohn and Tickle, 1999).

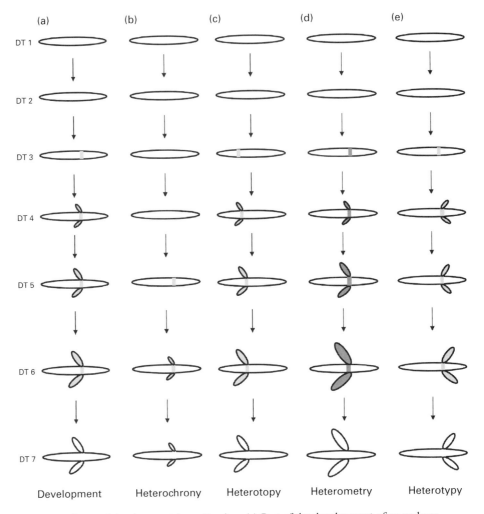

Figure 5.22 Cases of developmental repatterning. (a) Part of the development of an embryo, divided in time frames (DT: developmental time), is shown. At DT 3 a DNA sequence is expressed in the mid-anterior part and a protein is synthesized (gray region), which guides the development of limbs until DT 6. (b) Heterochrony: the DNA sequence starts being expressed at a later time (DT 5) and stops being expressed at DT 6. As a result the protein was produced for a shorter time and the limbs are shorter. (c) Heterotopy: At DT 3 a DNA sequence is expressed, but in the mid-posterior part, and a protein is synthesized (gray region) which guides the development of limbs until DT 6, but at another place of the embryo than before. (d) Heterometry: At DT 3 a DNA sequence is expressed in the mid-anterior part and a protein is synthesized, but in a greater amount (dark gray region), which guides the development of longer limbs until DT 6. (e) Heterotypy: At DT 3 a DNA sequence is expressed in the mid-anterior part and a protein is synthesized (gray region) which guides the development of limbs until DT 6, but this time limbs grow in a different direction than before.

developmental processes). One example of **heterochrony** is the change of relative timing of egg hatching and segment formation, which are two important events in arthropod development in centipedes. In members of the Order Geophilomorpha segment formation ends before egg hatching, whereas in members of the Order Lithobiomorpha it ends long after that. As a result, in the former group the hatching has the full number of segments, whereas in the latter group this number increases over the course of a year after (Arthur, 2011, pp. 95–97). An example of **heterotopy** is the unique morphology of flatfish, the head of which is partially rotated in relation to the rest of their body, resulting in both eyes being on the same side of the skull. Their embryos are bilaterally symmetrical but during development one of the eyes moves across the head and ends up on the same side as the other (Arthur, 2011, pp. 107–109). The increasing brain size in the human lineage compared to the chimp lineage in a case of **heterometry**. Humans are roughly 1.25 times larger than chimps, but our brains are 2.7 times larger than those of our close relatives (Arthur, 2011, pp. 122–123). Finally, the production of hemoglobin S (HbS) in humans instead of the normal hemoglobin A (HbA) is a case of **heterotypy** (Arthur, 2011, pp. 135–136).

Changes in the regulation of development can produce significant changes in body structure. One of the most stunning examples is the inversion of the dorso-ventral axis of arthropods and other **protostomes**[28] compared to vertebrates. In protostomes the central nervous system is closer to the ventral region and the digestive system is closer to the dorsal region, whereas in **deuterostomes** it is the other way around. This is due to an inversion in the expression of genes that determine the dorso-ventral axis. The same developmental mechanism guides body formation in both protostomes and deuterostomes, and this makes the ventral region of protostomes homologous to that of deuterostomes (De Robertis and Sasai, 1996). This is an amazing finding that shows how very different organisms can emerge from changes in the regulation of development. The DNA sequences that regulate development have been conserved evolutionarily, and the variety of animal species has been produced through mutations that provide the variation on which natural selection operates (De Robertis, 2008). The important point that should be made explicit here is that evolution does not proceed through the selection of structural mutations, i.e., mutations in coding sequences only. Other genetic changes can produce variation, even when the structural components do not change significantly. Protostomes and deuterostomes share important underlying similarities in their development (Figure 5.23) despite their enormous morphological differences (remember that pizzas and cookies differ in how they look and how they taste, although they are both made predominantly from dough).

The effects of changes in developmental processes on evolution were more or less ignored during the twentieth century (but not entirely ignored; see Minelli, 2010). The focus was on population genetics and phenotype/allele frequencies, and almost no attention was paid to how phenotypes develop under the influence of these alleles.

---

[28] Most animal phyla are bilaterians and they are divided into protostomes and deuterostomes. Protostomes are those in which the mouth develops close to the blastopore. In deuterostomes, the anus develops close to the blastopore and the mouth develops from a second opening (De Robertis, 2008, 2009; see also Figure 5.23).

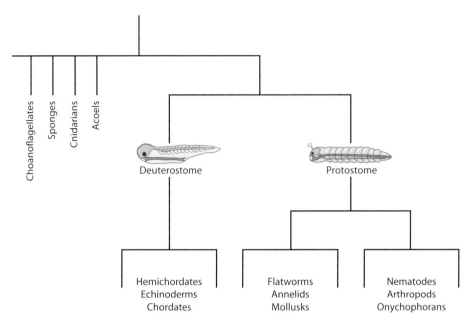

Figure 5.23 Evolutionary relationships among animals. Notice the "inversion" of systems between protostomes and deuterostomes. Protostomes have the central nervous system ventral to the digestive system, whereas deuterostomes have the central nervous system dorsal to the digestive system (dark gray: digestive system; light gray: nervous system; adapted from De Robertis, 2008, 2009). Image © Simon Tegg.

Population genetics and developmental genetics were concerned with different kinds of questions, but now a synthesis of these approaches seems to be taking place through evolutionary developmental biology (Gilbert and Burian, 2003a). There are both macroevolutionary and microevolutionary components in evolutionary developmental biology. The macroevolutionary component focuses on the mechanisms by which novel characters are brought about or, as it is often called, on the origins of **evolutionary novelty**. The microevolutionary component focuses on the processes that enable eggs, larvae, or embryos to survive in particular environments (Gilbert and Burian, 2003b). There is a lot of ongoing research on these topics. Here, I will focus on two examples for each case: **phenotypic/developmental plasticity** and the origin of evolutionary novelties.

At the microevolutionary level, one important phenomenon which is relevant and important to emphasize is phenotypic or **developmental plasticity**. It might simply be defined as the modifiability of the phenotype during development. This means that a genotype does not code for a fixed phenotype, but can produce a range of phenotypes depending on the environmental conditions. Depending on which one of a particular range of environments is experienced during development, an organism will come to possess one out of a specific range of phenotypes. For example, *Arabidopsis thaliana* plants with exactly the same genotype can be very different in morphology: If a plant is exposed to gentle mechanical stimulation during development, which might be caused by wind, rain, or herbivory in nature, it grows larger with a longer stem and with a different branching structure (Pigliucci, 2009). In other words, there is no point in distinguishing between any genetic and environmental contribution to the emergence of a phenotype, because both of them are important and any phenotype is the outcome of the interaction of the two. But is this enough to lead to evolutionary change? This is possible through phenotypic and eventually **genetic accommodation**. Developmental plasticity can provide the grounds for speciation through a specific sequence of events: (1) A population consists of organisms which are developmentally plastic, being capable of producing different phenotypic variants under the influence of inputs from the genome (mutations) or the environment (environmental change). (2) A new input may cause a reorganization of the phenotype, or "developmental recombination," which in turn produces novel variable phenotypes. Thus, **"phenotypic accommodation"** of the input takes place. (3) The new phenotypes provide material for selection, and if there is selection for them, they may become prevalent in the population. This leads to "genetic accommodation," that is, adaptive evolution that involves changes in the genetic structure of the population (see West-Eberhard, 2005; Pigliucci, 2009).

The phenomena of developmental plasticity are not in contrast or opposition, but rather are complementary with those that constitute **developmental robustness**. The latter can roughly be defined as the consistency of the phenotype despite environmental or genetic perturbation. In other words, robustness refers to the phenomenon that individuals of the same species develop the same general characters in much the same way, even if they live in very different environments. This is possible because many different processes at various levels, from the molecular to the behavioral, ensure that individuals of the same species develop to have the same general characters. One such example is insensitivity; if a developing organism cannot detect an environmental change because it lacks the necessary sensory organs, it cannot respond to it. In other cases, robustness is achieved through repair, the most impressive example being limb regeneration in salamanders. Eventually, the adult phenotype is the outcome of the interplay of the mechanisms generating robustness and plasticity (Bateson and Gluckman, 2011).

At the macroevolutionary level, explaining the origin of novel characters, or the origin of evolutionary novelties, is a very important task of evolutionary biology. However, accurately defining novelty seems to be difficult, especially if one accepts the continuum between homoplasy and homology that has been argued for in this chapter. If characters depend on deep, underlying homologies at the molecular developmental level, how novel can a new character really be? For example, should the digits of tetrapod limbs be considered as genuine novelties if it is eventually found out that this character is the outcome of rather minor rearrangements of ancestral developmental pathways and tissue interactions? Some would argue that it is not. Others would argue that a new morphological character is a novelty even if it shares some underlying molecular features with ancestral characters. Differences in definitions of evolutionary novelty could be due to the different epistemic role played by this concept, the structuring of a problem setting of an explanatory agenda, rather than its capacity to categorize characters as novel. However, what is important is explaining the origin of novel characters and it is possible to conduct research on novelties without any prior requirement to decide which characters count as "genuine" novelties. Identifying characters as novel or non-novel provides no explanation for their evolution, and it is the latter that is important (Brigandt and Love, 2010, 2012).

Now, if we insist on finding a criterion for identifying novelties, we might define a novelty as a character that has no homologous precursor. But then should all apomorphies count as novelties? And if not, what is an appropriate criterion to distinguish between those which should count as novelties and those which should not? One solution is to distinguish between different hierarchical levels (e.g., the structural and the developmental) and accept a new structure as a novelty, even if there is some underlying homology at the developmental level. The only mechanism that produces real novelties at the molecular level is horizontal DNA transfer: in this case DNA sequences can be transferred between organisms that are distantly related. It has already been mentioned that this is common among **microbes**, but it seems to be rare in animals. However, it may be the case that we are now starting to identify such cases. Such an example is the unique orange color polymorphism of pea aphids which, as revealed by genome sequencing, can be attributed to the acquisition of fungal carotenoid DNA sequences, which are not found in other animal genomes (Hall and Kerney, 2012).

Details notwithstanding, there is a lot that we still do not know, but there is also a lot that we are learning everyday through evolutionary developmental biology. Its findings not only provide further evidence for the common ancestry of all life on Earth, but also explain how divergence at the organismal level can be due to changes in regulatory DNA sequences. Thus, organisms which have many similar DNA sequences may differ significantly due to the different ways particular regulatory sequences are switched on and off. Most importantly, there is no paradox in how organisms may have evolved from ancestors from which they seem to be significantly different as seemingly large morphological changes may stem from minor changes at the molecular developmental level. This is very important given the widespread view that apparent extensive morphological differences can only result from corresponding extensive alterations of the respective DNA sequences.

Eventually, a main focus of evo-devo is **evolvability**, or the capacity or disposition of a population or lineage to evolve (Minelli, 2010). There actually exist three levels at which evolvability has been considered. The first level is within populations and refers to the available genetic variation and covariation, and determines the effect of natural selection within populations (e.g., in the case of moths and industrial melanism). The second level is within species and refers to variability, depends on genetic architecture and developmental **constraints**, and affects long-term adaptation (e.g., in the case of the adaptive radiation of the Galápagos finches). The third level is within clades and includes whatever exists within species, but also includes the capacity to overcome genetic and developmental constraints, generating major phenotypic novelties (e.g., the evolution of limbs in vertebrates) (Pigliucci, 2008). It also seems that evolvability itself evolves; however, how exactly this happens and whether this is an outcome or by-product of natural selection is not yet clear (see Pigliucci, 2008; Wagner and Draghi, 2009). Nevertheless, what is important for our discussion is that the capacity or disposition of organisms to evolve is a very important and distinctive property of organisms and life.

Conclusions

Common ancestry is an acceptable idea for many people, as long as humans are not considered. This has been the case since Darwin's time, and the problem still persists. What is missing is a proper understanding of what common ancestry entails. Sharing a vast amount of our DNA with other organisms does not mean we do not differ from them in other crucial, non-biological (even though biologically rooted or influenced) aspects of our life. Sharing a common ancestor with other organisms does not make us less important or unique, nor does it threaten our moral values and our social structure. Human uniqueness is not due to the fact that we are somehow biologically distinct from other organisms. Rather, our uniqueness is due to the fact that although we share a common ancestry and consequently crucial biological features with other organisms, we have nevertheless developed our own distinctive cultures, societies, and systems of moral values. We were neither specially created, nor created for some special purpose. However, we have evolved to occupy a distinct, unique place among all other organisms on Earth. Comparative genomics and evolutionary developmental biology, among other disciplines, show the deep underlying similarities among all organisms as diverse as *Drosophila* and humans. What is more important is that changes in developmental processes can produce morphologically different forms with significant corresponding differences in DNA sequences. This is crucial for understanding how evolution proceeds, because one does not need to assume extensive changes in DNA for the evolution of several different forms. Changes in coding and regulatory DNA sequences produce changes at the phenotypic or the developmental level, which in turn provide the raw material for evolutionary change. This is the topic of the last chapter of this book.

Further reading

Understanding concepts is a prerequisite for understanding a scientific theory. But the meanings of concepts cannot be fully given in simple definitions. Furthermore, concepts are historical entities and their meanings may change over time. Fortunately, there are two books with which one could start trying to understand evolutionary theory in more detail by better understanding its conceptual foundations: *Keywords in Evolutionary Biology*, edited by Evelyn Fox Keller and Elizabeth Lloyd, and *Keywords and Concepts in Evolutionary Developmental Biology*, edited by Brian Hall and Wendy Olson. For a contemporary view of microbiology, genomics, and more one could try the very interesting, but a bit technical, *The Logic of Chance: The Nature and Origin of Biological Evolution*, by Eugene Koonin. Lynn Margulis contributed a lot to our understanding of the origin of eukaryotes, and her books *Symbiotic Planet: A New Look at Evolution* and *Acquiring Genomes: A Theory of the Origins of Species* (the latter co-authored with her son, Dorion Sagan) are both interesting and readable. For a general account of microbes and their impact on life on Earth, one could read the *March of the Microbes: Sighting the Unseen* by John Ingraham. Regarding homology, homoplasy, and evo-devo there exist numerous books one could read, and indeed should read. Perhaps the best one to start with is Alessandro Minelli's *Forms of Becoming: The Evolutionary Biology of Development*. Then one could turn to *Biased Embryos and Evolution* by Wallace Arthur and *Endless Forms Most Beautiful: The New Science of Evo-Devo* by Sean B. Carroll. For more recent accounts one should start from the very informative and concise *Plasticity, Robustness, Development and Evolution* by Patrick Bateson and Peter Gluckman. Two more technical ones are *Evolution: A Developmental Approach*, by Wallace Arthur, and David Stern's *Evolution, Development, and the Predictable Genome*. Stephen Jay Gould's book *Ontogeny and Phylogeny* seems to have been quite influential on the development of evo-devo and it is an interesting read. Finally, for those interested in understanding phylogenetic trees, *Tree Thinking: An Introduction to Phylogenetic Biology* by David Baum and Stacey Smith is a highly recommended book.

6   Evolutionary change

In the previous chapter I explained how morphological, molecular, and fossil data consistently point to a common ancestry of all organisms on Earth. What is interesting is that organisms which may differ significantly in their morphology (e.g., *Drosophila* and humans) may share crucial similarities at the molecular developmental level. Although scientists are not yet aware of all the details, they can infer evolutionary relationships among species as these are related genealogically through common ancestry, having evolved from ancestral species through natural processes. Thus, the important question that will be the topic of this chapter is: How is evolutionary change possible? How do new species evolve from common ancestors? What causes this divergence that has produced the enormous diversity of life in our world?

Before turning to the details of these topics, I want to address some wider concerns and obstacles that are pertinent to understanding evolutionary change. It seems it is difficult for people to accept the idea of evolution through natural processes. I think this is due to the fact that it is rather difficult to understand because of two conceptual problems. The first problem is to understand how such different forms can have evolved from more primitive forms, or how organisms can have evolved from ancestors very different from them. This problem was addressed in the previous chapter, and it was explained how evolutionary modification of developmental processes can produce significant morphological changes with small changes at the molecular level (e.g., West-Eberhard, 2005). The second problem is to understand deep time: that evolution has taken place on Earth for an amount of time that is difficult for us to comprehend. It is also difficult to realize that during that time numerous life forms have emerged, but most of them have died out and only a small minority remains. One way to consider time spans is to compare actual time to a 24-hour equivalent. Table 6.1 presents some major events in evolution in actual timing and in a 24-hour equivalent. Table 6.2 presents the geologic time scale, i.e., the various periods in which the history of the Earth is divided.

The dates in Tables 6.1 and 6.2 are estimated by dating the rocks in which fossils are found. This is done by using the principles of radioactivity in what is called radiometric dating. Simply put, this technique is based on the comparison of the measured abundance of a naturally occurring radioactive isotope[1] and its decay products as well as its

---

[1] Isotopes are different forms of the same chemical element with slightly different masses, because the nuclei have the same number of protons but different numbers of neutrons, e.g., the radioactive isotope of carbon-12 is carbon-14 (see Macdougall, 2008).

Table 6.1 Some major events in evolution in actual time and in a 24-hour equivalent (dates from Benton, 2009; Knoll and Hewitt, 2011)

Evolutionary event	Actual time (million years ago)	24-hour equivalent (hours, minutes, seconds ago)
Formation of Earth	4600	24 hours
First prokaryotic cells	3600	18 hours, 47 minutes
Accumulation of oxygen in atmosphere	2400	12 hours, 31 minutes
First eukaryotic cells	1300	6 hours, 47 minutes
First (simple) multicellular organisms	1200	6 hours, 15 minutes
First (complex) multicellular organisms	600	3 hours ago
Dinosaurs	225	1 hour, 10 minutes
End-cretaceous extinction	65	20 minutes
Split of chimp and human lineages	6	2 minutes
Modern humans	0.2	4 seconds

Note that the period during which dinosaurs lived on Earth, which might seem to be a long time ago, is actually quite recent.

Table 6.2 The geologic time scale (based on Macdougall, 2008, p. 240; a complete version can be found in Ogg *et al.*, 2008 or at www.stratigraphy.org)

Eon	Era	Period	Epoch	Approximate beginning
Phanerozoic	Cenozoic	Neogene	Holocene	11 400 ya
			Pleistocene	1.8 mya
			Pliocene	5.3 mya
			Miocene	23.0 mya
		Paleogene	Oligocene	33.9 mya
			Eocene	55.8 mya
			Paleocene	65.5 mya
	Mesozoic	Cretaceous		146 mya
		Jurassic		200 mya
		Triassic		251 mya
	Paleozoic	Permian		299 mya
		Carboniferous		359 mya
		Devonian		416 mya
		Silurian		444 mya
		Ordovician		488 mya
		Cambrian		542 mya
Proterozoic		Ediacaran		630 mya
				2.5 bya
Archean				3.8 bya
Hadean				4.6 bya

Mya: million years ago; bya: billion years ago.

half-life, i.e., the time required for the decaying isotope to decay to half of its initial quantity. For example, when organisms die, carbon-14 decay begins and so the quantity of this isotope decreases exponentially. The half-life of carbon-14, i.e., the time required for half of its quantity to decay away, is 5730 years. Thus, by measuring the remaining amount of carbon-14 in some material, one is able to estimate its age. Other isotopes, such as uranium-238 with a half life of 4.7 billion years, can be used for this purpose (see Macdougall, 2008).[2]

It is evident from Table 6.1 that some events which we may consider as quite old, such as the extinction of dinosaurs, are not that old after all: just 20 minutes ago in the 24-hour scale. It seems that for the first five hours there was no life on Earth. Prokaryotic cells were the only form of life for the next 12 hours (half of the age of Earth!), and eukaryotic cells appeared less than seven hours ago. Dinosaurs only appeared about an hour ago and they disappeared along with the majority of life then on Earth just 20 minutes ago. Our lineage diverged from that of our closest relatives (chimpanzees) only two minutes ago and we have been here in our current form for less time than it took you to read this paragraph. These are important to keep in mind when wondering about evolution. The amount of time available for evolution is immense, and it is not easy for us to comprehend it.

Evolution of life on Earth has been a process of increasing complexity. This gives the impression of progress but, as Gould has explained, this is an illusion. Life started from a minimum of complexity and it could only increase during evolution (Gould, 1996). As already explained in Chapter 5, all organisms currently living on Earth have a common ancestry. Most life on Earth is microbial, but nevertheless some taxa have evolved forms of enormous complexity compared to microbes. In such cases, evolution has been characterized by significant changes in the way information is passed from one generation to the other and in the ways living systems are organized. These have been described as the major transitions in evolution (Maynard Smith and Szathmáry, 1995, 2000; see also Sterelny and Calcott, 2011).[3] These transitions are summarized in Table 6.3.

In this chapter I describe the main processes of evolution. I explain why natural selection is a very important process that can account for the origin of adaptations. However, **stochastic processes**, i.e., ones with unpredictable outcomes, are also very important in evolution and should be taken into account in evolutionary explanations. Most importantly, we should distinguish between microevolutionary processes – i.e., processes below the species level, which have been observed and studied in detail – and macroevolutionary processes, including speciation and other evolutionary transitions which have not been, and cannot be, directly observed and which are inferred from the study of fossils and molecular sequences. Finally, evolutionary explanations, both for microevolutionary but mostly for macroevolutionary processes, are historical in nature and thus have a distinctive structure and nature.

---

[2] It should be noted that fossil interpretation is neither simple nor straightforward. It can be the case, as in amphioxus, that decay of chordate characters is non-random and that apomorphic characters are prone to decay whereas plesiomorphic ones are decay resistant (see Sansom et al., 2010).

[3] The classic book is Maynard Smith and Szathmáry (1995). However, here I rely on Maynard Smith and Szathmáry (2000), which is a presentation of the same ideas to a wider audience.

Table 6.3 The major transitions in evolution (based on Maynard Smith and Szathmáry, 2000, pp. 16–19)

Transition	Main features
Replicating molecules → Populations of molecules in compartments	The first replicating molecules became informational when they started cooperating with others and assisting their replication. This was achieved when they were all enclosed within the same membrane.
Independent replicators → Chromosomes	When replicating molecules were linked, their replication was coordinated and this prevented competition and forced cooperation among them. This could lead to more complex properties (e.g., regulation of expression).
RNA as gene and enzyme → DNA and protein	Informational and catalytic properties were acquired by different molecules, nucleic acids, and proteins, respectively. This allowed for greater specialization of enzymes and the genetic code evolved.
Prokaryote →Eukaryote	This transition allowed for increased compartmentalization and complexity.
Asexual clones → Sexual populations	All prokaryotes and some eukaryotes reproduce asexually. This limits the origin of new variation to mutational events only. Sexual reproduction increases variation through meiosis and various possible combinations of male and female gametes.
Protists → Animals, plants, and fungi	Organisms with new properties emerged: multicellularity, cell differentiation and specialization, development.
Solitary individuals → Colonies	Some animals, insects mostly, lived together, forming a colony that resembled a super-organism where individual organisms performed different roles, as cells in multicellular organisms do.
Primate societies → Human societies and the origin of language	Language is a crucial, distinctive characteristic of humans. Language and DNA are two important natural systems of inheritance.

Adaptation and natural selection

In Chapter 2 I described Paley's approach to explaining adaptations (described as contrivances at the time) on the basis of intentional design; in Chapter 4 I described in detail the development of Darwin's alternative explanation, which relied mostly on natural selection. In this section I first describe the contemporary definitions of adaptation and then I distinguish between two conceptions of natural selection: selection *for* vs. selection *of* and selection *for* vs. selection *against*.[4] In the *Origin* Darwin described adaptation, in fact co-adaptation, as a kind of a special relation between organisms, crucial for their existence. He noted that:

A corollary of the highest importance may be deduced from the foregoing remarks, namely, that the structure of every organic being is related, in the most essential yet often hidden manner, to that of all other organic beings, with which it comes into competition for food or residence, or from which it has to escape, or on which it preys. (Darwin, 1859, p. 77)

---

[4] I will not get into all philosophical issues; Forber (2013) and Depew (2013) provide a concise overview of these.

But what exactly is adaptation? In the everyday use of the word, to *adapt* means to make something suitable for a new use or to adjust it to new conditions. Accordingly, adaptation may refer to the process of adapting something or of being adapted. A character that is the outcome of such a process might also be called an adaptation. Thus, based on these definitions and on everyday experience, one could infer that biological adaptation is the process by which populations become better suited to their environment, which consequently might mean that some of their characters change and become suitable for new roles. These new characters might be called adaptations as well.

These definitions, however, leave unanswered two questions: Why and how do some organisms come to possess adaptations? In addition, what is it that eventually becomes adapted to its environment: individual organisms or populations? To answer these questions one needs to describe the processes that produce adaptations. There has been much debate concerning the appropriate definition of adaptation (see Lewens, 2007 for an overview). The word adaptation has been used to refer either to a process or to a character. Two quite different definitions have been used for adaptation as a character: a historical definition, considering adaptation as a character that is the outcome of natural selection, and an ahistorical[5] definition, considering it as a character that contributes to the survival and reproduction of its possessors.[6]

George Williams defined adaptation as a character which is effective in performing a particular role, and which is the outcome of a selection process because of its effectiveness in this role (Williams, 1996, pp. 6, 212). In a later analysis, Elliott Sober argued that selection, and the historical process underlying it, are important for defining adaptation (Sober, 1993, pp. 203, 208). Robert Brandon also defined adaptation as the product of evolution by natural selection, and insisted that the criterion for considering a character as an adaptation should be its causal history (Brandon, 1990, pp. 40–41). This is often described as the historical definition of adaptation. A crucial distinction they all made was between a character being favorable to its possessors *and* being selected; a character is not an adaptation just because it confers some advantage to its possessors, but because their ancestors were selected due to this advantage. Thus, a character cannot be considered as an adaptation simply because it is beneficial, because this could be the result of chance (Williams, 1996, p. 12). A character can be considered as an adaptation if it spread in a population due to natural selection, even if it does not currently confer any advantage. In addition, not all characters that currently enhance their possessors' survival and reproduction are considered as adaptations if they have not evolved through selection (Brandon, 1990, p. 43; Sober, 1993, p. 196). Moreover, in order for a character to be considered as an adaptation, its prevalence does not only have to be the

---

[5] Meaning "non-historical"; "ahistorical" is more appropriate because it is closer to the etymological origins of the term.

[6] This is what is usually described as **fitness**. However, this is another ambiguous or at least potentially confusing term (see Ariew and Lewontin, 2004). Thus, in order to avoid dealing with this concept as well, I have refrained from using the term fitness in this book. Instead of describing adaptations as characters that contribute to increased fitness, I will refer to them as characters that contribute to the survival and reproduction of their possessors.

outcome of a selection process, but of selection for this particular character, or of this character *for* a role (Burian, 1992, p. 8). In short, according to the **historical definition** a character is an adaptation only if it is the outcome of natural selection, independently of whether it currently confers any advantage to its possessors, and only if it was selected *for* performing a particular role. Hence, natural selection for a character because of its role is a necessary and sufficient condition for being an adaptation, whereas contributing to current increased survival and reproduction is neither a necessary nor a sufficient condition for being an adaptation according to the historical definition.

Others seem to favor an ahistorical definition of adaptation. According to this, an adaptation can be defined as a character that can perform a biological role under specific circumstances (Bock, 1980, p. 221), or one that contributes to the survival and reproduction of its possessors among a specified set of variants in a given environment (Reeve and Sherman, 1993, p. 1; Mayr, 2002, p. 149). In this case, apparently, the emphasis is on the current contribution of the character and not on its history. Ernst Mayr argued that whether a character had the adaptive quality from the very beginning or not is irrelevant for its classification as an adaptation. A character should be considered as an adaptation only if it is currently favored by selection over alternative ones (Mayr, 2002, pp. 149–150). Consequently, the history of the character in general, and the selection process through which it was spread in a population or species in particular, do not have to be included in the definition of adaptation. A character is an adaptation if it is currently beneficial; whether it has evolved in order to be so is irrelevant (Bock, 1980, p. 224). What is important is the current advantageous contribution of the character, and not its selective history (Reeve and Sherman, 1993, p. 7). This, of course, means that selective history is not necessary, not that it is irrelevant (Mayr, 2002, p. 150). Reeve and Sherman further suggested that selective history may not be the only kind of history that results in adaptations, and that different kinds of historical processes may have produced current phenotypes (Reeve and Sherman, 1993, p. 14). In short, according to the **ahistorical definition of adaptation**, a character is an adaptation if it currently confers an advantage to its possessors, which gives them better chances of survival and reproduction among a specific set of alternative characters, and if it is consequently favored by selection over them. In this sense, the current contribution of a character to survival and reproduction is a necessary and sufficient condition for being an adaptation, whereas selection for a character because of its role is neither a necessary, nor a sufficient condition for being an adaptation.[7] The above definitions are summarized in Table 6.4 and are compared with an illustrative example in Figure 6.1.

---

[7] Different definitions of **adaptation as a process** also seem to exist, but they do not significantly differ from each other. Overall, adaptation has been defined as an evolutionary process for which natural selection seems to be an important factor; the differences among the various definitions have to do with how important natural selection is. For example, adaptation as a process has been defined as any evolutionary change in the form or the function of a feature which enables it to continue to perform a specific biological role under specific circumstances (Bock, 1980, p. 221), or as the evolutionary modification of a character under selection for advantageous functioning in a particular context (West-Eberhard, 1992, p. 13). Adaptation has also been defined as the evolutionary process during which the features or the capacities of organisms change in a way that enables them to overcome problems posed by their environment (Burian,

Table 6.4 Conditions in order for a character to be considered as an adaptation according to the historical and ahistorical definitions

Conditions	Historical definitions	Ahistorical definitions
Outcome of a selection process	• Necessary • Sufficient	• Unnecessary • Insufficient
Current contribution to increased survival and reproduction	• Unnecessary • Insufficient	• Necessary • Sufficient

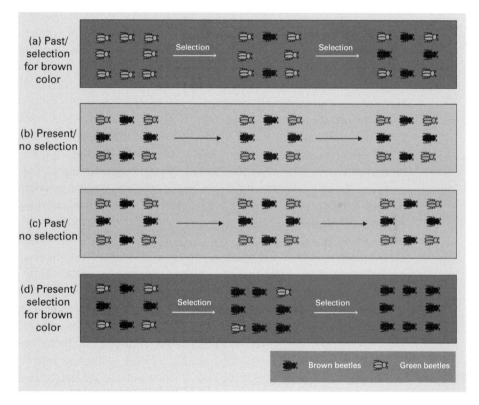

Figure 6.1 Historical and ahistorical definitions of adaptation. According to the historical definition of adaptation, the brown color of beetles is an adaptation only if (a) has taken place, independently of whether (b) or (d) are currently taking place. According to the ahistorical definition of adaptation, the brown color of beetles is an adaptation only if (d) is currently taking place, independently of whether (a) or (c) took place in the past (in all cases proportions, not actual numbers of the various types of individuals, are depicted).

1992, p. 7) or as the process during which different species become fitted to different environments by natural selection (Lewontin, 2001, p. 42). Interestingly enough, some do not provide a definition of adaptation as a process because they identify natural selection with the process of adaptation (Brandon, 1990, p. 40; Sober, 1993, p. 203). Others suggest that natural selection may not have produced, but simply maintained, a modification that arose by chance (Reeve and Sherman, 1993; Mayr, 2002, p. 148).

Figure 6.1 illustrates the differences between the historical and the ahistorical definitions of adaptation by using the beetles already introduced in Chapter 1. The beetles will be used to illustrate all evolutionary processes in this chapter.[8] There are four evolutionary processes illustrated (a–d). Two of them are evolutionary processes of the past (a, b) and two are evolutionary processes of the present (c, d). The processes are: (a) past selection for brown color; (b) no selection in the present; (c) no selection in the past; (d) present selection for brown color. The assumption made is that brown beetles have an advantage in the darker environment because they can conceal themselves, whereas there is no selective advantage for either type of beetle in the lighter colored environment. According to the historical definition of adaptation, the brown color of beetles is an adaptation only if past selection for brown color (a) has taken place, independently of whether selection is currently taking place or not (b or d). According to the ahistorical definition of adaptation, the brown color of beetles is an adaptation only if selection for brown color is currently taking place (d), independently of whether selection took or did not take place in the past (a or c).

Let me use another example in order to illustrate the differences between the two types of definitions. One may wonder under which conditions the white color of Arctic bears counts as an adaptation. According to the historical definition, it is an adaptation only if it has come to be the prevalent color in that population through natural selection. This means that there must have been selection *for* this color, e.g., because it facilitated its bearers' concealment in a snowy environment, and consequently contributed to their survival and reproduction as they could, e.g., attack their prey without being easily noticed. As a result, a process of differential survival and reproduction took place, the individuals with white color had an advantage compared to the others with different colors, and eventually the white color became the prevalent one in that population. The white color of the Arctic bears is considered an adaptation according to the historical definition because it was selected during a historical process. Whether it still is advantageous or not is irrelevant. On the other hand, according to the ahistorical definition, the white color of the Arctic bears can be considered an adaptation only if it currently provides its bearers with better chances of survival and reproduction compared to other individuals with different colors in the same environment, so that it will consequently be selected. It is not important how this character became prevalent in the particular population. What counts is that the color currently enhances the survival and reproduction of its bearers and that it is being selected. Whether or not it was selected in the past is not important; it may have been but it does not have to.

Note the major difference between the two definitions in the use of past and present tenses: in the historical definition there *was* selection, in the ahistorical one there *is* selection. One might wonder whether this difference is really of any importance. In both cases selection has a role. Whether it happened in the past or is happening now

---

[8] I made the choice of using the beetles in the illustrations of evolutionary processes in this book because Darwin was fond of them and because they are a nice example, easy to illustrate. No Beatles influence here!

may not make such a difference. However, a thought experiment may show that the difference is important. Imagine that an offspring of two brown bears, who live in a snowy environment, is accidentally born white. This could be, for instance, due to a mutation in one of its parents' alleles that was related to the production of the brown color. The mutation resulted in a defective product, which interrupted the process that produced the brown color. Given that the bears live in a snowy environment, the particular white offspring may now have an advantage in surviving compared to its siblings (because it may be more effective in attacking its prey, as they will not easily see it, being thus more capable of obtaining food) and in reproducing (because for the same reason it may be capable of feeding its own offspring more effectively, most of which will survive and reproduce as well). The white color is considered as an adaptation according to the ahistorical definition because it contributes to its bearer's survival and reproduction, and it will eventually be selected. But according to the historical definition the white color is not an adaptation because it was not selected for its role but just happened to emerge (due to a mutation) and to be inherited by the offspring.

In a similar sense imagine that the white color which was initially possessed only by that white bear is inherited by all its descendants, that it is selected for providing an advantage to its bearers, and that it eventually becomes the prevalent color in the bear population living in the snowy environment. After such a process the character can be considered an adaptation according to the historical definition because there was selection for it. But what if the climate suddenly changes and there is less snow in the environment than there used to be? The white color may not only stop conferring an advantage to its bearers, but it may also prove to be disadvantageous because, e.g., their prey will easily see the white bears in the new, darker environment. The white color will be considered as an adaptation according to the historical definition because it became prevalent due to selection, even if it does not currently confer any advantage to its bearers. However, it can no longer be considered as an adaptation according to the ahistorical definition if it does not contribute to the survival and reproduction of its bearers.

Apparently, it is not only selection that matters. Whether it happened in the past or is happening now makes a difference so that a character might be considered as an adaptation according to the one definition but not according to the other. It is important to note that in all cases it is populations, and not individuals, which adapt (in the process sense). The individual bear which was accidentally born white cannot be said to have adapted to its environment. Rather, it is the bear population that adapts because the phenotypic (and genotypic) make-up of its individuals and their frequencies change. If more white bears survive and reproduce in the particular environment compared to bears with other colors, after a number of generations the population may come to consist (mostly or exclusively) of white individuals which have an advantage in this particular environment. Thus, white color is an adaptation of the particular population of bears living in the particular environment. Whether adaptation is the outcome of a historical or contemporary process of selection, it is always a property of populations.

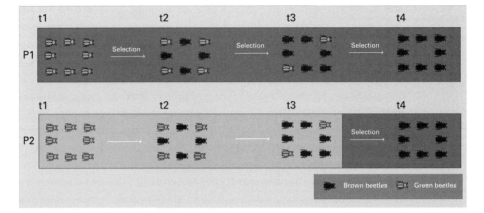

Figure 6.2 Adaptation and exaptation (t1–t4) are used to temporally divide processes P1 and P2. In P1, brown color in beetles is an adaptation because it was selected (from time t1 to time t4) for conferring an advantage (concealment) to its bearers in the dark environment. In P2, brown color became prevalent in the population of beetles up to time t2 without conferring an advantage, as the environment had a light color and so there was no advantage for beetles with brown color. Reasons other than selection for (e.g., more brown beetles happened to reproduce compared to green beetles) increased the frequency of brown beetles in the population. However, after an environmental change that made the environment darker, brown beetles had a concealment advantage. From t3 to t4 they were selected for this advantage and increased in frequency. In this case, brown color was "co-opted" and so it is an exaptation (in all cases proportions, not actual numbers of the various types of individuals, are depicted).

I think evolutionary history is important for understanding adaptation, no matter how one defines it (see Kampourakis, 2013b).[9] One might thus combine the two definitions and define adaptation as any character that enhances survival and reproduction and has become prevalent due to selection for its current role. However, a currently beneficial character may have evolved for some other role or no role at all, and may have been later "co-opted" for its current role. The increase of the frequency of a character in a population could be accidental and not the outcome of selection (see next section). Should this be called adaptation, too? It has been suggested that such a character should be distinguished from adaptations and be called an **exaptation** (Gould and Vrba, 1982; Gould, 2002, pp. 1232–1233). The distinction between adaptations and exaptations has undergone some criticism (Reeve and Sherman, 1993, pp. 3–4), but I think it is important and useful. Consider two different evolutionary processes, P1 and P2 (Figure 6.2). In P1 a brown color appears in a given environment and is selected

---

[9] I note that adaptation is a concept with a theological load. Just consider Paley's argument, already discussed in Chapter 2: "for, in the watch which we are examining, are seen contrivance, design; an end, a purpose; means for the end, adaptation to the purpose. And the question, which irresistibly presses upon our thoughts, is, Whence this contrivance and design? The thing required, is, the intending mind, the adapted hand, the intelligence by which that hand was directed" (Paley, 2006/1802, p. 14). Given this, I think that there are important semantic issues that should be taken into account in the public discourse of evolution: book titles such as *The Blind Watchmaker* (Dawkins, 2006a) or *The Tinkerer's Accomplice: How Design Emerges from Life Itself* (Turner, 2007) may be confusing.

because it increases the survival and reproduction of its possessors. Thus, brown color is an adaptation because it was selected for conferring an advantage (concealment) to its bearers in the dark environment. In P2 a brown color emerges and spreads in a population for reasons other than selection, conferring no advantage to its possessors. The environment had a light color and so there was no advantage for beetles with brown color. Then, an environmental change happens, and the character comes to play an important role because the environment became darker and so brown beetles gained a concealment advantage. According to Gould and Vrba (1982) such a character that initially spread for non-adaptive reasons and was later co-opted for a new role in the new environment is an exaptation.[10] The two processes are different. In P1 the character spreads in the population adaptively because it is advantageous in the particular environment. In P2 the character first spreads in the population non-adaptively, and becomes advantageous only after the environmental change. There is an important difference between these two processes and I think that the distinction between adaptation and exaptations makes it clear.[11]

You must have noted that so far I have been writing about *selection for* a character. It is important to distinguish between *selection for* and *selection of* (Sober, 1993, pp. 97–102). Sober distinguishes between *selection for properties* and *selection of objects*. In a toy in which balls which differ in size and color (e.g., white balls are large and black ones are small), only the small balls can pass through some holes when the toy is shaken, whereas the large ones cannot. Thus, there is *selection for size* (small balls pass; large balls do not pass). However, it is also the case that only black balls and no white balls pass, because all large balls are white whereas all small balls are black. In this case, there is no *selection for color*: whether a ball is black or white makes no difference on whether it can pass through the holes. However, there is selection *of* black balls because all small balls, which can pass through the holes, are also black. Sober notes that this points to the distinction between *selection for properties* (smallness is selected; largeness is not selected) and selection of objects (small balls, which also

---

[10] A character could also spread in the population due to selection for one kind of advantage and later be selected for a different kind of advantage. Exaptation does not require (though it can have) a non-adaptive origination step. However, to make the distinction clearer, I am using a non-adaptive first step in my example.

[11] A similar distinction is between adaptations, characters that exist as a consequence of natural selection for one or more of their effects, and adaptive characters which contribute to the survival and reproduction of their bearers whether they are the outcome of natural selection or not (Sterelny and Griffiths, 1999, pp. 217–220). According to this distinction, adaptations do not need to be currently adaptive, whereas adaptive characteristics may also be adaptations, but they may arise by chance as well. Thus, one might say that the historical definition refers to adaptations, whereas the ahistorical one refers to adaptive characteristics. Nevertheless, this distinction does not say much for the history of the character, whereas the exaptation/adaptation distinction does. Both adaptation and exaptation refer to history with the main difference being on whether the character initially evolved for its current role (adaptation) or for some other role (exaptation). Quoting Gould: "Thus, Vrba and I recommended that features crafted for current use continue to be called *adaptations* [. . .], and that features coopted for current use, following an origin for some other reason, be called *exaptations*. We would also prefer that biologists embrace 'aptation' rather than 'adaptation' as the general descriptive term for a character now contributing to fitness, with exaptation and adaptation defined as two subcategories of aptation, thus designated to recognize the crucial distinction between cooptation and direct shaping in the historical construction of characters" (Gould, 2002, p. 1233).

Figure 6.3 Selection *for* size and selection *of* color (based on Sober, 1993, p. 99). White balls are also the larger ones and do not pass through the holes. Light gray balls are smaller and pass to the next layer. Dark gray balls are even smaller and pass to the next layer. However, only black balls pass to the bottom layer because they are the smallest. In this case, there is selection for size (small balls pass to the bottom layer; large balls do not) and selection of color (black balls pass to the bottom layer because they are also small; white balls do not because they are also large).

happen to be black, are selected). Based on this, one could claim that there was *selection for* smallness but *selection of* blackness.[12] There was selection for one property (small size) because the objects with that property were "favored" (small balls could pass through the holes). However, there was also selection of another property (black color); this selection was incidental because the objects with that property (black balls) happened to be "favored" because they had another property (small size) for which they were selected (Figure 6.3).

Sober provides two examples on which I want to elaborate: **pleiotropy** and linkage. Let us start with pleiotropy, i.e., the phenomenon in which a DNA sequence affects two different characters. Imagine a DNA sequence G which is implicated in the development of two characters, A and B. If there is selection for A, the consequence will be that organisms with this character will survive and reproduce. Eventually, there will be selection of individuals with G for character A. However, these individuals with G will probably have character B, too. Thus, selection of B will be a consequence of selection for A. In the case of linkage, imagine that two linked DNA sequences, G_A and G_B, affect the development of characters A and B, respectively. If there is selection for A,

---

[12] Sober does not write about selection *of* objects *for* their size, but only about selection *for* size and selection *of* objects. In what follows I distinguish between two kinds of selection relevant to properties/characters: selection *for* a character if it is that character that makes the difference for selection, and selection *of* a character if this selection is incidental. In the next section I will explain that there is a stochastic component in the latter case. Nonetheless, it is a kind of selection.

the consequence will be that organisms with this character will survive and reproduce. Eventually, there will be selection of individuals with G_A for character A. However, these individuals with G_A will also have the sequence G_B, because the two sequences are linked, and also character B. Thus, selection of B will be a consequence of selection for A. In both of these cases, there is selection *of* character B but nevertheless this character is not considered an adaptation because there is no selection *for* it. Only if a character confers an advantage to its bearers, as happens with A in both cases, and it is because of this character that individuals survive and reproduce, can one consider it as an adaptation. In short, only when there is selection for a character, and not selection of it, is this character an adaptation.

But what does *selection for* consist of? David Depew (2013) explains that Darwin's view of natural selection was different from that of his supporters and successors. It was the latter's view that persisted until the **Modern Synthesis**, when a view similar to Darwin's became widely accepted again. What is more important is that these two views differ significantly and it is important not to conflate them. Let us explain the difference. Adaptations can evolve through natural selection only when the latter operates for many generations on variants of characters, which first arise independently of the advantage they may eventually confer to their bearers. This was Darwin's view in the *Origin*; he wrote about "the accumulation of slight but useful variations" (Darwin, 1859, p. 61; see Chapter 4). This is different from the idea of the "survival of the fittest," a term coined by Herbert Spencer and eventually adopted by Darwin in the third and subsequent editions of the *Origin*. In this case, there is no accumulation of slight, advantageous variations, but simply elimination of the individuals which bear the non-advantageous characters and survival of those individuals which possess the advantageous ones. There is an important difference between the two views: In the case of **selection for**, adaptation is the cumulative effect of selection operating on the available variation for many generations; in the case of **selection against**, variation is quickly eliminated. In other words, the first view is about gradual selection over trans-generational time *for* adaptive characters, whereas the second view is about selection *against* organisms that do not have such characters.

It may be the case that it is the *selection against* view that easily comes to mind, assisted by the slogan "survival of the fittest," so that it is difficult to understand the other view and the differences between them. This is a major conceptual issue, because it is difficult to understand how novel adaptations can evolve through selection if the latter just eliminates variants. Selection for a character does not simply mean that this character is preserved and all the others are eliminated. In contrast, selection for a character means that slight variations of this character provide an advantage and so those individuals bearing them survive and reproduce more efficiently than others. Gradually, over many generations, these variations of the character become prevalent in the population. Thus, selection is more than an eliminative process; it gradually drives a population to adaptation.

Let me give an example. Imagine a population consisting of dark green and dark brown beetles. In a dark brown environment selection can take place so that the population will eventually consist exclusively of the dark brown beetles. However, there are two ways in which this can happen. In the first case, there can be selection *against* the dark green beetles. This means that even within a generation these could be eliminated and only the dark brown ones could survive. In the second case, there could be selection *for* the dark brown color. This would be a gradual process that could span several generations. During this process, other variants could emerge (e.g., light green and light brown beetles). The varieties of green beetles could be eliminated during the selection processes, but the light brown beetles could survive along with the dark brown ones, although in a lower proportion. Eventually, the outcome of the *selection against* and the *selection for* processes is not exactly the same. In the first case, the population consists exclusively of dark brown beetles. That was an eliminative process that allowed only one variant to survive. In the second case, the process of selection spanned several generations and so a gradual accumulation of variations could take place. Eventually, the population now exhibits higher variation than in the first case as several varieties of brown beetles co-exist (Figure 6.4).

Let me illustrate this in a different way, using the dough example from the previous chapter. Imagine the cookie shop owner wants to make available those cookies which customers prefer most. One way to do this is to prepare four kinds of cookies: chocolate cookies, raisin cookies, almond cookies, and butter cookies. One Saturday morning she invites her customers to the shop, gives to each of them a cookie of each kind to taste, and asks them to decide which one they like more. Let's assume that most of them like the chocolate cookies more than the others. If the shop owner decides to stop producing the other kinds of cookies because her customers liked the chocolate ones more, this is *selection against* the other kinds of cookies. Now imagine that, in contrast, the shop owner keeps producing all kinds of cookies but then realizes that her customers like the chocolate cookies a bit more than the others. Thus, she starts asking them how she could improve those cookies more. One day some customers suggest that some more butter would make the chocolate cookies taste better. She does that and starts selling cookies with extra butter, in addition to the normal chocolate cookies. Some weeks later other customers suggest that more chocolate would make the cookies taste even better, so in addition to the normal and buttery cookies, she begins selling extra-chocolate cookies. A month after that another customer suggests that if the cookies were a bit softer, they would be great, so the owner brings out a fourth variation of the cookies. Thus, over the course of two months, chocolate cookies have been changing and customers have started preferring them more and more. In the same time, the owner produces more and more chocolate cookies of different kinds because more people ask for them as they hear that they have become better. Eventually, most people come to the shop to buy the more-butter and/or more-chocolate and/or softer chocolate cookies. Eventually, the shop owner stops producing the other kinds of cookies and keeps producing the various kinds of chocolate ones. In this case, *selection for* chocolate cookies has taken place.[13]

---

[13] I must note that in this example the "evolution" of cookies is driven by the intentions of the shop owner. There is no such intentionality in nature and this is important to keep in mind. Thus this example of selection is not equivalent to natural selection. It is only used in a metaphorical sense to illustrate the difference between selection *against* and selection *for*.

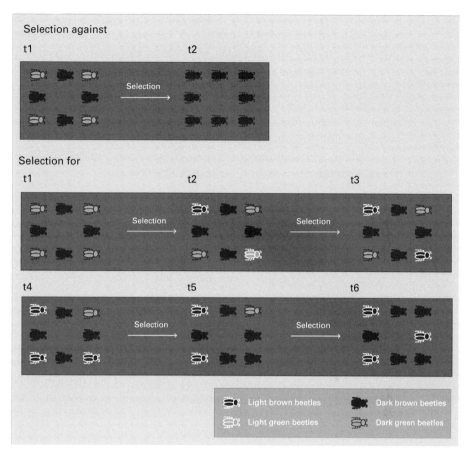

Figure 6.4 Selection *against* and selection *for*. In both cases dark green and dark brown beetles initially live in a dark brown environment. In the first case, only the dark brown beetles survive while the dark green ones are quickly eliminated; there is *selection against* the latter. In the second case, there is a gradual selection for brown color over many generations with a gradual accumulation of useful variations: not only dark brown beetles but also light brown ones can be adapted (in all cases proportions, not actual numbers of the various types of individuals, are depicted).

At the beginning of this section I described definitions of adaptation which are based on (past or present) selection. It is important to note that adaptations are the outcomes of "*selection for*" processes in populations. I do not want to favor one definition of adaptation over the other (historical or ahistorical), but I have already stressed the importance of history for understanding these processes. Evolutionary processes have a historical dimension and consequently there is a causal dependence on particular antecedent conditions or events; had these never existed a different evolutionary outcome would have come up (Beatty, 2006; see also the final section of this chapter). Consider the example in Figure 1.11 (evolution of multi-drug resistance). Drug resistance is an adaptation for bacteria, because it enhances their survival and reproduction in a particular environment. To understand how this happened, some explicit reference to history is required in order to explain that those bacteria who happened to have resistance survived whereas non-resistant bacteria died. Consequently, the bacterial population evolved drug resistance, which is an adaptation. To give another example which has already been discussed in Chapter 3, wings are adaptations for flying, at least for those birds which use them for this purpose (e.g., eagles – there is no need to discuss here whether or not it is an adaptation for ostriches). This character conferred an advantage to the ancestors of birds and it became prevalent through selection in that population that evolved, giving rise to birds with wings. But there is more than selection in evolution. Stochastic processes are also important and they will be the topic of the next section.

Stochastic events and processes in evolution

In 1979 Stephen J. Gould and Richard Lewontin published a now famous paper, often described as the Spandrels paper (Gould and Lewontin, 1979). There, they argued against the all-importance of natural selection and adaptation for evolution. As already explained in the previous section, some characters can become prevalent for reasons other than selection and eventually be "co-opted" for their current use (see also Pigliucci and Kaplan, 2006, pp. 112–129). Thus, the question is through which processes, except for selection, can a character become prevalent in a population? There are several undirected processes which can drive the evolution of a population without any kind of selection. These are generally described as stochastic processes: undirected processes with unpredictable outcomes in which chance seems to play a major role. Genetic drift, or the indiscriminate sampling of gametes in a population, is the exemplar of such processes. In this section I describe how **drift** has been conceptualized, especially with respect to natural selection. I also describe a related concept, called genetic **draft**. But before turning to these, let us first discuss what stochastic events and processes are.

In the previous chapter I mentioned that changes in DNA sequences can produce new characters. These include changes within DNA (coding or regulatory) sequences or larger rearrangements in chromosomes (insertions, duplications, deletions, etc.). All these are stochastic events, as chance plays a major role in their occurrence: Whether they will occur and what their outcome will be is a matter of chance, in the sense that it cannot be predetermined or predicted. What we usually describe as mutation in biology includes all of these phenomena.[14] For example, whether a nucleotide in DNA will change to another one or whether unequal crossing over will take place is not something that can be determined in advance. For example, whether a nucleotide change in a DNA sequence will lead to the production of a synonymous codon in mRNA and the same amino acid in the protein (silent mutation) or a codon that corresponds to a different amino acid is in large part a matter of chance. Similarly, whether unequal crossing over will result in the duplication of a DNA sequence without harmful effects or in the

---

[14] Mutation is a "bad" word outside biology, but the respective phenomena are actually one important source of variation.

disruption of a DNA sequence that is implicated in the production of an important peptide is also a matter of chance. In short, changes at the molecular level are stochastic events because we cannot tell in advance when they will take place or what their outcome will be (see Griffiths *et al.*, 2012 for the details of these and other similar phenomena).

Stochastic events also take place at the organismal level. One very important case is that of horizontal DNA transfer, already discussed in Chapter 5. To illustrate stochasticity at this level, I will use an example of horizontal DNA transfer in bacteria. Imagine that a bacterial strain (let us call it B) that infects humans (Bh) is not resistant to anti-B drugs, whereas a similar strain that infects birds is resistant to drugs (Bb-r). There are several mechanisms for drug resistance (Alekshun and Levy, 2007). Let's assume that in this case drug resistance is due to a DNA sequence, let us call it *RABD* (resistance to anti-bacterial drugs), located on a plasmid.[15] Infection of humans by Bh can effectively be fought by drugs, whereas infection by Bb-r has no harmful effects. Thus, a human infected simultaneously by Bh and Bb can take drugs and avoid suffering from the disease. However, it is possible that inside the human body bacteria of the two strains come close to each other and are connected through a cytoplasmic bridge. Once they are connected, transfer of the plasmid on which *RABD* is found from Bb-r to Bh is possible. If this happens, then it is possible that bacteria of the strain Bh will also become resistant to drugs (Bh-r). This may cause problems, even a pandemic, if the new transformed bacterium multiplies and is transmitted to other people. The disease that was so far controlled with drugs will no longer be possible to control if one is infected by the Bh-r, which is drug resistant. Whether such a DNA transfer from one bacterium to another will happen or not is a matter of chance, or in other words cannot be determined in advance. In this sense, this is a stochastic event.

Finally, stochastic events take place at the population level. The exemplar process of this kind is genetic drift or simply drift.[16] Drift can be defined as a process of indiscriminate (parent or gamete) sampling (Beatty, 1984), or as a natural process where differences in reproductive success are not due to differences between corresponding heritable characters (Millstein, 2002, 2005). This process can be better understood if it is contrasted to selection. Selection can be described as a process of discriminate sampling: Some characters confer an advantage to their possessors which in turn survive and reproduce better than other individuals that do not have these characters. Thus, there is selection for these characters and not for others, and in this sense (gamete or parent) sampling is discriminate (some characters are sampled through the reproduction of their possessors but not some others). In this case, differences in reproductive

---

[15] Plasmids are small, circular DNA molecules that many bacteria possess. These are distinct from their main (usually quite long) circular DNA molecule which makes up most of their genetic material.

[16] There is some disagreement about how drift should be conceptualized. Some argue that drift is a process (Millstein, 2002, 2005; Millstein *et al.*, 2009) whereas others deny this view (Matthen and Ariew, 2002; Walsh *et al.*, 2002; Pigliucci and Kaplan, 2006; Walsh, 2007). I want to refrain from getting into the complexities of this debate. In this book I consider drift as a process (following Millstein *et al.*, 2009), which is alternative to evolution by natural selection (Beatty, 1984). The reason for this is that different antecedent conditions may cause one or the other process (see the last section of this chapter).

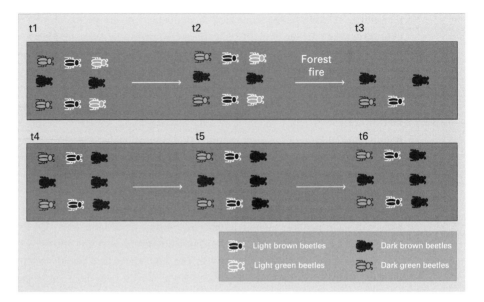

Figure 6.5 Drift due to indiscriminate parent sampling. Only those individuals that happened to survive produced offspring; because more brown rather than green beetles survived, the frequencies of the different varieties in the population changed (in all cases proportions, not actual numbers of the various types of individuals, are depicted).

success are due to differences between corresponding characters which facilitate the survival and reproduction of their bearers. Drift is a process of indiscriminate sampling in which reproductive success is not due to selection. Which characters are sampled is a matter of chance.[17] To illustrate drift as a process, I will describe two examples: a fire as a case of **indiscriminate parent sampling** and random mating as an example of **indiscriminate gamete sampling** (based on Beatty, 1984).

Indiscriminate parent sampling can be defined as the process which determines which organisms will have offspring, and how many offspring each parent will have. This sampling is indiscriminate in the sense that differences in offspring contribution are independent of any differences in characters among the individuals of the parental generation (Beatty, 1984). Imagine a forest in which green and brown beetles live and where there is no selection for color (Figure 6.5). This population consists of four different varieties of beetles (two varieties of green beetles – dark green and light green ones – and two varieties of brown beetles – dark brown and light brown ones). There is no selection for color in the particular environment and so each variety constitutes 25% of the beetle population. Assuming that frequencies are stable because there is no selection (this is not always the case as will be explained in the following example), up to t2 there is no change. However, a forest fire happens to kill more green beetles than brown ones. As a result, the brown varieties are now 75% of the remaining

---

[17] The **neutral theory of molecular evolution** suggests that drift has a major role at the molecular level (see Dietrich, 2013 for details).

population whereas the green ones are just 25%. Even when the population reaches the size before the fire, the frequencies are still the same and because there is no selection for color they might remain the same for some time. In this case, some individuals of one variety happen to die before reproducing. Thus, only the remaining individuals reproduced and the proportion of each variety in the population changed. There was a parent sampling (some beetles had offspring, others did not), but it was indiscriminate since it was purely contingent. The frequency of brown-colored beetles increased in the population but there was no such thing as selection *for* this character.

The relative frequencies of individuals with different characters may also change from generation to generation in cases of indiscriminate gamete sampling, defined as the process which determines which of the two genetically different types of gametes produced by a heterozygote parent actually contributed to each of its offspring. This sampling is indiscriminate in the sense that whether a particular offspring will stem from one or the other gamete of a heterozygote parent is independent of any differences between the gametes themselves (Beatty, 1984). Imagine a population consisting of four heterozygotes (G1G2), and four homozygotes of each kind (G1G1 and G2G2). Let us label them $G1G2_{(i)}$–$G1G2_{(iv)}$, $G1G1_{(i)}$–$G1G1_{(iv)}$, $G2G2_{(i)}$–$G2G2_{(iv)}$. Using high-school genetics, we can predict all possible crosses among G1G1, G1G2, and G2G2 individuals and the probability of each kind of offspring occurring:

I.   G1G1 × G1G1 → 100% G1G1
II. G1G1 × G1G2 → 50% G1G1, 50% G1G2
III. G1G1 × G2G2 → 100% G1G2
IV. G2G2 × G1G2 → 50% G2G2, 50% G1G2
V.   G2G2 × G2G2 → 100% G2G2
VI. G1G2 × G1G2 → 25% G1G1, 50% G1G2, 25% G2G2

It is evident that whenever a heterozygote is involved, there are different possible gamete contributions it can make to its offspring (crosses II, IV, VI). Let's assume now that each of these possible crosses takes place once in the population described above and that each cross yields one offspring only:

I. $G1G1_{(i)}$ × $G1G1_{(ii)}$ → G1G1
II. $G1G1_{(iii)}$ × $G1G2_{(i)}$ → G1G2
III. $G1G1_{(iv)}$ × $G2G2_{(i)}$ → G1G2
IV. $G2G2_{(ii)}$ × $G1G2_{(ii)}$ → G1G2
V. $G2G2_{(iii)}$ × $G2G2_{(iv)}$ → G2G2
VI. $G1G2_{(iii)}$ × $G1G2_{(iv)}$ → G1G2

If you count the offspring, you will realize that the genotypic ratio is G1G1:G1G2:G2G2 at 1:4:1, whereas in the parental generation it was 1:1:1. Although all possible crosses were done, crosses II, IV, and VI involve heterozygotes which can contribute different gametes to their offspring. When a heterozygote is crossed with another individual, although both its gametes have equal chances of contributing to the offspring, only one of them will be used. So, although all possible crosses can take place and all gametes have equal chances of contributing to the offspring, what actually happens during sexual reproduction is that some gametes but not others are used. This may cause fluctuation of frequencies from generation to generation. Gamete sampling may take place (some gametes of heterozygotes contribute more to the offspring than others), but it is indiscriminate (there is no reason other than chance that this is happening).[18]

As already mentioned, the main difference between the process of selection and the process of drift is direction. In both processes phenotypic or genotypic frequencies may change. However, in the case of selection there is direction toward the increase of the frequency of those organisms which can better survive and reproduce in the particular environment. In contrast, drift is undirected. Whether one or the other individual will contribute to offspring and whether it will contribute one or the other gamete is a matter of chance. However, not all stochastic processes are as undirected as drift. One interesting process is genetic draft (see Skipper, 2006). Draft is another name for linked selection, already discussed in the previous section. Imagine two linked DNA sequences, G_A and G_B, which affect the development of characters A and B, respectively. If there is selection for A, the consequence will be that organisms with this character will survive and reproduce. Eventually, individuals with this character will increase in proportion. Thus, character A may become prevalent in the population, but so will character B because it is linked to it.

Let me illustrate this process with an example. Imagine that a DNA sequence G_{A1} is implicated in brown coloration in beetles. Imagine also that a DNA sequence G_{B1} is implicated in the black coloration of the internal part of their wing blades. The respective alleles are G_{A2} (green coloration) and G_{B2} (white wing blades). Consequently, there are four possible combinations: brown color with black wing blades, green color with white wing blades, brown color with white wing blades, and green color with black wing blades. In an environment where brown coloration confers an advantage to beetles, compared to green coloration, there will be selection for brown color. However, there will also be selection of black or white wing blades as a result of the process of genetic draft. Black or white wing blades will become the prevalent character in the population but not because there was selection for them. Whether it will be one or the other character that will become prevalent is a matter of chance, depending on which of the alleles are linked (Figure 6.6).

These are some examples of stochastic events (mutation, horizontal DNA transfer) and processes (drift, draft) which can drive evolution in one or another direction unpredictably. A usual criticism against evolution is its "randomness"; critics state that it is impossible for complex systems to occur through random events. However,

---

[18] Note that it is possible that frequencies do not change in the next generation:

  I. $G1G1_{(i)} \times G1G1_{(ii)} \rightarrow G1G1$
  II. $G1G1_{(iii)} \times G1G2_{(i)} \rightarrow G1G1$
  III. $G1G1_{(iv)} \times G2G2_{(i)} \rightarrow G1G2$
  IV. $G2G2_{(ii)} \times G1G2_{(ii)} \rightarrow G2G2$
  V. $G2G2_{(iii)} \times G2G2_{(iv)} \rightarrow G2G2$
  VI. $G1G2_{(iii)} \times G1G2_{(iv)} \rightarrow G1G2$

In this case, the genotypic ratio G1G1:G1G2:G2G2 of the offspring is 1:1:1, similar to that of the parents.

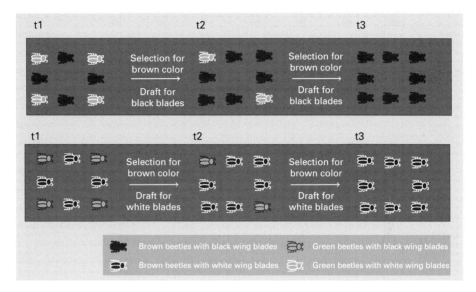

Figure 6.6 Genetic draft: there is selection for brown color, but also of black or white wing blades in beetles, depending on which of the alleles are linked. Whether black or white wing blades will become the prevalent character in the population is a matter of chance, depending on which of the two alleles is linked to the allele that affects the character which is being selected for (brown color) (in all cases proportions, not actual numbers of the various types of individuals, are depicted).

evolution is not a random process but particular events and processes can have a high degree of unpredictability. For example, whereas one can predict the outcome of selection, it is not possible most of the time to predict the outcome of drift because one cannot know in advance which of the events or processes such as those described above will take place and when. In a recent book (McShea and Brandon, 2010, pp. 2–3) it is argued that in the absence of processes such as natural selection, diversity and complexity can arise by the simple accumulation of accidents and eventually increase on average. It is also argued that this is actually the natural or background condition of evolving populations and organisms (what they call the "zero-force evolutionary law" or ZFEL). They illustrate this with the example of a picket fence. Such a fence may consist of pickets which initially are identical to each other. However, as time goes by different accidents can happen to different pickets (a pollen grain stains a picket; a passing animal knocks a chip of paint off another picket; the bottom of another one becomes moldy and crumbles where it touches the ground, etc.). As a result, the pickets become different from each other and this process can continue indefinitely. Eventually, there is an increase in the complexity and the diversity of the fence as it consists of pickets which over time become very different from each other. The important point here is that no external intervention or directed process is necessary. In this sense, undirected, unpredictable, stochastic processes can have dramatic effects.

One important concept that helps describe the implications of stochastic processes and events is the concept of **contingency**, proposed by Stephen Jay Gould, who argued that the history of life is not predictable as organisms have evolved through a series of contingent events. Gould defined contingency[19] as the "affirmation of control by immediate events over destiny" (Gould, 2000, p. 284) and argued that the world is largely a product of contingency. Gould illustrated the idea of contingency by the metaphor of the tape: "You press the rewind button and, making sure you thoroughly erase everything that actually happened, go back to any time and place in the past. [. . .] Then let the tape run again and see if the repetition looks at all like the original" (p. 48), "any replay of the tape would lead evolution down a pathway radically different from the road actually taken" (p. 51). Thus, the evolutionary contingency thesis suggests that the history of life on Earth has been determined by contingent events. For example, mutations are sources of contingency (Beatty, 1995). There are two versions of contingency: the unpredictability version and the causal dependence version (Beatty, 2006). There are several possible evolutionary paths (contingency); it is impossible to predict in advance which of them is going to actually be taken (unpredictability) and there are certain constraints in the possible outcomes once a specific pathway is taken (causal dependence).

Here is an example to illustrate this. Imagine a population consisting of equal numbers of different varieties of brown and green beetles, living in an environment where there is no selection for either color. How should one expect this population to evolve? It might remain as it is for years. However, if this population migrated to a brown environment (or if brown color somehow became the dominant one in their current environment, e.g., due to destruction of vegetation) then this population might evolve to one consisting of brown beetles only (outcome B in Figure 6.7). Similarly, if this population migrated to a green environment (or if green color somehow became the dominant one in their current environment, e.g., due to increase of vegetation) then this population might evolve to a population of green beetles only (outcome G in Figure 6.7). What is the most probable outcome? No one can tell in advance. Beetles might migrate or their environment might change, but this cannot be known in advance (unpredictability). Now if one of the two evolutionary paths is taken, this will determine the outcome of evolution (B or G). And the population will not be able to revert to its original condition if its genetic structure changes during evolution. If all green beetles die out, and a population of brown beetles evolves, there might never be any green beetle in that population again[20] (causal dependence). Finally, if outcome G was the actual outcome, Gould's tape metaphor suggests that if we could let that initial population evolve again antecedent conditions might be different and thus outcome B might be the result of evolution in that case. This is, of course, a hypothesis that we cannot test; but it highlights the importance of contingency in

---

[19] It should be noted here that the importance of contingency for evolution has been criticized on the basis of evidence for convergent evolution, the process of acquiring similar characters independently and not of their being derived from a common ancestor (Conway Morris, 2003). However, it has been argued that this is not enough to undermine the importance of contingency, even if it was not as high as Gould believed (Sober, 2003; Sterelny, 2005; Szathmàry, 2005).

[20] Although evolutionary developmental biology, discussed in the previous chapter, suggests that underlying homologies exist so there are ways that this could be possible.

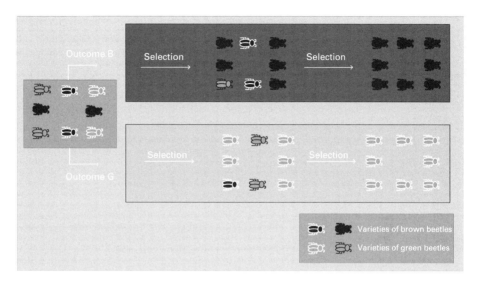

Figure 6.7 Contingency in evolution. Which of the two evolutionary paths will be taken is unpredictable. Once a path is taken there is a causal dependence of the outcome on the antecedent conditions (in all cases proportions, not actual numbers of the various types of individuals, are depicted).

evolution. Whether the environment in which the initial population of beetles lived would change to a green or a brown one is unpredictable.

Speciation, extinction, and macroevolution

The *Origin* is about how natural selection causes the gradual modification of populations so that they diverge enough to form new species which are related by common ancestry. However, Darwin refrained from providing a single definition for the term "species":

Nor shall I here discuss the various definitions which have been given of the term species. No one definition has as yet satisfied all naturalists; yet every naturalist knows vaguely what he means when he speaks of a species. Generally the term includes the unknown element of a distinct act of creation. The term "variety" is almost equally difficult to define; but here community of descent is almost universally implied, though it can rarely be proved. (Darwin, 1859, p. 44)

He also thought that it is really difficult to demarcate species from varieties, and he thought that their difference is actually a matter of degree: a well-marked variety could be called an incipient species:

Certainly no clear line of demarcation has as yet been drawn between species and sub-species – that is, the forms which in the opinion of some naturalists come very near to, but do not quite arrive at the rank of species; or, again, between sub-species and well-marked varieties, or between lesser varieties and individual differences. These differences blend into each other in an insensible series; and a series impresses the mind with the idea of an actual passage. Hence I look at individual differences, though of small interest to the systematist, as of high importance for us, as being the first step towards such slight varieties as are barely thought worth recording in works on natural history. And I look at varieties which are in any degree more distinct and permanent, as steps leading to more strongly marked and more permanent varieties; and at these latter, as leading to sub-species, and to species. The passage from one stage of difference to another and higher stage may be, in some cases, due merely to the long-continued action of different physical conditions in two different regions; but I have not much faith in this view; and I attribute the passage of a variety, from a state in which it differs very slightly from its parent to one in which it differs more, to the action of natural selection in accumulating (as will hereafter be more fully explained) differences of structure in certain definite directions. Hence I believe a well-marked variety may be justly called an incipient species; but whether this belief be justifiable must be judged of by the general weight of the several facts and views given throughout this work. (Darwin, 1859, pp. 51–52)

Darwin thought that those taxa naturalists called "species" really existed, but he doubted the existence of a well-defined "species" category. One important distinction thus made by Darwin is between the category of species, however defined, and the taxa that biologists identify as somehow distinct from others. It seems that even if the species category does not exist in nature, the taxa that biologists call "species" actually exist and the term "species" can be retained to describe them (Ereshefsky, 2010a). Darwin's view that there is a continuity between varieties and species seems to be supported by recent evidence (Mallet, 2008).

Whatever the case, we still need an, at least instrumental, definition for species. Since Darwin, many different definitions have been employed for different purposes (see Wilkins, 2009; Richards, 2010). Perhaps the most widely used definition is the one that is based on reproductive isolation. Based on this criterion, Coyne and Orr (2004, p. 30) consider groups of populations as constituting different species under two conditions: (1) if their genetic differences preclude them from living in the same area; or (2) if they inhabit the same area but their genetic differences make them unable to produce fertile hybrids. Coyne and Orr note that distinct species are characterized by substantial but not necessarily complete reproductive isolation; in other words, species may have some limited exchange of DNA sequences through reproduction with others living in the same area. This definition is actually problematic because it overlooks the facts and complexities of microbial life (Ereshefsky, 2010b; Duncan *et al.*, 2013). However, if one focuses on sexually reproducing organisms, as I have done in this book, this definition works well enough. If a species is defined as a group of potentially interbreeding natural populations which are reproductively isolated from other such groups, speciation can be defined as the evolution of new populations which are reproductively isolated from other populations (Coyne, 2009, p. 270). Two populations are reproductively isolated when their members cannot mate or cannot produce fertile offspring if they mate. There are several factors that act as barriers and thus cause reproductive isolation (see Table 6.5 for an overview of these).

But how does speciation occur? How do two populations which originally belonged to the same species come to be reproductively isolated from each other so that they can be regarded as distinct species? Roughly put, there are two main processes of speciation at the two extremes, and a continuum between them. On the one hand, speciation can take place when two populations are geographically isolated from each other,

Table 6.5 Barriers causing reproductive isolation (adapted from Coyne and Orr, 2004, pp. 28–29)

Stage of reproduction	Reproductive isolating barriers
Pre-mating (individuals of different populations do not mate)	(1) Behavioral isolation: individuals of different populations are not "attracted" to each other and do not mate. (2) Ecological isolation: individuals of different populations do not mate because of differences in habitat preference, timing of breeding, and pollinator interactions. (3) Mechanical isolation: individuals of different species do not mate because they have incompatible reproductive structures. (4) Mating system isolation: evolution of self-fertilization or of asexual production of offspring, which can result in the formation of new species.
Post-mating, pre-zygotic (individuals of different populations mate, but no zygote is formed)	(1) Copulatory behavioral isolation: individuals of different populations mate but behave in such a way during copulation that fertilization does not occur. (2) Gametic isolation: individuals of different populations mate but gametes cannot effectively cause fertilization.
Post-zygotic (individuals of different populations mate, a zygote is formed but the organism is either non-viable or sterile)	(1) Extrinsic barriers: sterility and non-viability are due to the external environment as, e.g., in the case of ecological non-viability (the ability of hybrids to survive and reproduce is low because they lack an appropriate niche) and behavioral sterility (hybrids cannot obtain mates). (2) Intrinsic barriers: sterility and non-viability are due to problems in development as, e.g., in the case of hybrid non-viability (problems in development cause full or partial lethality), hybrid sterility (which can be physiological sterility due to problems in the development of gametes or reproductive organs, or behavioral that leads to developmental problems causing hybrids to be incapable of successful mating).

e.g., because of a mountain or a river between the areas they live in. This process is called **allopatric speciation**. In this case, individuals of the two populations do not meet at all, and any new genetic variants are restricted to the population in which they occur and cannot be passed to the other. As a result, the two populations may evolve independently and eventually diverge from each other, developing isolating barriers which are the by-products of their divergent evolution. On the other hand, speciation can take place when a population evolves to two or more reproductively isolated groups

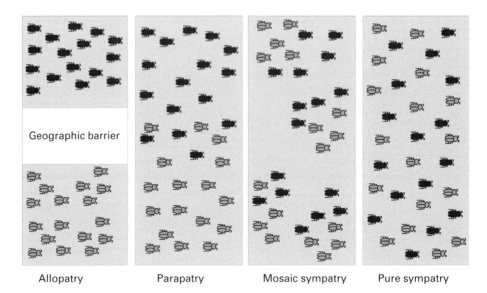

| Allopatry | Parapatry | Mosaic sympatry | Pure sympatry |

Figure 6.8 The continuum between allopatry and sympatry. At one extreme, the two populations are completely separated because of a geographic barrier (e.g., a river) they cannot cross (allopatry). At the other extreme, the two populations live together (pure sympatry). Between these two extremes, the two populations might interact slightly (parapatry) or quite extensively at specific locations (mosaic sympatry) (adapted from Mallet, 2008; Mallet *et al.*, 2009) (in all cases proportions, not actual numbers of the various types of individuals, are depicted).

which are not geographically isolated but live in the same area. This is the process of **sympatric speciation**, and in this case individuals encounter each other and are able to reproduce while they diverge. Between these two extremes one can identify cases of **parapatric speciation**, a process through which distinct species evolve from populations which are somehow, but not completely, isolated geographically.

   Figure 6.8 provides examples of different kinds of situations that can lead to speciation. In the case of allopatry, the two populations are kept apart from each other because of a geographic barrier. In the case of parapatry, the two populations interact at the edges of their habitats. In the case of mosaic sympatry, individuals from the two populations live together but in particular parts of the area they inhabit. Finally, in the case of pure sympatry, the individuals from the two populations live together throughout the whole area. What the figure shows is that a continuum from allopatry to pure sympatry can exist. Allopatry is the major condition leading to speciation and allopatric speciation is perhaps the most usual process. Under particular circumstances, allopatric speciation can lead to extensive diversification, as in the cases of adaptive radiation, when new species emerge from an initial one and adapt to previously unoccupied niches. The traditional example here is that of the Galápagos finches,[21] but there are other cases like this, such as the *Anolis* lizards (Losos, 2010). However, recent evidence

---

[21] Chapter 4 provides a detailed account of how Darwin came to realize that.

shows that sympatric speciation is indeed possible and perhaps more easy than commonly thought (e.g., Papadopulos *et al.*, 2011).

Whether in the same or different habitats, what is important is that significant diversification of populations and divergence from an initial state is possible. The environmental conditions under which speciation can take place are important, but what is more crucial to understand is how organisms diversify. Understanding this requires reference to phenomena already discussed in Chapter 5. Environmental conditions can cause selection for some character; different kinds of barriers causing reproductive isolation may promote divergence of two different populations to two distinct species. Divergence results from the accumulation of different genetic variants in either of the two populations because for some reason their individuals do not mate and so do not give rise to offspring with shared DNA sequences. Consequently, and depending on the genetic changes that will take place in the course of evolution, two populations may diverge significantly so as to end up being reproductively isolated. Allopatric speciation gives the clearest example of how this can happen, since geographically isolated populations do not mate at all (Figure 6.7 provides a simple illustration of how allopatric speciation might be initiated). In contrast, sympatric populations may not diverge significantly because they mate and produce offspring which have several combinations of DNA sequences which are eventually shared by several members of the two populations, although it has been shown that sympatric speciation is indeed possible.

But how does this divergence take place? Once again, evolutionary developmental biology provides important insights, although further research is necessary. Studies show that changes in developmental processes can be implicated in speciation events. For example, it has been found that DNA sequences involved in developmental signaling and regulation are significantly more likely to be evolutionarily retained in multiple copies after duplication than other DNA sequences, suggesting a role for developmental regulation in speciation. It has also been shown that amphibian and fish clades in which polyphenism, a form of phenotypic plasticity in which two or more distinct phenotypes are produced in different environments by the same genotype, has evolved are more species-rich than closely related clades without polyphenism. Another example is phenological isolation – isolation due to differences in the time of maturity or reproductive activity – which is one of the outcomes of changes in the timing of developmental processes (see Minelli and Fusco, 2012 for a review). The important point made here is that the morphological changes caused by changes in developmental processes, already discussed in the previous chapter, may be implicated in speciation because they produce significant reproductive barriers.

This brings us to an important conclusion: In the continuum from allopatry to pure sympatry, different kinds of barriers may cause populations of the same species to diverge significantly and become reproductively isolated. However, in many cases it is difficult to set strict limits on when this happens. A clear example of this is the so-called "ring species," as in the case of the salamander species *Ensatina eschscholtzii*. Ring species are considered to illustrate stages in the process of speciation because they include a full array of intermediate conditions between well-marked species and

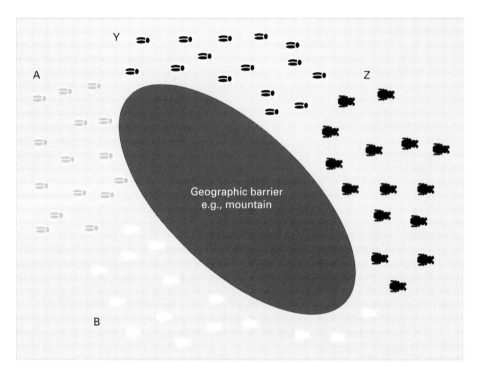

Figure 6.9 An illustration of the so-called ring species. Populations A and Y initially evolved from a common ancestral population. Individuals from populations living close to each other can interbreed successfully (B with A; A with Y; Y with Z), whereas those living further away cannot. Although populations B and Z live close to each other, they cannot interbreed successfully because they have diverged and currently are reproductively isolated. In this case, B and Z can be considered as distinct species. This could also be the case for A and Z, as well as for B and Y (in all cases proportions, not actual numbers of the various types of individuals, are depicted).

geographically variable populations (Wake, 1997). In such cases, neighboring populations can interbreed successfully, whereas those which are geographically more distant do not (Figure 6.9). This shows that the process of speciation should be better perceived as constituting a continuum, too. Two populations sharing a common ancestry may initially diverge, evolving different characters (**anagenesis**). Divergence may continue, or long periods without significant change (stasis) may occur. Such a divergence may eventually give rise to new species and the initial lineage may split into two or more lineages (**cladogenesis**). Throughout this process extinction is always a possibility. Hence, the important question in the study of speciation is to understand in which part of this continuum a species is actually found (Ptacek and Hankinson, 2009). Figure 6.10 provides an illustration of this continuum.

Extinction may be perceived as an exceptional case, but it is actually the rule in evolution. It is a fundamental process in nature as more than 99% of all species which have ever lived on Earth have gone extinct (Jablonski, 2004a). Although extinction events can take place here and there, of considerable interest are those massive events described as mass extinctions. Such events can have a profound impact as they have an

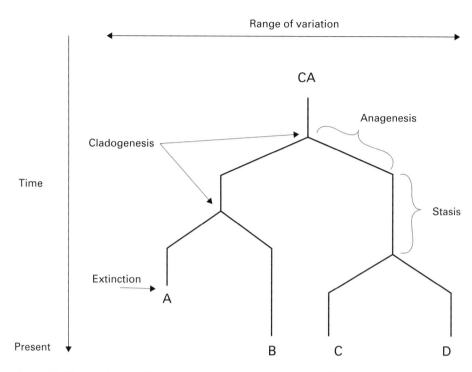

Figure 6.10 The continuum of speciation. Species may split to more lineages, and evolve by diverging, remaining the same or going extinct (adapted from Ptacek and Hankinson, 2009, p. 178). One problem is not so much the incompleteness of the fossil record, but the fact that one cannot deal with "species" in the same way when dealing with extant and extinct organisms. In particular, in the case of lineages where considerable anagenesis but no cladogenesis occurs, specimen samples, e.g., at ten million year intervals, can be so different as to invite recognition of separate species. However, these are not comparable to extant, distinct species derived from repeated cladogenesis from a common ancestor (CA: common ancestor; A–D: species).

effect on the spatial characteristics of the biodiversity that we currently observe (Jablonski, 2004b). One such example is the end-Cretaceous mass extinction, which took place 65 million years ago and famously caused the extinction of non-avian dinosaurs and of the majority of species then living on Earth. This mass extinction also largely determined the taxonomic and biogeographic characteristics of modern biota (Krug *et al.*, 2009). For example, the extinction of dinosaurs made possible the evolution and diversification of mammals among other taxa, something that probably was not possible before the extinction due to predation on mammals. Thus, extinction events can have a long-lasting impact on the extant biodiversity and consequently on evolution. Available data suggest that clades with a wide geographic range are more extinction resistant than other clades with narrow ranges. A probable explanation is that perturbations operate at a local scale so that those clades with a wide geographic range are less affected. However, this correlation changes in the case of mass extinctions, as events have a larger impact, and even clades with a wide geographic range are affected (Jablonski, 2007).

Speciation and extinction are included among those phenomena usually described as macroevolutionary. The distinction between micro- and **macroevolution** is useful because there are major differences between them: **Microevolution** encompasses phenomena of evolution within a species, whereas macroevolution encompasses phenomena across species. The important difference is that we can observe microevolutionary phenomena because in many cases (especially in microbes) they occur within a short time span. But such changes are also observable in multicellular organisms such as the Galápagos finches (Grant and Grant, 2002, 2008). In contrast, it is almost impossible to observe macroevolutionary phenomena because they are usually completed over very long time spans. Distinguishing between these two is important because they are often confused. Thus, unanswered questions about the latter are sometimes deliberately used by anti-evolutionists to question the foundations of the former. But this is entirely wrong. The fact that we do not know all details about a macroevolutionary process (e.g., the transition of lobe-finned fish to tetrapods discussed in Chapter 1) does not entail any criticism about the main processes of microevolution, such as natural selection or drift, for two reasons: (1) because microevolutionary processes can be (and actually have been) demonstrated in the lab or in the wild; and (2) because we may eventually come to know more about macroevolution. There is some disagreement about whether macroevolution and microevolution are governed by the same processes (Dietrich, 2010; Erwin, 2010). This does not raise any questions about whether these processes actually take place, but only about how exactly they do so.

One important conclusion is that, other phenomena notwithstanding, selection has a significant role in macroevolution. As I have already described, selection for some characters can drive the evolution of a population and eventually produce changes in its genetic and phenotypic structure. This is selection within the species level. However, there can also be selection at the species level, described as species selection. There are two senses of species selection: (1) a broad sense according to which speciation and extinction depend on characters at the organism level, such as body size and fecundity, and (2) a strict sense according to which speciation and extinction depend on characters which are emergent at the species level such as geographic range and population size (Jablonski, 2007). Table 6.6 presents some proposed species-level and organismal-level characters which are hypothesized to have an impact on rates of speciation and extinction. It is important to note that these characters may affect speciation and extinction in different ways and so should be studied carefully before conclusions are made. The important point here is that in order to understand speciation and extinction, we need to take into account all these different characters, their possible interactions, and their effects (Jablonski, 2008).

But how does macroevolution proceed? Figure 6.11 presents all possible combinations of the variation in rate of evolution between different lineages and over evolutionary time (tempo) and the mechanisms driving these varying rates of change (mode). All these combinations have been recorded from the fossil record. The mode of evolutionary change includes anagenesis and cladogenesis, already discussed above. Anagenesis is the evolutionary divergence of a lineage over time, whereas cladogenesis is the splitting of a lineage into two or more. The tempo of evolutionary

Table 6.6 Proposed species-level and organismal-level characters which are hypothesized to have an impact on rates of speciation and extinction (adapted from Jablonski, 2008, pp. 505–507)

Level	Characters
Organism	• Body size • Ecological specialization • Competitive ability • Host specificity • Intensity of sexual selection • Mating system • Generation time • Phenotypic plasticity • Trophic level
Species	• Geographic range • Genetic population structure • Sex ratio • Population size • Population density • Intraspecific variation • Evolvability • Social organization

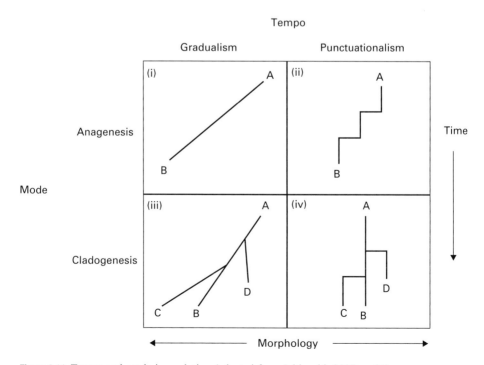

Figure 6.11 Tempo and mode in evolution (adapted from Jablonski, 2007, p. 91).

change includes gradualism and punctualism. The main difference between them is that in the first case all intermediate forms are found in the fossil record, whereas in the second they are not. According to Eldredge and Gould (1972), the evolutionary histories of most species display stasis, absence of significant evolutionary change, which is punctuated by rapid morphological evolution associated with cladogenesis. This model of punctuated equilibrium was offered as an alternative to gradualism: slow, continuous evolutionary change.

Interpreting macroevolutionary patterns and explaining macroevolutionary processes is conceptually challenging. Evolution within the species level can be understood more easily, because studies of natural populations and laboratory experiments are possible. In contrast, macroevolutionary processes require access to the deep past, which is difficult to achieve. Figure 6.11 illustrates the components which are crucial for such explanations. The challenge is to observe extant morphological diversity and combine such observation with fossil data in order to explain its evolution. The important component in this case is time. This is what makes evolutionary explanations, and particularly macroevolutionary explanations, distinctively historical. It is to this topic, the structure and the historical nature of evolutionary explanations, to which we now turn.

Evolutionary explanations and the historicity of nature

Explanation in biology is often characterized by pluralism. Some phenomena may require causal explanations, whereas others may require a subsumption under general patterns or laws. In all cases the context of explanation and the explanatory aims of scientists seem to be important (Brigandt, 2013; Potochnik, 2013). Many philosophers agree that identifying causes is important in the process of developing explanations. Thus, for example, the aim of an evolutionary biologist could be to identify the causes of (1) particular events/outcomes (e.g., end-Cretaceous extinction) or (2) general patterns (e.g., punctuated equilibrium). In this case, an explanation would be an answer to questions such as: (1) What caused the extinction of dinosaurs? Or (2) what causes the punctuated patterns in the fossil record (long periods of stasis punctuated by periods of rapid morphological change)? However, identifying causes is not simple and what is usually done is to rely on a piece or pieces of the complete explanation which are relevant to the occurrence of the explanandum.

The account of explanation that seems to be by far the more appropriate in evolutionary biology is the so-called "Inference to the Best Explanation" (IBE) (see Thagard, 1978; Lipton, 2004 for a detailed account; Lipton, 2008 for an overview). The central idea here is that explanatory considerations are a guide to inference: scientists make an inference from the available evidence to a hypothesis which would, if correct, best explain that evidence. According to IBE, hypotheses are supported by the available evidence they are supposed to explain; evidence supports the hypotheses precisely because they could explain it, and it is only by asking how well various alternative hypotheses could explain the available evidence that one can determine which hypotheses merit acceptance. A distinction that is important is that between potential and actual explanations. A potential explanation is one that satisfies all the conditions of actual explanation, with the possible exception of truth. According to IBE, we infer that what would best explain the available evidence is likely to be true. Thus, the best potential explanation is likely to be an actual explanation. There are two important advantages of IBE: (1) it is context dependent, i.e., a particular scientific hypothesis would, if true, explain particular observations; and (2) it discriminates between different hypotheses, all of which would explain the evidence, since it points to the hypothesis which would best explain it.

Based on these considerations, IBE could be defined as inference to the best of the available competing explanations, when the best one is sufficiently good. But how good is "sufficiently good"? Does this refer to the most probable explanation (Lipton calls it the likeliest explanation) or to the explanation that would, if correct, provide the greatest degree of understanding (Lipton calls it the loveliest explanation)? Let me clarify the difference with an example. We know that HIV causes death because of opportunistic infections. One explanation is that HIV causes deficiency of the immune system (likeliest explanation). This explanation is likely but provides no understanding. Another explanation is that HIV destroys T-cells (loveliest explanation). This explanation should be correct because it provides understanding, and as such it would also be the most likely one. Thus, IBE could be defined as inference to the loveliest explanation. The central claim is that scientists take loveliness as a guide to likeliness. As a result, the explanation that would, if correct, provide the most understanding is the explanation that is judged as the likeliest to be correct. There are particular criteria which can be applied in order to conclude which explanation is the best one. One approach is not only to consider the merits of a particular explanation, but also to contrast it to other alternative explanations. Thus, we should not simply ask "Why A?" but rather "Why A rather than B, C, etc.?" In this way, what would count as the best explanation would depend on both A and B, C, etc., and would identify a cause that made the difference between A and B, C, etc. (see Chapter 3 on causes and scientific explanations).

I will now describe how IBE applies to the case of historical explanations, using a case study: the end-Cretaceous mass extinction, already mentioned in the previous section. By doing so, I will also highlight the distinctive characteristics of historical explanations in order to conclude this section with a general description of the nature of such explanations. The analysis of this case study draws on previous philosophical work (Cleland, 2002, 2011; Forber and Griffith, 2011).[22] Before 1980, many competing explanations had been proposed by paleontologists for the end-Cretaceous (K-Pg)[23]

---

[22] There are some differences between the "smoking gun" approach of Cleland (2002, 2011) and the "consilience" approach of Forber and Griffith (2011). However, there is an agreement over the elements I will focus on here, namely, the importance of particular pieces of evidence for historical reconstruction and of relying on multiple independent lines of evidence. The account of how evidence in support of the asteroid impact event was accumulated that is presented here is brief; Cleland (2002, 2011) and Forber and Griffith (2011) provide detailed accounts of the discoveries.

[23] K is the abbreviation for Cretaceous period and the Tertiary period has been divided up into Paleogene and Neogene periods. Nowadays we are referring to the end-Creatacous extinction as the K-Pg extinction (see Table 6.2).

extinction, including changes in oceanographic, atmospheric, or climatic conditions, a magnetic reversal, a nearby supernova, volcanism, or the flooding of the ocean surface by water from a postulated arctic lake. However, the available evidence did not provide strong support for any of these alternative hypotheses. It was already assumed that the so-called K-Pg boundary might provide important information about this event. The K-Pg boundary is a 1 cm thick, distinct, thin layer of clay between two layers of limestone which are chemically similar to each other. The K-Pg boundary is found all over the world and marks the end of the Cretaceous and the beginning of the Paleogene periods (see Table 6.2). Walter Alvarez, a geologist, and his father Luis, a physicist, used the element iridium as a clock because it can be measured at low levels and because it mostly comes from meteoritic dust. They found that clays from the K-Pg boundary contained iridium levels more than 30 times higher than the limestones on either side, which was too much to be explained in terms of known geological processes (Alvarez *et al.*, 1980). Later studies confirmed the presence of an iridium anomaly in the K-Pg boundary more than 100 times higher compared to the background (Ganapathy, 1980; Kyte *et al.*, 1980; Smit and Hertogen, 1980).

The Earth's crust does not contain much iridium because it is a heavy element and most of it sank into the mantle and core during planet formation. Although not all meteorites are rich in iridium, asteroids and comets from the formation of the solar system usually have higher concentrations. However, volcanism also brings mantle material to the surface and so this was also a plausible explanation for the iridium anomalies (Officer and Drake, 1985). Meteorite impact and volcanism thus became the only two alternative explanations for the iridium anomaly, because none of the other competing hypotheses could explain it. However, further research supported meteorite impact over volcanism. Analysis of K-Pg boundary sediments showed large quantities of mineral grain, predominately quartz, exhibiting a highly unusual pattern of fractures. Sudden application of extremely high pressure is required to fracture minerals in this way. The observed mineralogical features were characteristic of shock metamorphism and formed evidence that the shocked grains were the product of a high-velocity impact between a large extraterrestrial body and the Earth (Bohor *et al.*, 1984). Further studies showed that lamellar deformation features in quartz from tectonic and explosive volcanic environments only superficially resemble features from known shock and/or impact environments (Alexopoulos *et al.*, 1988).

The excess iridium and shocked quartz in the K-Pg boundary was evidence suggesting that a huge meteorite hit the Earth 65 million years ago. However, this did not necessarily suggest that the mass extinctions were caused by the meteorite impact. In an analysis of fossil record data (accumulated over 12 years from seven measured sections in the Bay of Biscay, France, and Spain) of end-Cretaceous macroinvertebrates, 40 molluscan species were recovered. From these, only two seem to have survived into the Paleogene and three were excluded from the analysis because they are each known from a single fossil. Thus, the fossil records of the 28 species of ammonites and seven species of inoceramid bivalves were analyzed. Of these, six ammonite species appear to have become extinct before the K-Pg impact event, most likely due to background extinction processes. All but one of the inoceramid bivalve species became extinct well before the K-Pg impact event, perhaps due to global changes in deep-sea circulation. All the remaining species (22 ammonites and one inoceramid bivalve) are possible victims of the K-Pg mass extinction – however, the available evidence is not conclusive (Marshall and Ward, 1996). The point made here is not that the meteorite impact event is not the main cause of the mass extinction, but that we do not (and perhaps cannot) know all the details. The meteorite certainly had an impact but other processes may also have contributed to the K-Pg mass extinction.

The explanation of the K-Pg mass extinction discussed here exhibits two important characteristics: the overdetermination of causes by effects and the importance of obtaining evidence from multiple independent sources. Let's see them in some detail with an example (the discussion here draws on Cleland, 2002, 2011; Forber and Griffith, 2011). Imagine you throw a ball at a window. Under what conditions will a window break when a ball is thrown at it? As I have described, scientific explanations are often causal, so we may identify a moving ball as the cause and the broken window as the effect. The ball has some features with causal influence, i.e., due to which the effect takes place. Some of these are: the weight of the ball (a ball weighing 1 g might not break the window, whereas one weighing 1 kg probably would); the material of which the ball is made (a plastic ball might not break the window, whereas one made of iron probably would); the size of the ball (a larger ball of the same material, e.g., $10\,\mathrm{cm}^3$ instead of $1\,\mathrm{cm}^3$, would likely have a more significant effect); and the speed with which the ball falls on the window (a ball moving at higher speed, e.g., 60 km/h instead of 1 km/h, would likely have a more significant effect).

In this case there is underdetermination of effects by causes. The weight of the ball, the material of which the ball is made, the size of the ball, and the speed with which the ball falls on the window all have a causal influence on the effect. If one of these features is different, a different effect may occur. Several combinations of these features can have the same effect: a broken window. However, if one of them is different, the window may not break. For example, if a tennis ball is thrown at high speed at the window, the window will most likely break. But a table tennis ball thrown at high speed will not break the window. Thus, speed alone is not sufficient for the window to break; the material of which the ball is made matters, too. But neither is the material alone sufficient to break the window. A tennis ball thrown at low speed will probably not break the window, whereas if it is thrown at high speed it will. In this sense, causes or features with causal influence underdetermine their effects. This means that considered alone the features with the causal influence are not enough to guarantee that their effects will occur. However, even if we do not know all the details, we can sufficiently explain why a window broke. If we observe dispersed pieces of glass and a ball on the floor, these are traces of an effect which in this case is the breaking of the window. We do not need every piece of glass in order to explain how the window was broken. Nor do we need every detail of the causal history of the event, e.g., what the weight of the ball is, what exactly it is made of, what its speed was, etc. in order to conclude what happened. Thus, causes, or features with causal influence, are overdetermined by their effects. We can identify a cause by just observing traces of its effect, even if we are not aware of all the details.[24]

Evolutionary biology does involve experimentation (see Pigliucci, 2013) and this can yield important results and conclusions about how evolutionary processes take place (e.g., Desjardins, 2011). However, in many cases evolutionary biologists aim to explain either particular events in the history of the Earth (e.g., the extinction of non-avian dinosaurs as the outcome of a catastrophic event, caused by an asteroid that hit the Earth) or general patterns, and reveal generalizations which link prior causes to their present-day effects (speciation, extinction, or other macroevolutionary phenomena as the outcome of general principles and processes). To explain such occurrences, scientists study the available data and develop alternative explanations. Then they choose the one that best explains the available data, especially, when it is supported by distinct and independent bodies of evidence. In these cases, explanations take the form of a narrative (narrative explanation), which involves the construction of a story: a coherent, continuous, causal sequence of events that produced the phenomena (traces) under explanation. In some cases, the purpose is only to establish the plausibility that certain sorts of causal processes could have given rise to the phenomena observed (potential explanation). In other cases, however, the narrative is interpreted as showing how the phenomena actually came about (actual explanation) (Brandon, 1990, pp. 176–184; Forber, 2010).

In the example of the K-Pg mass extinction described above, not only traces of the effect were found, but these also constitute independent lines of evidence for such an effect. These include the existence of excess iridium and shocked quartz in a thin layer of sediment found all over the world, as well as the lack of particular ammonite fossils after that, and support the impact of a huge meteor as the best explanation for these observations. These independent observations provide the foundations both for the inference to the particular impact event and for selecting this one over competing hypotheses (e.g., volcanism, etc.). Trying to identify the details of an event which had a significant impact is actually the attempt to describe one important component of evolutionary explanations: the antecedent conditions (see Hull, 1992 for such a view). For example, in explaining adaptations emphasis has been given to the role of natural selection, perhaps overlooking the fact that natural selection takes place as long as some antecedent conditions exist. For instance, there will be no selection in a population that has no variation, because there is nothing to be "selected." It is antecedent conditions like variation in a particular environment that cause natural selection, which in turn brings evolutionary change and, perhaps, adaptations. And it is not only whether natural selection will take place, but also its direction that is affected by antecedent conditions (e.g., what kind of variation was available in a population, or in what kind of environment was this population living?).

---

[24] This is what Cleland (2002, 2011) describes as the asymmetry of overdetermination: effects are underdetermined by their causes (a single cause or causal property is not sufficient to bring about the effect) but causes are overdetermined by their effects (a single effect can be sufficient to explain what happened, i.e., identify causes).

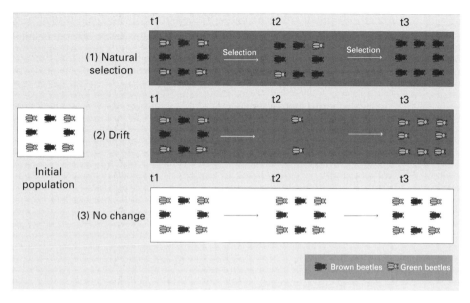

Figure 6.12 The importance of antecedent conditions in evolution (1) natural selection, (2) drift, (3) no change (in all cases proportions, not actual numbers of the various types of individuals, are depicted).

To illustrate the importance of antecedent conditions, I will use an example (Figure 6.12). A particular population of beetles might evolve in different ways, through different processes, or not evolve at all, depending on antecedent conditions. For example, a population consisting of 50% green and 50% brown beetles might: (1) evolve to a population consisting exclusively of brown beetles, through natural selection, if the brown beetles have an advantage in the particular environment; (2) evolve to a population consisting exclusively of green beetles, if during a fire only green beetles happen to survive from the initial population, even if the environment would have otherwise favored the survival of brown beetles (drift); and (3) remain as it was for numerous generations, as long as green and brown beetles have equal chances to arise through reproduction and neither of these two types has an advantage over the other in the particular environment. In all these cases, what is causally important is not only the process that does (or does not) take place, but also – and perhaps mostly – the antecedent conditions that cause each process or cause no process at all. In (1), the antecedent conditions are the variation within the population as well as the fact that brown color confers an advantage to its possessors in this particular environment (in another environment the outcome could have been different). In this case, it is the antecedents that cause natural selection. In (2), the antecedent conditions are the variation within the population, as well as the fire that killed all brown beetles. If there is no fire, no such change might take place. In this case, it is the antecedent conditions that cause drift. Finally, in (3), the antecedent conditions do not cause any process of change. Although variation exists within the population, it is nevertheless not enough to cause evolutionary change as long as green and brown beetles have equal chances to arise through reproduction, and neither of them has any adaptive advantage in the particular environment. As is obvious from Figure 6.12, different outcomes are possible: the population may evolve to one consisting exclusively of brown beetles, or green beetles, or not change at all.

Conclusions

Natural selection is one very important but not the only important process in evolution. It certainly is the process through which adaptations emerge. However, other stochastic processes are important in driving the evolution of life on Earth. I have shown that a character might become prevalent through selection for it. However, it could also become prevalent through draft, i.e., because another character to which it is linked is being selected for. Or a character could simply become prevalent in a population through drift, a process of indiscriminate sampling: some individuals but not others happened to survive and reproduce. Eventually, what matters most for evolution are the antecedent conditions (see Figure 6.12). Whether a population will change or not and what kind of change it will undergo depends largely on its inherent characteristics (variation, developmental mechanisms, and more, already discussed in Chapter 5), but also on the particular antecedent conditions, as the same population could evolve to different conditions depending on what they were. Now, in the long run populations may change significantly to give rise to new species and so speciation may occur. This usually requires some kind of isolation between varieties of the same species which may eventually evolve differently and become distinct species. One main problem that scientists have is that they cannot directly observe these processes because they take place over huge time spans. Thus, whereas one might observe microevolutionary phenomena, e.g., changes within a population over a few years due to some dramatic environmental change, it is very difficult to directly observe speciation, especially in the case of multicellular organisms. Consequently, what scientists can do is rely on their limited epistemic access to the past and use traces of past changes to infer what actually happened. This is not the same as direct observation, but it can nevertheless yield significant knowledge and understanding of the past.

Further reading

A good book about natural selection to start with is Elliott Sober's *The Nature of Selection: Evolutionary Theory in Philosophical Focus*. *Making Sense of Evolution* by Massino Pigliucci and Jonathan Kaplan offers a nice overview of several core concepts of evolutionary theory. Adaptation has often been linked to natural design. Two interesting books on this topic are *The Blind Watchmaker* by Richard Dawkins and *Not by Design: Retiring Darwin's Watchmaker* by John Reiss. An account of the evolution of the Galápagos finches is given in *How and Why Species Multiply: The*

Radiation of Darwin's Finches by Peter and Rosemary Grant. The work of Stephen J. Gould was cited several times in this chapter and his writings are worth reading. All of his popular books have recently been re-published by Harvard University Press (see Ruse, 2012). Gould's *The Structure of Evolutionary Theory*, his *magnus opus*, is an interesting book but quite hard to follow. In contrast, all of his popular books are highly recommended, particularly *Wonderful Life* and *Full House: The Spread of Excellence from Plato to Darwin*. For a very interesting discussion of Darwinism that requires close reading, the book *Darwinism Evolving: Systems Dynamics and the Genealogy of Natural Selection* by David Depew and Bruce Weber is highly recommended. The species concept has attracted the attention of philosophers for some time. Two book-length philosophical discussions of species are *Species: A History of the Idea* by John Wilkins and *The Species Problem: A Philosophical Analysis* by Richard Richards. Finally, *Biology's First Law: The Tendency for Diversity and Complexity to Increase in Evolutionary Systems* by Dan McShea and Robert Brandon provides an interesting account of the importance of stochastic processes in evolution.

Concluding remarks

In the previous chapters I first explained why, religious resistance aside, particular teleological and essentialist intuitions make evolution seem counter-intuitive. Then I explained what conceptual change in evolution consists of and I described Darwin's conceptual change as an exemplar case. After that, I presented some of the central concepts of evolutionary theory in the light of the conceptual obstacles already discussed. I hope it is clear that evolutionary theory is a powerful theory which answers many questions about the living world around us and which still has many questions to answer in the future. This is the nature of all scientific theories and this is what makes science exciting and interesting. In this concluding chapter I am explicit and specific about why evolutionary theory is a good scientific theory. Finally, I describe which questions I think this scientific theory cannot answer and what implications it has.

The virtues of evolutionary theory

What constitutes a good scientific theory? This is a difficult question to answer, but Ernan McMullin (2008) has provided a useful list of the virtues of a good scientific theory: (1) empirical fit (support by data); (2) internal consistency (no contradictions); (3) internal coherence (no additional assumptions); (4) simplicity (testability and applicability); (5) external consistency (consonance with other theories); (6) optimality (comparative success over other theories); (7) fertility (novel predictions, anomalies, change); (8) consilience (unification); and (9) durability (survival over tests). A final virtue is explanatory power, which is actually a consequence of all the other virtues. Evolutionary theory is the theory that explains the origin of species on Earth in general and of their particular characters in particular. More generally, evolutionary theory explains the unity and the diversity of life on Earth. It comprises several propositions, principles, and models – as any other valid scientific theory. However, evolutionary theory is an interestingly special case because it has often been described as "only a theory." This reflects a misunderstanding of science and a misuse of the colloquial meaning of the word theory, often synonymous with speculation or hypothesis (e.g., "I have a theory about this"). The statement that evolutionary theory is "only a theory" is nevertheless appropriate if it means to suggest that it is a scientific theory the same as many others (relativity theory, atomic theory, plate tectonics) and not a secular religion or anything else beyond the realm of science. Evolutionary theory is a valid scientific theory because of its many virtues, which are described below.

Empirical fit or support by data is a prerequisite in order for any theory to be considered as scientific. The main propositions of evolutionary theory, that all organisms on Earth share a common ancestry and that they all have evolved from pre-existing ones through natural processes, are supported by all available evidence from such diverse disciplines as paleontology, biogeography, molecular biology, cellular biology, and developmental biology. In Chapter 5 I explained that organisms that are very different in terms of structure and functions, such as bacteria and eukaryotes, share fundamental similarities at the cellular and the molecular levels (Tables 5.1 and 5.2). These similarities are easily explained with common ancestry: if two taxa T1 and T2 have evolved from the same ancestor A, it is anticipated that some of the characters of A will be found in T1 and T2. Differences are also explained by assuming that T1 and T2 evolved from A by divergence through natural processes such as natural selection or drift, described in Chapter 6. Over millions of years evolutionary processes can produce very different life forms which nevertheless share crucial similarities. Studies of the development, the genomes, the geographical distribution, and the ecologies of contemporary species, as well as of fossils when they are available, point to the conclusion that evolution has occurred as described by evolutionary theory. The fact that we do not know some details yet, as well as that we may never know all the details, does not undermine how strongly evolutionary theory is supported by empirical data.

What is more important, and this is another virtue of evolutionary theory, is that it exhibits internal consistency and no contradictions among its propositions. All data accumulated from different fields of research point to the same conclusions consistently. Organisms living in neighboring areas are found to be more closely related to each other – genetically speaking – than others living in more remote areas, even if their environments are similar. Similar structures may evolve in otherwise very different organisms – remember the wings of bats and birds (Figure 5.16); fossils exhibit similarities with extant species; embryos of closely related organisms, e.g., vertebrates are very similar; and so on. Figure 5.19 shows the similarities in terms of DNA sequences between (a) humans, chimpanzees, and gorillas and (b) chordates, arthropods, and nematodes. It is no surprise that, despite the difficulties identifying homoplasy and its effects, there are more similarities among taxa in (a) than among taxa in (b). This coincides with what one would expect from comparing the structures of these taxa, and it is sufficiently explained by the fact that the time of divergence of taxa in (a) from their common ancestor is more recent than the respective time for taxa in (b).

Another virtue of evolutionary theory is its internal coherence, meaning that no additional assumptions are required. The principles and models of evolutionary theory can account for all observed phenomena, although it is not always possible to explain everything. What is important is that no additional arguments are required, such as the "God of the gaps" argument discussed in Chapter 2. An important distinction here is between what the theory cannot explain and what the theory has not explained yet. There is nothing about the living world that evolutionary theory cannot explain in principle, so that additional arguments would be required. Of course, that evolutionary theory can in principle explain everything about the living world does not entail that it will actually explain everything. But the inability to explain every single aspect of the living world has to do with the lack of data or with human perception, not with the theory itself. As was explained in Chapter 6, achieving epistemic access to the deep past is very difficult; only traces of past events are available and these may not always adequately represent the actual events. However, if we somehow had access to all relevant data, then the models and propositions of evolutionary theory would be adequate to explain what we currently observe as the outcome of natural (evolutionary) processes.

Simplicity is another important virtue of evolutionary theory, and it relates to its testability and applicability. Evolutionary theory can be easily tested (e.g., by making predictions about the distributions of species and comparing it to biogeographical data). In Chapter 1 I described in some detail the discovery of *Tiktaalik*. This discovery was made as the result of a successful prediction: looking for particular fossils at rocks which were accessible and were of the appropriate age. In Chapter 1 I also explained how AIDS or multi-drug resistance to tuberculosis develops. By applying evolutionary theory to understand these phenomena, it is then possible to develop strategies for fighting disease and eliminating the responsible pathogens. This is not always success-ful, but the evolution of pathogens gives important insights about the onset and the progression of infectious disease. This, in turn, may allow for the development of successful therapies. Evolutionary theory can account for such phenomena and can also provide important insights.

Evolutionary theory also exhibits external consistency, which practically means it exhibits consonance with other theories, such as those of chemistry and physics. As already explained in Chapter 6, evolutionary processes, such as natural selection and drift, take place even at the molecular level. All propositions made about how DNA sequences change and eventually evolve do not violate any of the laws/principles of physics and chemistry. At more complex levels of organizations, no matter if it is about how cellular processes, organismal characters, or population properties evolve, what-ever evolutionary theory entails is in accordance with, e.g., the conservation of energy, the entropy of systems, or how chemical reactions take place. Again, we may not always be able to explain everything. However, none of the explanations provided contradicts the fundamental laws of physics and chemistry and what we know about how atoms and molecules, or other components of complex systems, interact.

A particularly interesting aspect of evolutionary theory is its consilience: the enormous potential for unification. Evolutionary theory brings together and explains different kinds of data (fossils, biogeography, morphology, genomics) which become evidence for the common ancestry of all life on Earth, discussed in Chapter 5, and of its evolution through natural processes, discussed in Chapter 6. The fossil evidence of extinct species, the distribution of extant species, as well as their morphology and DNA sequences, can all be explained by evolutionary theory. Populations are more closely related – genetically, morphologically, ecologically – to those living in neighboring areas, no matter how different their habitat is, than to those living in distant areas even in identical environments. Darwin's study of the Galápagos finches discussed in

Chapter 4 or contemporary studies of the so-called "ring species" described in Chapter 6 point consistently to the conclusion that two populations are more closely related to each other the closer they live and the more they interbreed. This is explained as the outcome of evolution by bringing together different kinds of data that all support this explanation.

Another virtue of evolutionary theory is its fertility. Evolutionary theory has made novel predictions and it has also been modified in the light of anomalies. For some reason, some people perceive the potential of a scientific theory to change in the light of anomalies as a weakness; however, this is a real strength. The critical point is what kind of change the theory undergoes. All scientific theories depend on auxiliary hypotheses. It is these hypotheses that often change, not the core of the theory. The opposite would be problematic because if the core of a theory changed, the theory would no longer exist. Darwin's theory, the development of which was described in detail in Chapter 3, can be summarized in the phrase "descent with modification": new taxa evolve through the modification of older taxa and as a result all of them share a common descent. This was the core proposition of Darwin's theory and this still is the core proposition of contemporary evolutionary theory, despite the advancement of our understanding of genetics, development, and evolutionary processes over the past 155 years since the *Origin* was published. During that time, evolutionary scientists have obtained a better understanding of evolution and they have also been accumulating evidence that supports evolutionary theory. This theory has of course itself evolved, as new pieces of evidence came in, anomalies arose, and modifications to address them took place; however, the core of the theory is still fundamentally the same.

A relevant virtue is the durability of evolutionary theory: its ability to surpass all tests. The legend has it that John Burdon Sanderson Haldane, a major contributor to twentieth-century evolutionary theory, once said that he would give up his belief in evolution if someone found a fossil rabbit in the Precambrian. This meant that one should not expect to find a mammalian fossil in rocks a few hundred million years older than the common ancestors of all vertebrates. In other words, given our knowledge of how the various taxa have evolved, we should not find any of them in strata and rocks older than we would expect. The cartoon *The Flintstones* depicted humans and dinosaurs living together, something entirely impossible given what we know about the extinction of dinosaurs 65 million years ago and the evolution of our species less than 200 000 years ago (see Table 6.1). Finding the fossil skeleton of Fred Flintstone next to the fossil skeleton of Dino, the family's pet dinosaur, would be a real problem for evolutionary theory.

Another virtue of evolutionary theory is its optimality: its comparative success over other theories. Indeed, evolutionary theory offers the best explanations available for the characters of organisms by relying exclusively on natural processes. The only alternative is Intelligent Design, which is insufficient and inaccurate. To put it simply, evolutionary theory provides the simplest, most coherent, and most unifying explanations for what we observe. On the one hand, all similarities between organisms can be explained with reference to shared DNA sequences, a consequence of common descent (homologies) or of the evolution under similar conditions (homoplasies) (see Chapter 5). On the other hand, all differences observed between organisms can be explained with reference to different DNA sequences, a consequence of evolutionary processes such as natural selection and drift which caused divergence of subgroups of what was initially the same population.

Perhaps the greatest virtue of evolutionary theory, which is actually a consequence of all the other virtues discussed above, is its enormous explanatory power. This does not mean that evolution can explain everything about the world; quite the contrary. No scientific theory can explain everything, and no "theory of everything" exists. Nor can evolutionary theory explain everything about life and organisms. We do not always have all the required evidence to answer all kinds of questions related to the processes that give rise to organisms and their distinctive features or the patterns produced by these processes in DNA, fossils, and biogeography. But no scientific theory can explain everything in its domain anyway. Why, then, is evolutionary theory an explanatorily powerful one? The answer to this question is that evolutionary theory can explain a wide variety of phenomena based on a small number of propositions and models. It can explain the peculiar features and properties of organisms; the vast variety of life forms on Earth; the similarities of the DNA sequences of organisms which are morphologically very different; and much more. To summarize, evolutionary theory can explain both the unity and the diversity of life. How this is done was described in Chapters 5 and 6, respectively. As all scientific theories, evolutionary theory provides answers to questions. But, at the same time, as with all scientific theories, evolutionary theory cannot answer all kinds of questions. It is to these questions that I now turn.

Questions not answered by evolutionary theory

We must accept the limits of our cognitive abilities; we must learn to live with ignorance. In this final section I discuss the implications which I think evolutionary theory has and does not have for human life. At this point, my account will become more personal, and I am going to be explicit about how I personally understand the world around us. I believe that the so-called evolution wars and the supposed evolution–religion controversy do not have much to do with science, but rather with politics. What is a requirement for science education and the public understanding of science is to make clear why evolutionary theory is a good scientific theory and why we should not expect from it more that it can provide.

In a recent book, Stuart Firestein (2012) elegantly describes that it is ignorance and not knowledge that actually drives science. Firestein argues that it is ignorance that follows knowledge, not the other way around. And this "is knowledgeable ignorance, perceptive ignorance, insightful ignorance. It leads us to frame better questions, the first step to getting better answers. It is the most important resource we scientists have, and using it correctly is the most important thing a scientist does" (p. 7). We live in a world of ignorance; we humans have evolved to be consciously aware of this and to want to know more. Science is one way of knowing; Audi (2011, chapter 12) explains that we may have distinct kinds of knowledge: scientific, religious, and moral knowledge.

Perhaps there can be more. What seems certain to me is that scientific knowledge, or if you prefer scientific ignorance, cannot bring an end to our questions, worries, and concerns by answering everything. Scientific knowledge has specific characteristics, its most important being that it has limits that correspond to the limits of human cognitive abilities. As a consequence, there will always be questions that scientific theories will not be able to answer, perhaps not even to address.

This book is about evolution, so I will explain which questions, I think, cannot be answered by evolutionary theory. This is particularly important because in my view public resistance to evolution is due to two main reasons. One has to do with the misuse of the theory in trying to answer questions it cannot actually answer, already addressed in Chapter 2, and the other with deep human intuitions already addressed in Chapter 3. I think science and philosophy can help humans derive a meaningful life (see Pigliucci, 2012 for a book-length discussion of this topic). But at the same time, I also think we should be able to distinguish between the less subjective approach of science and the more subjective approach of philosophy. This is not to deny that scientific endeavor exhibits some elements of subjectivity; quite the contrary. However, the subjective elements in philosophy are more predominant and less easy to clarify.

In Chapter 2 I argued that scientists may have very different religious views and I described the views of Richard Dawkins, an atheist, Stephen Jay Gould, an agnostic, and Simon Conway Morris, a believer. There I argued that their religious views notwithstanding, all three of them have relied on science to answer questions that fall outside its realm. This is not a very bad thing to do. My concern is that evolutionary theory is unfairly blamed when it is being misused to answer questions it cannot actually answer, and it is actually blamed for too much. In order to explain why this is the case without unnecessarily raising religious concerns, I will describe how I see the relationship between science and religion[1] in analogy with the relationship between science and ethics. In particular, I will explain what implications biological science has for ethics and I will argue that the implications that evolutionary theory has for religion are not more significant than that.

Generally speaking, science has implications both for ethics and for religion. However, moral behavior and religious attitudes cannot be guided by the rational tools of science. The important point here is that although one could be informed by science to make appropriate decisions, one cannot base these decisions solely on science. Ethics and religion also have to do with worldviews and philosophical perspectives. These can be enriched by science in various ways, but science cannot guide them because decisions about what is bad or wrong, as well as about whether human life has an inherent purpose or not, are made on a subjective basis. For example, we might decide that humans only or all organisms should be included in the moral community. There are several arguments for and against each choice (see Millstein, 2013). Knowledge of which beings are sentient or not might be useful here but it could not alone point to any decision about this. Other arguments, provided by the various ethical theories, are also

---

[1] As in Chapter 2, it is Christian religion I am referring to because evolutionary theory was developed within cultures in which Christianity was considered the dominant religion.

important to consider. Eventually, there will always be a counter-argument about what we decide, and whether we will take it into account or ignore it is quite subjective.

Morality is not threatened by science in any way, because science can only inform but not guide moral choices and decisions. Scientific knowledge may make people question ethical norms, but this does not mean the latter will persist only if the former is rejected. For example, clinical trials could be considered immoral if healthy human subjects were given experimental drugs without their consent. However, this does not mean we should reject clinical trials altogether as a practice. Rather, we should be careful in their implementation and ensure it is done appropriately and with respect for participants. Ethical decisions should be made after having obtained a good understanding of the respective scientific knowledge, but the final decision about what is moral or not requires more than scientific knowledge alone. In other words, science can make important contributions to decisions about ethics, but it cannot guide them. Some people might acquire the contemporary scientific knowledge, e.g., that double-blind clinical trials might provide us with important conclusions about drugs and their effects, but nevertheless decide not to perform them because of, e.g., their particular cultural characteristics. It is their right to do so, and they should not blame other cultures for adopting them, as long as their implementation respects human life. Of course, such decisions at the community level can only be made if a consensus is reached.

Religion is perhaps crucial at a more personal rather than a community level, but besides that people could make decisions in a similar manner. People should be aware of contemporary scientific knowledge and then decide if it has implications for their religion and how serious these implications are. Some people might decide to change their religious views, some others might not. No matter what happens to one's religious beliefs, science can inform but not guide such a decision. Biologists Francisco Ayala and Ken Miller, already presented in Chapter 2, are examples of devout believers whose religious views were not affected by their scientific knowledge. One might decide to sustain his/her religious beliefs, even despite an apparent incompatibility with science. What is important is that he/she is aware of this knowledge; then a conscious decision can be made.

Science per se is not religious or atheistic, moral or immoral. Scientists can be either of these. Therefore, it is scientists that should be criticized about how they act and what conclusions they make. And if we disagree with a religious person or with an atheist, we should neither reject nor blame science for this. Actually, disagreement is a healthy endeavor. If I use science to justify my religion or to justify my atheism, you should not blame science for my views, but me. Similarly, if I use science to support or reject the use of stem cells, I am solely responsible for this. A fertilized ovum will not develop to an embryo unless it is implanted in a uterus. This is a fact of life. Now, whether I consider a fertilized egg as a human being because it has the potential to develop to an adult once it is implanted, or as just a bunch of cells with some potential that will never develop to an adult if it is not implanted is, in my view, a subjective decision. More generally, how one relies on science to make philosophical conclusions cannot have implications for science.

To make my argument explicit, that Richard Dawkins relies on evolutionary theory to promote militant atheism should not have implications for the theory. In my view people like Dawkins use science to draw conclusions about issues that fall outside the realm of science. I would argue exactly the same about Simon Conway Morris, who argues that our scientific knowledge of nature indicates that there is much more out there and that consequently we need theology to get the whole picture. My disagreement with Gould's view that we cannot really know would seem a minor one, but this is not a claim I make either. None of these conclusions is justified in my view as stemming from scientific knowledge. Rather, it stems from an interpretation of scientific knowledge within different contexts. Dawkins and Conway Morris could argue that, e.g., "based on our knowledge of evolutionary biology, I infer that there is no God" or "I infer that God exists." However, this is a personal worldview and not a scientific inference, i.e., one based on concrete evidence. Thus, since they both agree on the basics of evolutionary theory, but disagree on implications they draw, it should be made clear that criticism should be relevant to the conclusions they make based on their scientific knowledge and not to the knowledge itself.

I must also note that I think that Dawkins (2006b) is right in much of what he writes about how religious belief can influence human actions; but again, is it religion that should be blamed or religious fundamentalists? When people are killing each other in the name of their religion, why is religion to blame, and not people themselves? It might make sense to claim that if there were no cars, there would be no traffic accidents – and thousands die every year in traffic accidents. However, in most such cases the drivers, and not the cars, are to blame. The point I want to make is that all of us are responsible for what we say, write, or do. Whether or not we have free will is another philosophical question that is perhaps difficult to answer. But whether or not I have free will, I am responsible for many of my actions. If I drink at a party, I can decide not to drive my family home. If I drink so much that I am not able to make such a decision, this is my fault – I cannot blame my favorite red wine for this. Similarly, if my scientific knowledge makes me believe that God does or does not exist, I am responsible for such a view. I have the right to defend it, but I must take responsibility for it; no one should blame scientific knowledge for the conclusions I reach. If you disagree with what I am writing here and with the conclusions I will soon make from evolutionary theory, you have the right to reject them. There is no point in rejecting evolutionary theory because of the conclusions I draw. Evolutionary theory per se is not responsible for my atheism, agnosticism, or religiosity. This is my decision; my choice; my responsibility.

I am trying to respect those who disagree with me. We do not always do this, and I believe that misallodoxy,[2] hating other people because they hold different beliefs than I do, is a major problem in human societies and in part responsible for the so-called "evolution wars." Interestingly enough, misallodoxy could be explained in evolutionary terms, e.g., as a consequence of group selection – hating those with different beliefs makes you support those who hold the same beliefs as you, i.e., your own group – but I do not want to get into that discussion now. The problem in the case of evolution and religion is that we fail to respect the views of those who disagree with us. Militant

---

[2] This is a Greek word: misos = hate, allo = other, doxa = belief.

atheists fail to respect the decision of religious people to believe; religious fundamentalists fail to respect the decision of irreligious people or atheists not to believe. Militant atheists blame religion and religious fundamentalists blame science. I think they are all wrong. Writing on the science side, I want to conclude that evolutionary theory influences but does not guide atheism, as well as that it has implications for religion but does not hurt it. Evolutionary theory provides a deep, coherent understanding of our natural world; it has nothing to say about supernaturalism. As all science, evolutionary theory is a human construct and indeed a successful one given how many questions it can answer and how many applications it has (see, for example, Mindell, 2007; Poiani, 2012). How we use it to make claims about anything beyond the natural world is our own problem, and not one of the theory itself. Personally, I have made the choice to worry about whatever my human brain can process. Science is knowledge of this kind, and philosophy contributes a lot to this knowledge. Whether God exists or not, I feel that I do not have much to say. Let me again quote Charles Darwin on this: "I feel most deeply that the whole subject is too profound for the human intellect. A dog might as well speculate on the mind of Newton. Let each man hope and believe what he can" (Darwin, 1995/ 1902, p. 236).

Given these considerations, is there any purpose and meaning in our evolving world? I take evolutionary theory to suggest that there is no inherent purpose and meaning in our world. But note: This is my view, and if it is wrong this is my fault; it has no consequences for evolutionary theory. But even if you agree with me that this is the case, I do not think this means one cannot find purpose and meaning in life in this evolving world. In contrast, I see our ability to consider such questions as an evolutionary outcome – some might say a triumph of evolution; even our closest relatives do not seem to have such concerns. Evolutionary theory does not deprive life of meaning; in contrast, it shows us that we are fortunate to be able to feel happiness, satisfaction, to set goals and try to fulfill them. Most animals just kill other organisms to feed on their tissues, reproduce, and die. We do that too, but we are also in a position to be aware of that as well as that there is more we can do: think, communicate, and philosophize. Biologically speaking, we are primates; but perhaps in contrast with most primates and other animals, we have a wider scope of experiences that stem from our biological hypostasis but which also go beyond that. We have culture, morals, and much more. In many cases, but unfortunately not for most people on Earth, we do not just survive, we live.

I also think that evolutionary theory has another important implication for human life – again, this is my view, and if it is wrong there are no consequences for the theory. I think that those who understand evolutionary theory realize that we humans should not be arrogant. I take evolutionary theory to suggest that we have no special place in this world and thus no justification to believe we can rule it in any way we like. We are not a special or select species, but just one species among so many; we are a short and recent branch in the evolutionary network of life. We have no right to change the world any way we like, just because (we think) we can. Most interestingly, we do not have the power we think we have in order to do this. Although the human population at the global level increases everyday, millions of people still die from infectious disease. Consequently, we are not all-powerful – just in case we think we are. Yet we cause enormous change in the natural world. This is not necessarily a bad thing to do; photosynthetic organisms transformed the atmosphere of the Earth by producing molecular oxygen without which perhaps we would never have evolved. I do not know if the changes we cause will have a good or a bad outcome; history will tell. But we have no reason and no rational justification to believe we can change the world any way we like just because this is our world. It is not! We are its component, and actually a very minor one. Therefore, we should not be arrogant; we should be modest.

In accepting that there is no inherent purpose in the world and that we should not be so arrogant to believe we have any special place reserved in it, we can decide how to live our lives, and decide what matters most. I strongly believe we can find meaning and purpose in life. I believe that life becomes meaningful through sentiments. Given the violence that is inherent in our world, I am happy to be able not to worry about my survival and be able to write these lines. I find enormous meaning in my life in doing everyday things with my wife and our children. I feel extremely rich of sentiments when they all express their love for me. I find meaning in writing this book and sharing my thoughts and understanding of the world with people all over the globe. These different kinds of meaning contribute to the purpose I find in life: live long, love people, enjoy life, and feel full of sentiments and happiness when I get older and the final countdown begins. Evolutionary theory has nothing to say about all this, but these are implications I have drawn from my own understanding of evolution. So if I may make a suggestion, it is this: Try to understand evolutionary theory and then draw your own implications for your life and find your own meaning and purpose.

Glossary

adaptation, ahistorical definition of: A character that currently confers an advantage to its possessors which as a consequence survive and reproduce more successfully than others in a particular environment.

adaptation, historical definition of: A character which has become prevalent in a population or species as a result of a history of selection for this character because it has conferred an advantage to its possessors, which as a consequence have survived and reproduced more successfully than others in a particular environment.

adaptation, process definition of: The evolutionary process by which populations become adapted to their environment.

allele: One of several variants of a particular DNA sequence (gene) that "encodes" a particular protein or RNA molecule and thus affects a particular biological process. Alleles are identified with particular parts of chromosomes which are described as loci (sing. locus).

anagenesis: The process through which two populations sharing a common ancestry may initially diverge, evolving different characters (evolution within a lineage).

apomorphy: A character in a set of homologous ones which is derived (apomorphic), i.e., in an innovative condition with respect to the condition in a reference ancestor.

argument from design: The argument suggesting that organisms are the products of intentional design and therefore are evidence for the existence of God as the Designer.

artifact: Any object intentionally created by humans which exhibits properties designed to serve a particular purpose.

belief: A mental state in which one accepts a particular proposition to be true.

blind watchmaker: Refers to natural selection considered as the natural equivalent, and eventually the alternative, to Paley's Divine Designer. In this view there is design in nature but it is natural; it is neither purposeful, nor intentional. Thus, there is a designer in nature but it is a blind and unconscious one: natural selection.

character (biological): Any recognizable feature of an organism that can exist in a variety of character states and at several levels from the molecular to the organismal.

clade: Hierarchically nested groups which include a common ancestor and all its descendants.

cladogenesis: The process through which a lineage splits into two or more and new species eventually give rise to new lineages.

concepts: Mental representations of the world, in terms of which our knowledge and understanding of the world is formulated.

conceptual change: In this book, the change of conceptions in the wider sense (including the change in the meaning of concepts or the change in the relation among concepts within an explanatory scheme or model) as a result of conceptual conflict, i.e., the realization that prior conceptions are wrong or explanatory insufficient.

conceptual obstacles: Conceptions (beliefs or ideas) which are strongly held and resistant to change, and which thus impede understanding and acquisition of correct concepts.

constraints: Any properties or processes that limit or facilitate evolutionary change by biasing what is or is not possible. They can be interpreted in terms of adaptation (e.g., constraints prevent optimal phenotypes from evolving) or development (e.g., constraints facilitate specific types of variation available for natural selection).

contingency, evolutionary: The view that the history of life on Earth has been determined by unpredictable events and that there are no inevitable outcomes in evolution.

convergence, evolutionary: The emergence of the same character through independent evolution, i.e., the evolution of the same character from two different ancestral ones in two different lineages.

Creation: The belief that everything that exists does so due ultimately to God as the ground of all being and existence.

Creationism: The belief that God created the universe, including Earth and humans, through a series of miracles.

deductive argument: An argument in which if the premises are true, then the conclusion must be true. This is a form of reasoning from generalizations to particulars.

Deism: The idea that God creates the physical laws needed for the universe to exist and function, but otherwise does not interact with the universe.

design, intentional: Pattern or arrangement arising by the actions of an intelligent, conscious agent (e.g., a human in the case of artifacts).

design, natural: Pattern or arrangement arising by natural processes (e.g., adaptations evolving by natural selection).

design teleology: A mode of teleological explanation which suggests that a feature exists for some purpose because it is intentionally designed to fulfill it. Teleological explanations based on design are appropriate for artifacts but not for organisms or non-living natural objects (e.g., stones, clouds).

deuterostomes: Those bilaterian animals in which the anus develops close to the blastopore and the mouth develops from a second opening (such as echinoderms, hemichordates, chordates).

development: The processes of growth, change, and transformation that organisms undergo in their life cycle, e.g., from a fertilized egg to a sexually mature adult. It is also called ontogeny. Development includes changes that entail increasing order, via the differentiation and integration of specialized parts.

developmental (phenotypic) plasticity: Modifiability of the phenotype during development. Individuals of the same species with the same genotype may exhibit phenotypic variation depending on local conditions.

DNA information: It is often stated that DNA "encodes," "contains," "stores," or "transmits" information. DNA plays an important role in certain bioinformational relationships, usually as a message. It should be noted that bioinformation is not a property of DNA, but a complex relation in which DNA has an important role.

domains: Superkingdoms of life; the most fundamental division of living entities into the three groups of Archaea, Bacteria, and Eukarya.

domestication: The process of controlled breeding of a species in a way that it is useful to humans.

draft: Another name for the process of linked selection. In this case a character B becomes prevalent in a population although there is no selection for it, because there is selection for another character A to which B is linked (i.e., the DNA sequences of A and B are located on the same chromosome).

drift: An indiscriminate sampling process that typically produces a pattern of random variability.

endemic species: Species which are only found in a particular place and not in other places in the world.

epistasis: The phenomenon during which the effect of an allele at one locus may hide the effect of an allele at another locus.

epistemology: The branch of philosophy that addresses questions having to do with the nature of knowledge and rational belief.

essence: The necessary properties of a thing that make it the kind of thing it is, often contrasted to "accidental" or contingent properties. These are the properties that all members of a kind must have, and the combination of which only members of this kind do, in fact, have.

essentialism: The idea that classes or kinds must have jointly necessary and severally sufficient conditions. That is, a general term like "animal" must be definable in terms of properties that only animals jointly have.

essentialism, psychological: The idea according to which certain categories are real rather than human constructions and they possess an underlying causal force, the essence, which is responsible for why category members are the way they are and share so many properties.

eukaryote: A macro- or microorganism that is not in Archaea or Bacteria, possessing well-defined cellular compartments, such as the nucleus.

evidence: Whatever can make a belief rational, such as experimental data, mathematical proofs, perceptual experiences, or memory.

evo-devo (evolutionary developmental biology): A constellation of biological disciplines that investigate the evolution of development (how developmental processes vary and change over time) and the developmental basis of evolution (how developmental processes causally impact the evolution of organismal characters).

evolution: The natural process by which new species emerge as the modified descendants of pre-existing ones. It accounts for both the unity and the diversity of life.

evolutionary theory: The scientific theory that explains how evolution has taken and still takes place on Earth, with reference to particular, old and current, aspects of life on Earth and to particular episodes of its history.

evolvability: The capacity or disposition to evolve, usually ascribed to a group of organisms (e.g., a population or lineage). Evolvability is often described as depending on other properties, such as modularity (e.g., increased modularity leads to increased evolvability).

exaptation: An adaptive character which originally evolved for reasons unrelated to its current biological role.

fitness: A measure of evolutionary success, often broken down into two components: survival and reproduction. It is usually stated that a character may increase the fitness of individuals. In this book, this has been described as the contribution of a character to the survival and reproduction of its possessors.

function: The role of a component in the organization of a system. The functions of the parts and activities of artifacts are "artifact functions" (e.g., the function of the wings of airplanes is enabling flight). The functions of the parts and activities of organisms in enabling their continued existence are "biological functions" or "biological roles" (the function of the wings of eagles is enabling flight, but the function of the wings of penguins is enabling swimming).

functional explanations: In biology, functional explanations answer the question of why particular organisms have a particular character by pointing out that the latter confers an advantage to those organisms because it efficiently performs a role.

genetic accommodation: Gene frequency change due to selection on variation in the regulation, form, or side effect of novel characters in the sub-population of individuals that express them.

genetic material: Any nucleic acid with the propensity to be inherited and to interact with other cellular components as a source of sequence information, eventually affecting or being implicated in cellular processes with local or extended impact.

genomics: Biological research that focuses on whole genomes, i.e., the base sequence of the genetic material of organisms.

genotype: Which alleles, related to a particular character, an individual carries.

God of the gaps argument: The argument that whenever there is a "gap" in the explanatory potential of science, this is filled by assuming that God intervened and so His intervention stands as the explanation for whatever cannot be explained otherwise.

heredity: The transmission of material from ancestors to descendants.

heterochrony: The differences in time or speed, i.e., *when* a particular structure is formed, during the development of two organisms under comparison.

heterometry: The differences in the amount of activity, i.e., *how much* of a molecule is produced, in developmental events during the development of two organisms under comparison.

heterotopy: The differences in the spatial location of developmental events, i.e., *where* a particular structure is formed, during the development of two organisms under comparison.

heterotypy: The differences in the type of developmental events, i.e., *what kind of* molecule is produced or structure is formed, during the development of two organisms under comparison.

heterozygote: An individual that carries two different alleles related to a particular character.

homologous DNA sequences: Sequences derived from some common ancestor. There are three different types: orthologous ones have evolved from a common ancestral DNA sequence through speciation events; paralogous ones have evolved from a common ancestral DNA sequence through duplication events; and xenologous ones have emerged through horizontal transfer of DNA sequences between different species.

homology: A relation of sameness between two or more characters in two or more organisms, or within the same organism, in an evolutionary context. There must be some connection through common ancestry in order for two characters to be considered as homologous. Thus, homologous characters are those that derive from the same character in the most recent common ancestor of those organisms. There exist specific concepts of homology such as serial homology (repetitive structures of the same individual, e.g., vertebrae), positional homology (different, non-homologous structures localized in homologous positions in individuals of two species), and special homology (the same homologous structure is localized in non-homologous positions in individuals of two species).

homoplasy: A relation of similarity between two characters in two or more organisms that do not derive from the same character in the most recent common ancestor of those organisms, but due to independent evolution (e.g., convergence).

homozygote: An individual that carries the same allele related to a particular character on both homologous chromosomes.

horizontal gene transfer: Exchange of genetic material between cells during their life cycle.

indiscriminate gamete sampling: The process which determines which of the two genetically different types of gametes produced by a heterozygote parent is actually contributed to each of its offspring.

indiscriminate parent sampling: The process which determines which organisms will have offspring, and how many offspring each parent will have.

inductive argument: An argument in which the premises support the conclusion but do not guarantee its truth. This is a form of reasoning from particulars to generalizations.

inference: An inference is made when a conclusion is drawn from a set of premises, and includes both the psychological process of drawing conclusions and the rules that justify drawing conclusions from these premises.

Intelligent Design: The idea that some intelligent, supernatural agent has influenced the history of organismal life on Earth. Intelligent Design proponents claim that it is possible to infer the past action of an intelligent designer from some features of extant organisms.

irreducible complexity: A subsystem of an organism (e.g., a molecular mechanism) consisting of parts that interact so as to fulfill a function, is irreducibly complex when the removal of any part leads to the system no longer performing the function.

knowledge: One knows a proposition when: (1) the proposition is true, (2) one believes the proposition, (3) one's belief in the proposition is based on sufficiently strong evidence, and (4) one satisfies whatever condition is required to handle the Gettier problem.

levels of selection: Different levels (such as genes, cells, individuals, or groups) at which natural selection can operate in a biological hierarchy.

macroevolution: The evolutionary processes above the species level, including speciation and other evolutionary transitions.

microbe: A microscopic organism or virus.

microevolution: The evolutionary processes within the species level, i.e., the evolution of different populations of the same species.

missing links: Taxa intermediate between major groups of organisms, such as between apes and humans, which have not yet been discovered.

modularity: The property of being a module or a partially independent, distinguishable unit (e.g., a segment), behaving in a quasi-autonomous fashion. Modularity allows for evolutionary change to occur in one character without affecting another character or the entire organism.

molecular clock: The hypothesis that the rate of molecular evolution is approximately constant for each different type of molecule.

molecular evolution: The study of the evolutionary patterns and processes of biological macromolecules.

monophyletic group: A group, the members of which are derived from a common ancestor.

natural selection: An evolutionary process that occurs when heritable variation in characters of organisms in a population produces difference in reproductive success. The result is the differential reproduction of organisms with different characters.

natural selection, creative view of: In this view natural selection evolves functional adaptive characters only by working over much time and many generations on small variants in characters that first arise independently of the utility they subsequently acquire. This was Charles Darwin's view.

natural selection, eliminative view of: In this view, natural selection is either favoring or eliminating organisms whose characters are or are not adapted *from the outset*. Selection is thus conceived as an eliminative force that discriminates among whole organisms rather than the slightly variant characters they bear and that adaptations are nothing but retained accidents. This is the view of natural selection summarized in the phrase "survival of the fittest," a phrase coined by Herbert Spencer and adopted by Charles Darwin.

natural theology: Systematic arguments purporting to demonstrate the existence and attributes of God based on the features of the natural world.

naturalism, metaphysical: The claim that the material phenomena studied by science are all that exist. This view denies that supernatural phenomena exist.

naturalism, methodological: The claim that science studies natural (as opposed to supernatural) phenomena only. It does not deny that supernatural phenomena exist, it only asserts that science does not and cannot study the supernatural.

neutral theory of molecular evolution: A theory of molecular evolution that claims that the majority of observed changes in biological macromolecules (DNA, RNA, and proteins) are neutral or nearly neutral, which means that their behavior is dictated by random drift, rather than selection. The neutral theory combines both drift and selection, since selection is presumed to operate on a number of molecular changes.

niche: Either the role that a species plays in the overall community dynamics, or the distinct portions of the abiotic environment in which species exist and reproduce.

non-genetic (epigenetic) inheritance: Refers to the many different mechanisms in addition to the transfer of DNA by which the parental phenotypes (or more remote ancestors) affect the development of their offspring. It includes cellular epigenetic inheritance, but it can also involve other mechanisms, including behavioral interactions between parents and offspring.

non-living natural object: Any natural object other than organisms (e.g., clouds, rocks).

novelty (evolutionary): A character which has no obvious homology with any other character in another organism or the same organism, and whose origin cannot be easily traced back to a modification of a body structure already existing in the ancestral lineage leading to that organism.

origin of life: The transition from a non-living suite of chemicals to a living system that occurred on Earth approximately four billion years ago and gave rise to all known life.

orthology: A relation between homologous DNA sequences which are present in different organisms and have evolved from a common ancestral DNA sequence through speciation events.

parallelism: The evolution of the same character from the same ancestral character in two different lineages. It can be considered as a special case of convergence.

paralogy: A relation between homologous DNA sequences which are present in the same organism or in different organisms and have evolved from a common ancestral DNA sequence through duplication events.

phenotype: The outcome of the expression of the alleles related to a particular character.

phenotypic accommodation: The capacity of organisms for mutual adjustment of different parts during development to produce a functional phenotype even when perturbed by genetic or environmental input.

plasticity, developmental or phenotypic: The potential of organisms with the same genotype to produce different phenotypes during development as a response to genetic or environmental perturbations.

pleiotropy: The phenomenon during which the effect of an allele at one locus affects multiple phenomena within the organism.

plesiomorphy: A character in a set of homologous ones which is primitive (plesiomorphic), i.e., in the same condition with respect to the condition in a reference ancestor.

prokaryote: A microorganism that belongs to Archaea or Bacteria, possessing cellular structures that are less obviously compartmentalized than in cells of non-prokaryotes.

protostomes: Those bilaterian animals in which the mouth develops close to the blastopore (such as flatworms, annelids, mollusks, nematods, arthropods, onychophorans).

regulatory DNA sequences: DNA sequences which are not transcribed to mRNA like protein-coding sequences, but affect their expression. Particular molecules, such as transcription factors, can bind on regulatory sequences and influence the transcription of protein-coding sequences. Regulatory sequences can thus act as switches that regulate protein synthesis.

repatterning, developmental: Changes in the timing, positioning, amount, or type of expression of a DNA sequence, implicated in some developmental process, which produce novel characters.

robustness, developmental: The consistency of the phenotype of individuals despite genetic or environmental perturbation. Individuals exhibit the general characteristics of a species irrespective of the environment they live in.

scientific concepts: Systematic representations of the world through which explanations of and predictions about phenomena are possible.

selection teleology: A mode of teleological explanation which suggests that a feature exists in a population because it is being selected for its beneficial consequences to its bearers. Teleological explanations based on natural selection are appropriate for organisms but not for artifacts or for non-living natural objects (e.g., stones, clouds).

small probability argument: The argument that the origination of complex biological features (e.g., anatomical structures, cells, or genetic information) by means of Darwinian evolution is too unlikely to be credible.

speciation: The evolution of new populations which are reproductively isolated from other populations (on the assumption that species is defined as a group of potentially interbreeding populations which are reproductively isolated from other such groups).

speciation, allopatric: The process of speciation that takes place when two populations are geographically isolated from each other, e.g., because of a mountain or a river between them. In this case, individuals of the two populations do not meet at all, any new genetic changes are restricted to the population in which they occur, and cannot be passed to the other.

speciation, parapatric: A process through which distinct species evolved from populations which are somehow but not completely isolated geographically.

speciation, sympatric: The process of speciation that takes place when a population evolves to two or more reproductively isolated groups which are not geographically isolated but live in the same area. In this case, individuals encounter each other and are able to reproduce while they diverge.

species: In this book, this generally refers to a group of individuals which are reproductively isolated from other groups and/or genetically distinct. For sexually reproducing organisms, a species is usually defined as a number of, usually similar, organisms that can interbreed and produce fertile offspring.

stochastic processes: Undirected evolutionary processes with unpredictable outcomes in which chance seems to play a major role.

symbiosis: A broad term that covers parasitic, mutualist, and commensal interactions between biological entities; interactions may be obligatory or facultative; endosymbiosis refers to symbioses that take place within cells (as opposed to between cells or organisms).

symplesiomorphy: A plesiomorphy shared by members of a taxon.

synapomorphy: An apomorphy shared by members of a taxon.

Synthesis, Extended: A proposed extension of the Modern Synthesis to take into account a broader range of biological phenomena (e.g., phenotypic plasticity), to incorporate new disciplines (e.g., evo-devo, genomics), and to introduce new concepts (e.g., evolvability).

Synthesis, Modern: The standard theoretical framework in evolutionary biology, a synthesis of various disciplines, achieved from the 1920s through the 1940s.

systematics: The scientific practice of classifying objects, usually biological organisms, by the relations between them. It is similar to but not identical with taxonomy.

taxon: The term *taxon* (plural *taxa*) is used to refer to a taxonomic group of organisms such as a phylum, a class, a species, etc. (phylum, class, species are categories; individual phyla, classes, and species are taxa).

taxonomy: In biology the discipline of identifying and describing species and sub-specific kinds. It is the basis on which systematics is undertaken.

teleology (teleological explanation): A mode of explanation in which some property, process, or entity is explained by appealing to a particular result or consequence that it brings about. There are two distinct types of teleological explanations: teleological explanations based on design and teleological explanations based on natural selection.

Theism: The idea of a personal creator God who interacts with his creation and answers prayer.

transmutation: This is how the emergence of a new species from a pre-existing one was called in Darwin's time. The word "evolution" at the time referred to progress and development rather than the process we nowadays refer to.

xenology: A relation between homologous DNA sequences that are present in different organisms because of horizontal transfer of genetic material.

References

Akey, J. M., Ruhe, A. L., Akey, D. T., *et al.* (2010). Tracking footprints of artificial selection in the dog genome. *Proceedings of the National Academy of Sciences USA* **107**(3), 1160–1165.

Albert, F. W., Carlborg, O., Plyusnina, I., *et al.* (2009). Genetic architecture of tameness in a rat model of animal domestication. *Genetics* **182**, 541–554.

Alekshun, M. N. and Levy, S. B. (2007). Molecular mechanisms of antibacterial multidrug resistance. *Cell* **128**(6), 1037–1050.

Alexander, D. R. (2013). The implications of evolutionary biology for religious belief. In K. Kampourakis (Ed.) *The Philosophy of Biology: A Companion for Educators*. Dordrecht: Springer, 179–204.

Alexopoulos, J. S., Grievel, R. A. F., and Robertson, P. B. (1988). Microscopic lamellar deformation features in quartz: discriminative characteristics of shock-generated varieties. *Geology* **16**(9), 796–779.

Alizon, S., Luciani, F., and Regoes, R. R. (2011). Epidemiological and clinical consequences of within-host evolution. *Trends in Microbiology* **19**(1), 24–32.

Allchin, D. (2005). The dilemma of dominance. *Biology and Philosophy* **20**, 427–451.

Alvarez, L. W., Alvarez, W., Asaro, F., and Michel, H. V. (1980). Extraterrestrial cause for the Cretaceous–Tertiary extinction. *Science* **208**(4448), 1095–1108.

Appel, T. A. (1987). *The Cuvier–Geoffroy Debate: French Biology in the Decades Before Darwin*. Oxford: Oxford University Press.

Arabatzis, T. and Kindi, V. (2008). The problem of conceptual change in the philosophy and history of science. In S. Vosniadou (Ed.) *International Handbook of Research on Conceptual Change*. New York and London: Routledge, 345–373.

Ariew, A. (2003). Ernst Mayr's "Ultimate/Proximate" distinction reconsidered and reconstructed. *Biology & Philosophy* **18**, 553–565.

Ariew, A. (2007). Teleology. In D. Hull and M. Ruse (Eds.) *The Cambridge Companion to the Philosophy of Biology*. Cambridge: Cambridge University Press, 160–181.

Ariew, A. and Lewontin, R. C. (2004). The confusions of fitness. *British Journal for the Philosophy of Science* **55**, 347–363.

Arthur, W. (2002). The emerging conceptual framework of evolutionary developmental biology. *Nature* **415**, 757–764.

Arthur, W. (2004). *Biased Embryos and Evolution*. New York: Cambridge University Press.

Arthur, W. (2011). *Evolution: A Developmental Approach*. Oxford: Wiley-Blackwell.

Asher, Y. M. and Kemler-Nelson, D. G. (2008). Was it designed to do that? Children's focus on intended function in their conceptualization of artifacts. *Cognition* **106**, 474–483.

Athanasiou, K. and Papadopoulou, P. (2011). Conceptual ecology of the evolution acceptance among Greek education students: knowledge, religious practices and social influences. *International Journal of Science Education*, DOI:10.1080/09500693.2011.586072.

Atran, S. (2004). *In Gods We Trust: The Evolutionary Landscape of Religion*. Oxford: Oxford University Press.

Audi, R. (2009). Religion and the politics of science: can evolutionary biology be religiously neutral? *Philosophy & Social Criticism* **35**(1–20), 23–50.

Audi, R. (2011). *Epistemology: A Contemporary Introduction to the Theory of Knowledge* (3rd edn.). New York: Routledge.

Avise, J. C. (2010). *Inside the Human Genome: A Case for Non-Intelligent Design*. Oxford: Oxford University Press.

Ayala, F. J. (2006). *Darwin and Intelligent Design*. Minneapolis, MN: Fortress Press.

Ayala, F. J. (2007). *Darwin's Gift to Science and Religion*. Washington, DC: Joseph Henry Press.

Ayala, F. J. (2009). Molecular evolution. In M. Ruse and J. Travis (Eds.) *Evolution: The First Four Billion Years*. Cambridge, MA: Belknap Press, 132–151.

Ayala, F. J. (2010a). *Am I a Monkey?* Baltimore, MD: The Johns Hopkins University Press.

Ayala, F. J. (2010b). The difference of being human: morality. *Proceedings of the National Academy of Sciences USA* **107**(2), 9015–9022.

Ayala, F. J. (2013). Biology and religion: the case for evolution. In K. Kampourakis (Ed.) *The Philosophy of Biology: A Companion for Educators*. Dordrecht: Springer, 161–178.

Barlow, N. (Ed.) (2005) [1958]. *The Autobiography of Charles Darwin 1809–1882*. New York: W.W. Norton.

Bateson, P. and Gluckman, P. (2011). *Plasticity, Robustness, Development and Evolution*. Cambridge: Cambridge University Press.

Baum, D. A. and Offner, S. (2008). Phylogenies and tree-thinking. *The American Biology Teacher* **70**(4), 222–229.

Baum, D. and Smith, S. (2013). *Tree Thinking: An Introduction to Phylogenetic Biology*. Greenwood Village, CO: Roberts and Company Publishers.

Baum, D. A., DeWitt Smith, S., and Donovan, S. S. (2005). The tree thinking challenge. *Science* **310**, 979–980.

Beatty, J. (1984). Chance and natural selection. *Philosophy of Science* **51**(2), 183–211.

Beatty, J. (1994). The proximate/ultimate distinction in the multiple careers of Ernst Mayr. *Biology & Philosophy* **9**, 333–356.

Beatty, J. (1995). The evolutionary contingency thesis. In G. Wolters and J. G. Lennox (Eds.) *Concepts, Theories, and Rationality in the Biological Sciences: The Second Pittsburgh-Konstanz Colloquium in the Philosophy of Science*. Pittsburgh, PA: University of Pittsburgh Press, 45–81.

Beatty, J. (2006). Replaying life's tape. *Journal of Philosophy* **CIII**(7), 336–362.

Beatty, J. (2010). Reconsidering the importance of chance variation. In G. Müller and M. Pigliucci (Eds.) *Evolution: The Extended Synthesis*. Cambridge, MA: MIT Press, 21–44.

Bechtel, W. (2013). Understanding biological mechanisms: using illustrations from circadian rhythm research. In K. Kampourakis (Ed.) *Philosophical Issues in Biology Education*. Springer: Dordrecht, 487–510.

Bechtel, W. and Abrahamsen, A. (2005). Explanation: a mechanistic alternative. *Studies in History and Philosophy of the Biological and Biomedical Sciences* **36**, 421–441.

Belyaev, D. K. (1979). Destabilizing selection as a factor in domestication. *Journal of Heredity* **70**, 301–308.

Benton, M. (2009). Paleontology and the history of life. In M. Ruse and J. Travis (Eds.) *Evolution: The First Four Billion Years*. Cambridge, MA: Belknap Press of Harvard University Press, 80–104.

Benton, M. J. and Donoghue, P. C. J. (2007). Paleontological evidence to date the tree of life. *Molecular Biology and Evolution* **24**, 26–53.

Berger, L. R., de Ruiter, D. J., Churchill, S. E., *et al.* (2010). *Australopithecus sediba:* a new species of *Homo*-like Australopith from South Africa. *Science* **328**, 195–204.

Berkman, M. and Plutzer, E. (2010). *Evolutionism, Creationism, and the Battle to Control America's Schools*. New York: Cambridge University Press.

Berkman, M. and Plutzer, E. (2011). Defeating Creationism in the courtroom, but not in the classroom. *Science* **331**(6016), 404–405.

Berkman, M. and Plutzer, E. (2012). An evolving controversy: the struggle to teach science in science classes. *The American Educator* **Summer,** 12–23.

Blancke, S., Hjermitslev, H. H., Braeckman, J., and Kjærgaard, P. C. (2013). Creationism in Europe: facts, gaps and prospects. *Journal of the American Academy of Religion*, DOI:10.1093/jaarel/lft034.

Bloom, P. (2004). *Descartes' Baby: How the Science of Child Development Explains What Makes Us Human*. New York: Basic Books.

Bloom, P. and Weisberg, D. S. (2007). Childhood origins of adult resistance to science. *Science* **316**, 996–997.

Bock, W. J. (1980). The definition and recognition of biological adaptation. *American Zoologist* **20**(1), 217–227.

Bohor, B. F., Foord, E. E., Modreski, P. J., and Triplehorn, D. M. (1984). Mineralogic evidence for an impact event at the Cretaceous–Tertiary boundary. *Science* **224**(4651), 867–869.

Bowler, P. J. (1975). The changing meaning of evolution. *Journal of the History of Ideas* **36**(1), 95–114.

Bowler, P. J. (1976). Malthus, Darwin and the concept of struggle. *Journal of the History of Ideas* **37**(4), 631–650.

Bowler, P. J. (1983). *The Eclipse of Darwinism: Anti-Darwinian Evolution Theories in the Decades around 1900*. Baltimore, MD: The Johns Hopkins University Press.

Bowler, P. J. (2005). Revisiting the eclipse of Darwinism. *Journal of the History of Biology* **38**, 19–32.

Bowler, P. J. (2007). *Monkey Trials and Gorilla Sermons: Evolution and Christianity from Darwin to Intelligent Design*. Cambridge, MA: Harvard University Press.

Bowler, P. J. (2009a). *Evolution: The History of an Idea*. Berkeley, CA: University of California Press.

Bowler, P. J. (2009b). Geographical distribution in the *Origin of Species*. In M. Ruse and R. J. Richards (Eds.) *The Cambridge Companion to the "Origin of Species."* Cambridge: Cambridge University Press, 153–172.

Bowler, P. J. (2013). *Darwin Deleted: Imagining a World Without Darwin*. Chicago, IL: University of Chicago Press.

Bowler, P. J. and Morus, I. R. (2005). *Making Modern Science: A Historical Survey*. Chicago, IL: The Continuum University of Chicago Press.

Boyer, P. (2001). *Religion Explained: The Evolutionary Origins of Religious Thought*. New York: Basic Books.

Branch, G. and Scott, E. C. (2009). The latest face of Creationism. *Scientific American* **300**(1), 92–99.

Brandon, R. N. (1990). *Adaptation and Environment*. Princeton, NJ: Princeton University Press.

Brem, S. K., Ranney, M., and Schindel, J. (2003). Perceived consequences of evolution: college students perceive negative personal and social impact in evolutionary theory. *Science Education* **87**(2), 181–206.

Brigandt, I. (2013). Intelligent Design and the nature of science: philosophical and pedagogical points. In K. Kampourakis (Ed.) *The Philosophy of Biology: A Companion for Educators*. Dordrecht: Springer, 205–238.

Brigandt, I. and Love, A. C. (2010). Evolutionary novelty and the evo-devo synthesis: field notes. *Evolutionary Biology* **37**, 93–99.

Brigandt, I. and Love, A. C. (2012). Conceptualizing evolutionary novelty: moving beyond definitional debates. *Journal of Experimental Zoology (Mol. Dev. Evol.)* **318B**, 417–427.

Brooke, J. H. (1991). *Science and Religion: Some Historical Perspectives*. Cambridge: Cambridge University Press.

Brooke, J. H. (2001). The Wilberforce–Huxley debate: why did it happen? *Science & Christian Belief* **13**(2), 127–141.

Brooke, J. H. (2009a). Darwin and Victorian Christianity. In J. Hodge and G. Radick (Eds.) *The Cambridge Companion to Darwin* (2nd edn.). Cambridge: Cambridge University Press, 197–218.

Brooke, J. H. (2009b). "Laws impressed on matter by the creator"? The Origin and the question of religion. In M. Ruse and R. J. Richards (Eds.) *The Cambridge Companion to the "Origin of Species."* Cambridge: Cambridge University Press, 256–274.

Brooke, J. H. and Numbers, R. L. (2010). *Science and Religion Around the World*. Oxford: Oxford University Press.

Browne, J. (2003a) [1995]. *Charles Darwin: Voyaging*. London: Pimlico.

Browne, J. (2003b) [2002]. *Charles Darwin: The Power of Place*. London: Pimlico.

Browne, J. (2006). *Darwin's "Origin of Species": A Biography*. London: Atlantic Books.

Burian, R. M. (1992). Adaptation: historical perspectives. In E. F. Keller and E. A. Lloyd (Eds.) *Keywords in Evolutionary Biology*. Cambridge, MA and London: Harvard University Press, 7–12.

Burian, R. M. (2013). On concepts that change with the advance of science: comments on Lennox. In A. Gotthelf and J. G. Lennox (Eds.) *Concepts and their Role in Knowledge: Reflections on Objectivist Epistemology*. Pittsburgh, PA: University of Pittsburgh Press, 185–199.

Burian, R. M. and Kampourakis, K. (2013). Against "genes for": could an inclusive concept of genetic material effectively replace gene concepts? In K. Kampourakis (Ed.) *The Philosophy of Biology: A Companion for Educators*. Dordrecht: Springer, 597–628.

Carey, S. (1985). *Conceptual Change in Childhood*. Cambridge, MA and London: MIT Press.

Carey, S. (2009). *The Origin of Concepts*. Oxford: Oxford University Press.

Carroll, S. B. (2003). Genetics and the making of *Homo sapiens*. *Nature* **422**, 849–857.

Carroll, S. B. (2005a). *Endless Forms Most Beautiful: The New Science of Evo-Devo*. New York: W.W. Norton.

Carroll, S. B. (2005b). Evolution at two levels: on genes and form. *PLoS Biology* **3**, 1159–1166.

Cavicchioli, R. (2011). Archaea: timeline of the third domain. *Nature Reviews Microbiology* **9**, 51–61.

Chi, M. T. H. (2008). Three types of conceptual change: belief revision, mental model transformation and categorical shift. In S. Vosniadou (Ed.) *International Handbook of Research on Conceptual Change*. New York and London: Routledge, 61–82.

Ciccarelli, F. D., Doerks, T., Von Mering, C., Creevey, C. J., Snel, B., and Bork, P. (2006). Toward automatic reconstruction of a highly resolved tree of life. *Science* **311**, 1283–1287.

Cleland, C. E. (2002). Methodological and epistemic differences between historical science and experimental science. *Philosophy of Science* **69**, 474–496.

Cleland, C. E. (2011). Prediction and explanation in historical natural science. *British Journal of Philosophy of Science* **62**, 551–582.

Cleland, C. E. and Zerella, M. (2013). What is life? In K. Kampourakis (Ed.) *The Philosophy of Biology: A Companion for Educators*. Dordrecht: Springer, 31–48.

Cohn, M. J. and Tickle, C. (1999). Developmental basis of limblessness and axial regionalization in snakes. *Nature* **399**, 474–479.

Conway Morris, S. (2003). *Life's Solution: Inevitable Humans in a Lonely Universe*. Cambridge: Cambridge University Press.

Coppinger, R. and Coppinger, L. (2001). *Dogs: A Startling New Understanding of Canine Origin, Behaviour and Evolution*. New York: Scribner.

Cornell, J. F. (1984). Analogy and technology in Darwin's vision of nature. *Journal of the History of Biology* **17**(3), 303–344.

Corsi, P. (2005). Before Darwin: transformist concepts in European natural history. *Journal of the History of Biology* **38**, 67–83.

Coyne, J. A. (2009). *Why Evolution is True*. Oxford: Oxford University Press.

Coyne, J. A. (2012). Science, religion, and society: the problem of evolution in America. *Evolution* **66**(8), 2654–2663.

Coyne, J. and Orr, H. (2004). *Speciation*. Sunderland, MA: Sinauer.

Curry, A. (2009). Creationist beliefs persist in Europe. *Science* **323**, 1159.

Cyranoski, D., Grimme, S., and Watt, F. (2010). A global survey of the scientifically literate public reveals a Pacific divide on key issues in science. *Nature* **467**, 388–389.

Daeschler, E. B., Shubin, N. H., and Jenkins, F. A., Jr. (2006). A Devonian tetrapod-like fish and the evolution of the tetrapod body plan. *Nature* **440**, 757–763.

Danielson, D. R. (2009). That the Copernican System demoted humans from the center of the cosmos. In R. Numbers (Eds.) *Galileo Goes to Jail, and Other Myths About Science and Religion*. Cambridge, MA: Harvard University Press, 50–58.

Darwin, C. (1859). *On the Origin of Species by Means of Natural Selection, or the Preservation of Favoured Races in the Struggle for Life* (1st edn.). London: John Murray.

Darwin, C. (1868). *The Variation of Animals and Plants Under Domestication*, Vol. **2**. London: John Murray.

Darwin, C. (1871). *The Descent of Man, and Selection in Relation to Sex* (1st edn.). London: John Murray.

Darwin, C. (1964) [1859]. *On the Origin of Species: A Facsimile of the First Edition*. Cambridge, MA: Harvard University Press.

Darwin, C. R. and Wallace, A. R. (1858). On the tendency of species to form varieties; and on the perpetuation of varieties and species by natural means of selection. *Journal of the Proceedings of the Linnean Society of London. Zoology* **3**, 45–50.

Darwin, F. (1995) [1902]. *The Life of Charles Darwin*. London: Senate.

Darwin, F. (Ed.) (1909). *The Foundations of the Origin of Species: Two Essays Written in 1842 and 1844*. Cambridge: Cambridge University Press.

Dawkins, R. (2006a) [1986]. *The Blind Watchmaker*. London: Penguin Books.

Dawkins, R. (2006b). *The God Delusion*. London: Bantam Press.

Dawkins, R. (2009). *The Greatest Show on Earth: The Evidence for Evolution*. London: Bantam Press.

De Robertis, E. M. (2008). Evo-devo: variations on ancestral themes. *Cell* **132**, 185–195.

De Robertis, E. M. (2009). Spemann's organizer and the self-regulation of embryonic fields. *Mechanisms of Development* **126**, 925–941.

De Robertis, E. M. and Sasai, Y. (1996). A common plan for dorsoventral patterning in Bilateria. *Nature* **380**, 37–40.

Deniz, H., Donnelly, L. A., and Yilmaz, I. (2008). Exploring the factors related to acceptance of evolutionary theory among Turkish preservice biology teachers: toward a more informative conceptual ecology for biological evolution. *Journal of Research in Science Teaching* **45**(4), 420–443.

Dennett, D. (2006). *Breaking the Spell: Religion as a Natural Phenomenon*. New York: Viking.

Depew, D. (2008). Consequence etiology and biological teleology in Aristotle and Darwin. *Studies in the History and Philosophy of Biological and Biomedical Sciences* **39**, 379–390.

Depew, D. (2013). Conceptual change and the rhetoric of evolutionary theory: "Force talk" as a case study and challenge for science pedagogy. In K. Kampourakis (Ed.) *The Philosophy of Biology: A Companion for Educators*. Dordrecht: Springer, 121–144.

Depew, D. J. and Weber, B. H. (1995). *Darwinism Evolving: Systems Dynamics and the Genealogy of Natural Selection*. Cambridge, MA: MIT Press.

Desjardins, E. (2011). Historicity and experimental evolution. *Biology & Philosophy* **26**, 339–364.

Desmond, A. and Moore, J. (1994) [1991]. *Darwin: The Life of a Tormented Evolutionist*. New York: W.W. Norton.

Desmond, A. and Moore, J. (2009). *Darwin's Sacred Cause: How a Hatred of Slavery Shaped Darwin's Views on Human Evolution*. New York: Houghton Mifflin Harcourt.

Desmond, D., Moore, J., and Browne, J. (2007). *Charles Darwin*. Oxford and New York: Oxford University Press.

Diamond, J. M. (1997). *Guns Germs and Steel: The Fate of Human Societies*. New York: W.W. Norton.

Diamond, J. (2002). Evolution, consequences and future of plant and animal domestication. *Nature* **418**, 700–707.

Diedrich, C. R. and Flynn, J. L. (2011). HIV-1/*Mycobacterium tuberculosis* coinfection immunology: how does HIV-1 exacerbate tuberculosis? *Infection and Immunity* **79**(4), 1407–1417.

Dietrich, M. (2010). Microevolution and macroevolution are governed by the same processes. In F. J. Ayala and R. Arp (Eds.) *Contemporary Debates in Philosophy of Biology*. Oxford: Wiley-Blackwell, 169–179.

Dietrich, M. (2013). Molecular evolution. In K. Kampourakis (Ed.) *The Philosophy of Biology: A Companion for Educators*. Dordrecht: Springer, 239–248.

Dixon, T., Cantor, G., and Pumfrey, S. (Eds.) (2011). *Science and Religion: New Historical Perspectives*. Cambridge: Cambridge University Press.

Dobzhansky, T. (1973). Nothing in biology makes sense except in the light of evolution. *The American Biology Teacher* **35**(3), 125–129.

Donoghue, P. C. J. and Benton, M. J. (2007). Rocks and clocks: calibrating the tree of life using fossils and molecules. *Trends in Ecology and Evolution* **22**, 424–431.

Doolittle, W. F. (2000). Uprooting the tree of life. *Scientific American* **282**(2), 90–95.

Duit, R. and Treagust, D. F. (2012). How can conceptual change contribute to theory and practice in science education? In B.J. Fraser, K. G. Tobin, and C. J. McRobbie (Eds.) *Second International Handbook of Science Education*. Dordrecht: Springer, 107–118.

Duncan, M. J., Bourrat, P., DeBerardinis, J., and O' Malley, M. (2013). Small things, big consequences: microbiological perspectives on biology. In K. Kampourakis (Ed.) *The Philosophy of Biology: A Companion for Educators*. Dordrecht: Springer, 373–394.

Dupré, J. (2003). *Darwin's Legacy: What Evolution Means Today*. Oxford: Oxford University Press.

Dye, C. and Williams, B. (2010). The population dynamics and control of tuberculosis. *Science* **328**, 856–861.

Ecklund, E. H. (2010). *Science vs. Religion: What Scientists Really Think*. Oxford: Oxford University Press.

Eldredge, N. (2000). *The Triumph of Evolution and the Failure of Creationism*. New York: W.H. Freeman and Co.

Eldredge, N. and Gould, S. J. (1972). Punctuated equilibria: an alternative to phyletic gradualism. In T. J. M. Schopf (Ed.) *Models in Paleobiology*. San Francisco, CA: Freeman, Cooper, 82–115.

Endersby, J. (2008). *Imperial Nature: Joseph Hooker and the Practices of Victorian Science*. Chicago, IL: University of Chicago Press.

Endersby, J. (Ed.) (2009). *Charles Darwin: On the Origin of Species*. Cambridge: Cambridge University Press.

Ereshefsky, M. (2010a). Darwin's solution to the species problem. *Synthese* **175**, 405–425.

Ereshefsky, M. (2010b). Microbiology and the species problem. *Biology & Philosophy* **25**, 553–568.

Erwin, D. H. (2010). Microevolution and macroevolution are not governed by the same processes. In F. J. Ayala and R. Arp (Eds.) *Contemporary Debates in Philosophy of Biology*. Oxford: Wiley-Blackwell, 180–193.

Evans, L. T. (1984). Darwin's use of the analogy between artificial and natural selection. *Journal of the History of Biology* **17**(1), 113–140.

Evans, M. E. (2001). Cognitive and contextual factors in the emergence of diverse belief systems: creation versus evolution. *Cognitive Psychology* **42**, 217–266.

Evans, M. E. (2008). Conceptual change and evolutionary biology: a developmental analysis. In S. Vosniadou (Ed.) *International Handbook of Research in Conceptual Change*. New York and London: Routledge, 263–294.

Fabre, P. H., Rodrigues, A., and Douzery, E. J. P. (2009). Patterns of macroevolution among primates inferred from a supermatrix of mitochondrial and nuclear DNA. *Molecular Phylogenetics and Evolution* **53**, 808–825.

Firestein, S. (2012). *Ignorance: How it Drives Science*. Oxford: Oxford University Press.

Fitch, W. M. (2000). Homology: a personal view on some of the problems. *Trends in Genetics* **16**, 227–231.

Fong, J. J., Brown, J. M., Fujita, M. K., and Boussau, B. (2012). A phylogenomic approach to vertebrate phylogeny supports a turtle-archosaur affinity and a possible paraphyletic lissamphibia. *PLoS ONE* **7**(11): e48990, DOI:10.1371/journal.pone.0048990.

Forber, P. (2010). Confirmation and explaining how possible. *Studies in the History and Philosophy of Biological and Biomedical Sciences* **41**, 32–40.

Forber, P. (2013). Debating the power and scope of adaptation. In K. Kampourakis (Ed.) *The Philosophy of Biology: A Companion for Educators*. Dordrecht: Springer, 145–160.

Forber, P. and Griffith, E. (2011). Historical reconstruction: gaining epistemic access to the deep past. *Philosophy and Theory in Biology* **3**, e203, DOI:10:3998/ptb.6959004.0003.003

Frigg, R. and Hartmann, S. (2012). Models in science. In E. N. Zalta (Ed.) *The Stanford Encyclopedia of Philosophy*. http://plato.stanford.edu/archives/spr2012/entries/models-science.

Ganapathy, R. (1980). A major meteorite impact on the Earth 65 million years ago: evidence from the Cretaceous–Tertiary boundary clay. *Science* **209**, 921–923.

Gandhi, N. R., Moll, A., Sturm, A. W., *et al.* (2006). Extensively drug-resistant tuberculosis as a cause of death in patients co-infected with tuberculosis and HIV in a rural area of South Africa. *Lancet* **368**, 1575–1580.

Gandhi, N. R., Nunn, P., Dheda, K., *et al.* (2010). Multidrug-resistant and extensively drug-resistant tuberculosis: a threat to global control of tuberculosis. *Lancet* **375**, 1830–1843.

Gass, G. L. and Bolker, J. A. (2003). Modularity. In B. K. Hall and W. M. Olson (Eds.) *Keywords and Concepts in Evolutionary Developmental Biology.* Cambridge, MA: Harvard University Press, 260–267.

Gelman, S. A. (2003). *The Essential Child: Origins of Essentialism in Everyday Thought.* Oxford: Oxford University Press.

Gelman, S. A. (2004). Psychological essentialism in children. *Trends in Cognitive Sciences* **8**(9), 404–409.

Gelman, S. A. and Bloom, P. (2000). Young children are sensitive to how an object was created when deciding what to name it. *Cognition* **76**, 91–103.

Gelman, S. A. and Ebeling, K. S. (1998). Shape and representational status in children's early naming. *Cognition* **66**, 35–47.

Gelman, S. A. and Hirschfeld, L. A. (1999). How biological is essentialism? In D. L. Medin and S. Atran (Eds.) *Folkbiology.* Cambridge, MA: MIT Press, 403–446.

Gelman, S. A. and Markman, E. M. (1986). Categories and induction in young children. *Cognition* **23**, 183–209.

Gelman, S. A. and Markman, E. M. (1987). Young children's inductions from natural kinds: the role of categories and appearances. *Child Development* **58**, 1532–1541.

Gelman, S. A. and Rhodes, M. (2012). "Two-thousand years of stasis": how psychological essentialism impedes evolutionary understanding. In K. Rosengren, S. Brem, E. M. Evans, and G. M. Sinatra (Eds.) *Evolution Challenges: Integrating Research and Practice in Teaching and Learning About Evolution.* Oxford: Oxford University Press, 3–21.

Gelman, S. A. and Wellman, H. M. (1991). Insides and essences: early understandings of the nonobvious. *Cognition* **38**, 213–244.

Giere, R. N. (2006). *Scientific Perspectivism.* Chicago, IL and London: University of Chicago Press.

Gilbert, S. F. and Burian, R. M. (2003a). Developmental genetics. In B. K. Hall and W. M. Olson (Eds.) *Keywords and Concepts in Evolutionary Developmental Biology.* Cambridge, MA: Harvard University Press, 68–74.

Gilbert, S. F. and Burian, R. M. (2003b). Development, evolution and evolutionary developmental biology. In B. K. Hall and W. M. Olson (Eds.) *Keywords and Concepts in Evolutionary Developmental Biology.* Cambridge, MA: Harvard University Press, 61–68.

Gingerich, O. (1985). Did Copernicus owe a debt to Aristarchus? *Journal for the History of Astronomy* **16**(1), 37–42.

Gingerich, O. (2005). Copernicus Nicholas. In J. L. Heilbron (Ed.) *The Oxford Guide to the History of Physics and Astronomy.* Oxford: Oxford University Press, 73–74.

Glick, T. F. and Kohn, D. (1996). *Charles Darwin on Evolution: The Development of the Theory of Natural Selection.* Indianapolis, IN: Hackett.

Godfrey-Smith, P. (2003). *Theory and Reality: An Introduction to the Philosophy of Science.* Chicago, IL: University of Chicago Press.

Godfrey-Smith, P. (2009). *Darwinian Populations and Natural Selection.* Oxford: Oxford University Press.

Goldberg, R. F. and Thompson-Schill, S. L. (2009). Developmental "roots" in mature biological knowledge. *Psychological Science* **20**(4), 480–487.

Goldstein, B. R. (2002). Copernicus and the origins of his heliocentric system. *Journal for the History of Astronomy* **33**, 219–235.

Gould, S. J. (1977). *Ontogeny and Phylogeny*. Cambridge, MA: Belknap Press of Harvard University Press.

Gould, S. J. (1996). *Full House: The Spread of Excellence from Plato to Darwin*. New York: Harmony Books.

Gould, S. J. (1999). *Rocks of Ages: Science and Religion in the Fullness of Life*. New York: Ballantine Books.

Gould, S. J. (2000) [1989]. *Wonderful Life: The Burgess Shale and the Nature of History*. London: Vintage.

Gould, S. J. (2002). *The Structure of Evolutionary Theory*. Cambridge, MA and London: Belknap Press of Harvard University Press.

Gould, S. J. and Lewontin, R. C. (1979). The spandrels of San Marco and the Panglossian paradigm: a critique of the adaptationist programme. *Proceedings of the Royal Society London, Series B, Biological Sciences* **205**, 581–598.

Gould, S. J. and Vrba, E. S. (1982). Exaptation: a missing term in the science of form. *Paleobiology* **8**(1), 4–15.

Graebsch, A. and Schiermeier, Q. (2006). Anti-evolutionists raise their profile in Europe. *Nature* **444**, 406–407.

Grant, P. R. and Grant, B. R. (2002). Unpredictable evolution in a 30-year study of Darwin's finches. *Science* **296**, 707–711.

Grant, P. R. and Grant, B. R. (2008). *How and Why Species Multiply: The Radiation of Darwin's Finches*. Princeton, NJ: Princeton University Press.

Gregory, T. R. (2008). Understanding evolutionary trees. *Evolution: Education and Outreach* **1**, 121–137.

Grehan, J. R. and Schwartz, J. H. (2009). Evolution of the second orangutan: phylogeny and biogeography of hominid origins. *Journal of Biogeography* **36**(10), 1823–1844.

Greif, M., Kemler-Nelson, D., Keil, F. C., and Guiterrez, F. (2006). What do children want to know about animals and artifacts? Domain-specific requests for information. *Psychological Science* **17**(6), 455–459.

Greiffenhagen, C. and Sherman, W. (2008). Kuhn and conceptual change: on the analogy between conceptual change in science and children. *Science & Education* **17**, 1–26.

Grene, M. and Depew, D. (2004). *The Philosophy of Biology: An Episodic History*. Cambridge: Cambridge University Press.

Gribaldo, S., Poole, A. M., Daubin, V., Forterre, P., and Brochier-Armanet, C., (2010). The origin of eukaryotes and their relationship with the archaea: are we at a phylogenomic impasse? *Nature Reviews Microbiology* **8**, 743e752, DOI:10.1038/nrmicro2426

Gribbin, J. (2003). *Science: A History*. London: Penguin Books.

Griffiths, A. J. F., Wessler, S., Carroll, S. B., and Doebley, J. (2012). *Introduction to Genetic Analysis*. New York: W.H. Freeman.

Hall, B. K. (2003). Descent with modification: the unity underlying homology and homoplasy as seen through an analysis of development and evolution. *Biological Reviews of the Cambridge Philosophical Society* **78**, 409–433.

Hall, B. K. (2012). Parallelism, deep homology and evo-devo. *Evolution and Development* **14**(1), 29–33.

Hall, B. K. and Kerney, R. (2012). Levels of biological organization and the origin of novelty. *Journal of Experimental Zoology (Mol. Dev. Evol.)* **318B**, 428–437.

Hall, B. K. and Olson, W. M. (Eds.) (2003). *Keywords and Concepts in Evolutionary Developmental Biology*. Cambridge, MA: Harvard University Press.

Hameed, S. (2008). Bracing for Islamic Creationism. *Science* **322**, 1637.

Hare, B., Brown, M., Williamson, C., and Tomasello, M. (2002).The domestication of social cognition in dogs. *Science* **298**, 1634.

Harrison, T. (2010). Apes among the tangled branches of human origins. *Science* **327**, 532–534.

Heeney, J. L., Dalgleish, A. G., and Weiss, R. A. (2006). Origins of HIV and the evolution of resistance to AIDS. *Science* **313**, 462–466.

Heilbron, J. L. (2010). *Galileo*. Oxford: Oxford University Press.

Hempel, C. (1965). *Aspects of Scientific Explanation*. New York: Free Press.

Hempel, C. and Oppenheim, P. (1948). Studies in the logic of explanation. *Philosophy of Science* **15**, 135–175.

Herbert, S. (2005). *Charles Darwin, Geologist*. Ithaca, NY: Cornell University Press.

Hilpinen, R. (2011). Artifact. In E. N. Zalta (Ed.) *The Stanford Encyclopedia of Philosophy*. http://plato.stanford.edu/archives/win2011/entries/artifact.

Hitchcock, C. (2008). Causation. In S. Psillos and M. Curd (Eds.) *The Routledge Companion to Philosophy of Science*. New York: Routledge, 317–326.

Hodge, J. (2009). The notebook programmes and projects of Darwin's London years. In J. Hodge and G. Radick (Eds.) *Cambridge Companion to Darwin* (2nd edn.). Cambridge: Cambridge University Press, 44–72.

Hodge, J. (2010). Darwin, the Galápagos and his changing thoughts about species origins: 1835–1837. *Proceedings of the California Academy of Sciences* **61**(II, 7), 89–106.

Hodge, J. and Radick, G. (Eds.) (2009a). *The Cambridge Companion to Darwin* (2nd edn.). Cambridge: Cambridge University Press.

Hodge, J. and Radick, G. (2009b). Darwin's theories in the intellectual long run. In J. Hodge and G. Radick (Eds.) *The Cambridge Companion to Darwin* (2nd edn.). Cambridge: Cambridge University Press, 277–301.

Hodge, M. J. S. (1977). The structure and strategy of Darwin's "long argument." *British Journal for the History of Science* **10**, 237–246.

Hodge, M. J. S. (1992). Darwin's argument in the Origin. *Philosophy of Science* **59**(3), 461–464.

Hodge, M. J. S. and Kohn, D. (1985). The immediate origins of natural selection. In D. Kohn (Ed.) *The Darwinian Heritage*. Princeton, NJ: Princeton University Press, 185–206.

Hoekstra, H. E. and Coyne, J. A. (2007). The locus of evolution: evo devo and the genetics of adaptation. *Evolution* **61**(5), 995–1016.

Hokayem, H. and Boujaoude, S. (2008). College students' perceptions of the theory of evolution. *Journal of Research in Science Teaching* **45**(4), 395–419.

Hoyningen-Huene, P. (2013). *Systematicity: The Nature of Science*. Oxford: Oxford University Press.

Hull, D. (1992). The particular circumstance model of scientific explanation. In M. Nitecki and D. Nitecki (Eds.) *History and Evolution*. Albany, NY: SUNY Press, 69–80.

Hull, D. L. (2005). Deconstructing Darwin: evolutionary theory in context. *Journal of the History of Biology* **38**(1), 137–152.

Hull, D. L. (2009). Darwin's science and Victorian philosophy of science. In J. Hodge and G. Radick (Eds.) *The Cambridge Companion to Darwin* (2nd edn.). Cambridge: Cambridge University Press, 173–196.

Hume, D. (1993) [1779/1777]. *Dialogues Concerning Natural Religion and Natural History of Religion*. Oxford and New York: Oxford University Press.

Huxley, J. (1942). *Evolution: The Modern Synthesis*. London: Allen and Unwin.

Huxley, T. H. (1860). [Review of] The origin of Species. *Westminster Review* **17** (n.s.), 541–570.

Huxley, T. H. (1863). *Evidence as to Man's Place in Nature*. London: Williams and Norgate.

Inagaki, K. and Hatano, G. (2002). *Young Children's Naive Thinking About the Biological World*. New York: Psychology Press.

Ingraham, J. (2010). *March of the Microbes: Sighting the Unseen*. Cambridge, MA: Harvard University Press.

Ingram, E. L. and Nelson, C. E. (2006). Relationship between achievement and students' acceptance of evolution or creation in an upper-level evolution course. *Journal of Research in Science Teaching* **43**(1), 7–24.

Jablonski, D. (2004a). Extinction: past and present. *Nature* **427**, 589.

Jablonski, D. (2004b). Extinction and the spatial dynamics of biodiversity. *Proceedings of the National Academy of Sciences USA* **105**, 11535–11538.

Jablonski, D. (2007). Scale and hierarchy in macroevolution. *Palaeontology* **50**, 87–109.

Jablonski, D. (2008). Species selection: theory and data. *Annual Review of Ecology, Evolution and Systematics* **39**, 501–524.

Jamieson, A. and Radick, G. (2013). Putting Mendel in his place: how curriculum reform in genetics and counterfactual history of science can work together. In K. Kampourakis (Ed.) *The Philosophy of Biology: A Companion for Educators*. Dordrecht: Springer, 577–596.

Kampourakis, K. (Ed.) (2013a). *The Philosophy of Biology: A Companion for Educators*. Dordrecht: Springer.

Kampourakis, K. (2013b). Teaching about adaptation: why evolutionary history matters. *Science & Education* **22**(2), 173–188.

Kampourakis, K. and McComas, W. F. (2010). Charles Darwin and evolution: illustrating human aspects of science. *Science & Education* **19**, 637–654.

Kampourakis, K. and Nehm, R. (2014). History and philosophy of science and the teaching of evolution: students' conceptions and explanations. In M. R. Matthews (Ed.) *International Handbook of Research in History, Philosophy and Science Teaching*. Dordrecht: Springer.

Kampourakis, K. and Zogza, V. (2007). Students' preconceptions about evolution: how accurate is the characterization as "Lamarckian" when considering the history of evolutionary thought? *Science & Education* **16**(3–5), 393–422.

Kampourakis, K. and Zogza, V. (2008). Students' intuitive explanations of the causes of homologies and adaptations. *Science & Education* **17**(1), 27–47.

Kampourakis, K. and Zogza, V. (2009). Preliminary evolutionary explanations: a basic framework for conceptual change and explanatory coherence in evolution. *Science & Education* **18**(10), 1313–1340.

Kampourakis, K., Pavlidi, V., Papadopoulou, M., and Palaiokrassa, E. (2012a). Children's teleological intuitions: what kind of explanations do 7–8 year olds give for the features of organisms, artifacts and natural objects? *Research in Science Education* **42**(4), 651–671.

Kampourakis, K., Palaiokrassa, E., Papadopoulou, M., Pavlidi, V., and Argyropoulou, M. (2012b). Children's intuitive teleology: shifting the focus of evolution education research. *Evolution Education and Outreach* **5**(2), 279–291.

Keil, F. C. (1989). *Concepts, Kinds and Cognitive Development*. Cambridge, MA and London: MIT Press.

Keil, F. C. (1992). The origins of an autonomous biology. In M. R. Gunnar and M. Maratsos (Eds.) *Modularity and Constraints in Language and Cognition: Minnesota Symposium on Child Psychology*, Vol. **25**. Hillsdale, NJ: Erlbaum, 103–138.

Keil, F. C. (1994). The birth and nurturance concepts by domains: the origins of concepts of living things. In L. A. Hirschfeld and S. Gelman (Eds.) *Mapping the Mind: Domain Specificity in Cognition and Culture*. Cambridge: Cambridge University Press, 234–254.

Keil, F. C. (1995). The growth of causal understanding of natural kinds. In D. Sperber, D. Premack, and A. J. Premack (Eds.) *Causal Cognition: A Multi-Disciplinary Debate*. Oxford: Clarendon Press, 234–262.

Keil, F. C. and Newman, G. E. (2008). Two tales of conceptual change: what changes and what remains the same. In S. Vosniadou (Ed.) *International Handbook of Research on Conceptual Change*. New York and London: Routledge, 83–101.

Keil, F. C. and Wilson, R. A. (Eds.) (2000). *Explanation and Cognition*. Cambridge, MA: MIT Press.

Keil, F. C., Greif, M. A., and Kerner, R. S. (2007). A world apart: how concepts of the constructed world are different in representation and in development. In E. Margolis and S. Laurence (Eds.) *Creations of the Mind: Essays on Artifacts and their Representation*. Oxford: Oxford University Press, 231–247.

Kelemen, D. (1999a). Function, goals and intention: children's teleological reasoning about objects. *Trends in Cognitive Sciences* **3**(12), 461–468.

Kelemen, D. (1999b). The scope of teleological thinking in preschool children. *Cognition* **70**, 241–272.

Kelemen, D. (1999c). Why are rocks pointy? Children's preference for teleological explanations of the natural world. *Developmental Psychology* **35**, 1440–1452.

Kelemen, D. (2003). British and American children's preferences for teleo-functional explanations of the natural world. *Cognition* **88**, 201–221.

Kelemen, D. (2004). Are children "Intuitive Theists"? Reasoning about purpose and design in nature. *Psychological Science* **15**(5), 295–301.

Kelemen, D. (2012). Teleological minds: how natural intuitions about agency and purpose influence learning about evolution. In K. Rosengren, S. Brem, E. M. Evans, and G. M. Sinatra (Eds.) *Evolution Challenges: Integrating Research and Practice in Teaching and Learning About Evolution*. Oxford: Oxford University Press, 66–92.

Kelemen, D. and Carey, S. (2007). The essence of artifacts: developing the design stance. In E. Margolis and S. Laurence (Eds.) *Creations of the Mind: Theories of Artifacts and their Representation*. New York: Oxford University Press, 212–230.

Kelemen, D. and DiYanni, C. (2005). Intuitions about origins: purpose and intelligent design in children's reasoning about nature. *Journal of Cognition and Development* **6**(1), 3–31.

Kelemen, D. and Rosset, E. (2009). The human function compunction: teleological explanation in adults. *Cognition* **111**, 138–143.

Keller, E. F. (2010). *The Mirage of a Space Between Nature and Nurture*. Durham, NC: Duke University Press.

Keller, E. F. and Lloyd, E. A. (Eds.) (1992). *Keywords in Evolutionary Biology*. Cambridge, MA: Harvard University Press.

Keynes, R. (2001). *Darwin, His Daughter and Human Evolution*. New York: Riverhead Books.

Kitcher, P. (1981). Explanatory unification. *Philosophy of Science* **48**(4), 507–531.

Kitcher, P. (1989). Explanatory unification and the causal structure of the world. In P. Kitcher and W. C. Salmon (Eds.) *Scientific Explanation*. Minneapolis, MN: University of Minnesota Press, 410–505.

Kitcher, P. (2007). *Living with Darwin*. New York: Oxford University Press.

Kitcher, P. (2011). Militant modern atheism. *Journal of Applied Philosophy* **28**(1), 1–13.

Knoll, A. H. (2011). The multiple origins of complex multicellularity. *Annual Review of Earth and Planetary Sciences* **39**, 217–239.

Knoll, A. H. and Hewitt, D. (2011). Complex multicellularity: phylogenetic, functional and geological perspectives. In K. Sterelny and B. Calcott (Eds.) *The Major Transitions in Evolution Revisited*. Cambridge, MA: MIT Press, 251–270.

Kohn, D. (Ed.) (1985a). *The Darwinian Heritage*. Princeton, NJ: Princeton University Press.

Kohn, D. (1985b). Darwin's principle of divergence as internal dialogue. In D. Kohn (Ed.) *The Darwinian Heritage*. Princeton, NJ: Princeton University Press, 245–257.

Kohn, D. (2009). Darwin's keystone: the principle of divergence. In M. Ruse and R. J. Richards (Eds.) *The Cambridge Companion to the "Origin of Species."* Cambridge: Cambridge University Press, 87–108.

Koonin, E. V. (2010a). The two empires and three domains of life in the postgenomic age. *Nature Education* **3**(9), 27.

Koonin, E. V. (2010b). The origin and early evolution of eukaryotes in the light of phylogenomics. *Genome Biology* **11**, 209.

Koonin, E. V. (2011). *The Logic of Chance: The Nature and Origin of Biological Evolution*. Upper Saddle River, NJ: FT Press.

Krug, A. Z., Jablonski, D., and Valentine, J. W. (2009). Signature of the end-Cretaceous mass extinction in the modern biota. *Science* **323**, 767–771.

Kuhn, T. S. (1996) [1962]. *The Structure of Scientific Revolutions* (3rd edn.). Chicago, IL and London: University of Chicago Press.

Kyte, F. T., Zhou, Z., and Wasson, J. T. (1980). Siderophile-enriched sediments from the Cretaceous–Tertiary boundary. *Nature* **288**, 651–656.

Laland, K. N., Sterelny, K., Odling-Smee, F. J., Hoppitt, W., and Uller, T. (2011). Cause and effect in biology revisited: is Mayr's proximate-ultimate dichotomy still useful? *Science* **334**, 1512–1516.

Laland, K. N., Odling-Smee, F. J., Hoppitt, W., and Uller, T. (2012). More on how and why: cause and effect in biology revisited. *Biology & Philosophy*, DOI 10.1007/s10539-012-9335-1.

Largent, M. A. (2009). Darwin's analogy between artificial and natural selection in the *Origin of Species*. In M. Ruse and R. J. Richards (Eds.) *The Cambridge Companion to the "Origin of Species."* Cambridge: Cambridge University Press, 14–29.

Larson, E. J. (2004). *Evolution: The Remarkable History of a Scientific Theory*. New York: Modern Library.

Laudan, R. (1987). *From Mineralogy to Geology: The Foundation of a Science, 1650–1830*. Chicago, IL: University of Chicago Press.

Lennox, J. G. (1992). Teleology. In E. Lloyd and E.F. Keller (Eds.) *Keywords in Evolutionary Biology*. Cambridge, MA: Harvard University Press, 122–127.

Lennox, J. G. (2005). Darwin's methodological evolution. *Journal of the History of Biology* **38**(1), 85–99.

Lennox, J. G. (2010). The Darwin/Gray correspondence 1857–1869: an intelligent discussion about chance and design. *Perspectives on Science* **18**(4), 456–479.

Lennox, J. G. (2013a). Concepts, context and the advance of science. In A. Gotthelf and J.G. Lennox (Eds.) *Concepts and their Role in Knowledge: Reflections on Objectivist Epistemology*. Pittsburgh, PA: University of Pittsburgh Press, 112–133.

Lennox, J. G. (2013b). Conceptual development versus conceptual change: response to Burian. In A. Gotthelf and J.G. Lennox (Eds.) *Concepts and their Role in Knowledge: Reflections on Objectivist Epistemology*. Pittsburgh, PA: University of Pittsburgh Press, 201–211.

Lennox, J. G. and Kampourakis, K. (2013). Biological teleology: the need for history. In K. Kampourakis (Ed.) *The Philosophy of Biology: A Companion for Educators*. Dordrecht: Springer, 421–454.

Lennox, J. G. and Wilson, B. E. (1994). Natural selection and the struggle for existence. *Studies in the History and Philosophy of Science* **25**(1), 65–80.

Levine, A. T. (2000). Which way is up? Thomas S. Kuhn's analogy to conceptual development in childhood. *Science & Education* **9**, 107–122.

Lewens, T. (2004). *Organisms and Artifacts: Design in Nature and Elsewhere*. Cambridge, MA and London: MIT Press.

Lewens, T. (2007). Adaptation. In D. L. Hull and M. Ruse (Eds.) *Cambridge Companion to the Philosophy of Biology*. Cambridge: Cambridge University Press, 1–21.

Lewontin, R. C. (2001). *The Triple Helix: Gene, Organism and Environment*. Cambridge, MA: Harvard University Press.

Lewontin, R. C., Rose, S., and Kamin, L. J. (1984). *Not in Our Genes: Biology, Ideology, and Human Nature*. New York: Pantheon Books.

Lienau, E. K., DeSalle, R., Allard, M., *et al.* (2011). The mega-matrix tree of life: using genome-scale horizontal gene transfer and sequence evolution data as information about the vertical history of life. *Cladistics* **27**, 417–427.

Lindberg, D. C. (2007). *The Beginnings of Western Science: The European Scientific Tradition in Philosophical, Religious, and Institutional Context, Prehistory to A.D. 1450* (2nd edn.). Chicago, IL: University of Chicago Press.

Lipton, P. (2004). *Inference to the Best Explanation*. London: Routledge.

Lipton, P. (2008). Inference to the best explanation. In S. Psillos and M. Curd (Eds.) *The Routledge Companion to Philosophy of Science*. New York: Routledge, 193–202.

Livingstone, D. (2003). *Putting Science in its Place: Geographies of Scientific Knowledge*. Chicago, IL: University of Chicago Press.

Livingstone, D. N. (2009). That Darwin destroyed natural theology. In R. L. Numbers (Ed.) *Galileo Goes to Jail and Other Myths About Science and Religion*. Cambridge, MA: Harvard University Press, 152–160.

Lombrozo, T. and Carey, S. (2006). Functional explanation and the function of explanation. *Cognition* **99**, 167–204.

Long, J. (2012). Evolution, missing links and climate change: recent advances in understanding transformational macroevolution. In A. Poiani (Ed.) *Pragmatic Evolution: Applications of Evolutionary Theory*. Cambridge: Cambridge University Press, 23–36.

Long, J. A., Young, G. C., Holland, T., Senden, T. J., and Fitzgerald, E. M. (2006). An exceptional Devonian fish from Australia sheds light on tetrapod origins. *Nature* **444**, 199–202.

Losos, J. B. (2010). A tale of two radiations: similarities and differences in the evolutionary diversification of Darwin's finches and Greater Antillean Anolis lizards. In P. R. Grant and B. R. Grant (Eds.) *In Search of the Causes of Evolution: From Field Observations to Mechanisms*. Princeton, NJ: Princeton University Press, 309–331.

Love, A. C. (2002). Darwin and *Cirripedia* prior to 1846: exploring the origins of the barnacle research. *Journal of the History of Biology* **35**, 251–289.

Love, A. C. (2013). Teaching evolutionary developmental biology: concepts, problems, and controversy. In K. Kampourakis (Ed.) *The Philosophy of Biology: A Companion for Educators*. Dordrecht: Springer, 323–342.

Lucas, J. R. (1979). Wilberforce and Huxley: a legendary encounter. *The Historical Journal* **22**(2), 313–330.

Lustig, A. J. (2009). Darwin's difficulties. In M. Ruse and R. J. Richards (Eds.) *The Cambridge Companion to the "Origin of Species."* Cambridge: Cambridge University Press, 109–128.

Lyell, C. (1832). *Principles of Geology*, Vol. **2**. London: John Murray.

Macdougall, D. (2008). *Nature's Clocks: How Scientists Measure the Age of Almost Everything.* San Diego, CA: University of California Press.

Machamer, P., Darden, L., and Craver, C. F. (2000). Thinking about mechanisms. *Philosophy of Science* **67**(1), 1–25.

Mallet, J. (2008). Hybridization, ecological races, and the nature of species: empirical evidence for the ease of speciation. *Philosophical Transactions of the Royal Society of London, B Biological Sciences* **363**, 2971–2986.

Mallet, J., Meyer, A., Nosil, P., and Feder, J. L. (2009). Space, sympatry and speciation. *Journal of Evolutionary Biology* **22**, 2332–2341.

Margulis, L. (1998). *Symbiotic Planet: A New Look at Evolution*. London: Basic Books.

Margulis, L. and Sagan, D. (2002). *Acquiring Genomes: A Theory of the Origins of Species.* New York: Basic Books.

Margulis, L. and Sagan, D. (2009). Endosymbiotic origin of eukaryotes. In M. Ruse and J. Travis (Eds.) *Evolution: The First Four Billion Years*. Cambridge, MA: Harvard University Press, 534–541.

Marshall, C. R. and Ward, P. D. (1996). Sudden and gradual molluscan extinctions in the latest Cretaceous of western European Tethys. *Science* **274**(5291), 1360–1363.

Matthen, M. and Ariew, A. (2002). Two ways of thinking about fitness and natural selection. *Journal of Philosophy* **99**, 55–83.

Maynard Smith, J. and Szathmáry, E. (1995). *The Major Transitions in Evolution*. Oxford: Oxford University Press.

Maynard Smith, J. and Szathmáry, E. (2000). *The Origins of Life: From the Birth of Life to the Origin of Language*. Oxford: Oxford University Press.

Mayr, E. (1961). Cause and effect in biology. *Science* **131**, 1501–1506.

Mayr, E. (1982). *The Growth of Biological Thought: Diversity, Evolution and Inheritance.* Cambridge, MA: Belknap Press of Harvard University Press.

Mayr, E. (2002). *What Evolution Is*. London: Weidenfeld and Nicolson.

McGhee, G. R. (2011). *Convergent Evolution: Limited Forms Most Beautiful*. Cambridge, MA: MIT Press.

McLaughlin, P. (2001). *What Functions Explain: Functional Explanation and Self-Reproducing Systems*. Cambridge: Cambridge University Press.

McMullin, E. (2008). The virtues of a good theory. In S. Psillos and M. Curd (Eds.) *The Routledge Companion to Philosophy of Science*. New York: Routledge, 498–508.

McShea, D. W. and Brandon, R. N. (2010). *Biology's First Law: The Tendency for Diversity and Complexity to Increase in Evolutionary Systems*. Chicago, IL and London: University of Chicago Press.

Miller, J. D., Scott, E. C. and Okamoto, S. (2006). Public acceptance of evolution. *Science* **313**(5788), 765–766.

Miller, K. (2007). *Finding Darwin's God: A Scientist's Search for Common Ground Between God and Evolution*. New York: Harper Perennial.

Millstein, R. L. (2002). Are random drift and natural selection conceptually distinct? *Biology & Philosophy* **17** (1), 33–53.

Millstein, R. L. (2005). Selection vs. drift: a response to Brandon's reply. *Biology & Philosophy* **20**(1), 171–175.

Millstein, R. L. (2013). Environmental ethics. In K. Kampourakis (Ed.) *The Philosophy of Biology: A Companion for Educators*. Dordrecht: Springer, 723–744.

Millstein, R. L., Skipper, R. A., and Dietrich, M. R. (2009). (Mis)interpreting mathematical models: drift as a physical process. *Philosophy & Theory in Biology* **1**, e002, DOI:10.3998/ptb.6959004.0001.002.

Mindell, D. P. (2007). *The Evolving World: Evolution in Everyday Life*. Cambridge, MA: Harvard University Press.

Minelli, A. (2009). *Forms of Becoming: The Evolutionary Biology of Development*. Princeton, NJ: Princeton University Press.

Minelli, A. (2010). Evolutionary developmental biology does not offer a significant challenge to the neo-Darwinian paradigm. In F. J. Ayala and R. Arp (Eds.) *Contemporary Debates in Philosophy of Biology*. Oxford: Wiley-Blackwell, 213–226.

Minelli, A. (2011). Development, an open-ended segment of life. *Biological Theory* **6**, 4–15.

Minelli, A. and Fusco, G. (2012). On the evolutionary developmental biology of speciation. *Evolutionary Biology* **39**, 242–254.

Minelli, A. and Fusco, G. (2013). Homology. In K. Kampourakis (Ed.) *The Philosophy of Biology: A Companion for Educators*. Dordrecht: Springer, 289–322.

Moore, D. S. (2002). *The Dependent Gene: The Fallacy of "Nature vs. Nurture."* New York: Times Books/Henry Holt and Co.

Moore, D. S. (2013). Current thinking about nature and nurture. In K. Kampourakis (Ed.) *The Philosophy of Biology: A Companion for Educators*. Dordrecht: Springer, 629–652.

Moore, J. (2009). That evolution destroyed Charles Darwin's faith in Christianity – until he reconverted on his deathbed. In, R. L. Numbers (Ed.) *Galileo Goes to Jail and Other Myths About Science and Religion*. Cambridge, MA: Harvard University Press, 142–151.

Moreira, D. and López-García, P. (2009). Ten reasons to exclude viruses from the tree of life. *Nature Reviews Microbiology* **7**, 306–311.

Müller, G. B. (2007). Evo-devo: extending the evolutionary synthesis. *Nature Reviews Genetics* **8**, 943–949.

Müller, G. B. (2008). Evo-devo as a discipline. In A. Minelli and G. Fusco (Eds.) *Evolving Pathways: Key Themes in Evolutionary Developmental Biology*. Cambridge: Cambridge University Press, 3–29.

Müller, G. and Pigliucci, M. (Eds.) (2010). *Evolution: The Extended Synthesis*. Cambridge, MA: MIT Press.

Nersessian, N. J. (2008). *Creating Scientific Concepts*. Cambridge, MA: MIT Press.

Numbers, R. (2006). *The Creationists: From Scientific Creationism to Intelligent Design*. Cambridge, MA: Harvard University Press.

Numbers, R. (2009a). That Creationism is a uniquely American phenomenon. In R. Numbers (Ed.) *Galileo Goes to Jail, and Other Myths About Science and Religion*. Cambridge, MA: Harvard University Press, 215–223.

Numbers, R. (Ed.) (2009b). *Galileo Goes to Jail, and Other Myths About Science and Religion*. Cambridge, MA: Harvard University Press.

Numbers, R. (2011). Clarifying Creationism: five common myths. *History and Philosophy of the Life Sciences* **33**, 129–139.

O'Malley, M. A. (2012). When integration fails: prokaryote phylogeny and the tree of life. *Studies in History and Philosophy of Biological and Biomedical Sciences*, DOI: 10.1016/j.shpsc.2012.10.003.

Officer, C. B. and Drake, C. L. (1985). Terminal cretaceous environmental events. *Science* **227**(4691), 1161–1167.

Ogg, J. G., Ogg, G., and Gradstein, F. M. (2008). *The Concise Geologic Time Scale*. Cambridge: Cambridge University Press.

Okasha, S. (2002). *Philosophy of Science: A Very Short Introduction*. Oxford: Oxford University Press.

Okasha, S. (2006). *Evolution and the Levels of Selection*. Oxford: Oxford University Press.

Oppy, G. (1996). Hume and the argument for biological design. *Biology and Philosophy* **11**, 519–534.

Ospovat, D. (1981). *The Development of Darwin's Theory: Natural History, Natural Theology and Natural Selection, 1838–1859*. Cambridge: Cambridge University Press.

Owen, R. (1860). [Review of Origin and other works.] *Edinburgh Review* **111**, 487–532.

Pace, N. R. (2009). Mapping the tree of life: progress and prospects. *Microbiology and Molecular Biology Reviews* **73**(4), 565–576.

Paley, W. (2006) [1802]. *Natural Theology or Evidence of the Existence and Attributes of the Deity, Collected from the Appearances of Nature*. Oxford and New York: Oxford University Press.

Papadopulos, A. S. T., Baker, W. J., Crayn, D., *et al.* (2011). Speciation with gene flow on Lord Howe Island. *Proceedings of the National Academy of Sciences USA* **108**, 13188–13193.

Peeters, M., Courgnaud, V., Abela, B., *et al.* (2002). Risk to human health from a plethora of simian immunodeficiency viruses in primate bushmeat. *Emerging Infectious Diseases* **8**, 451–457.

Pennock, R. T. (2000). *The Tower of Babel: The Evidence Against the New Creationism*. Cambridge, MA: MIT Press.

Piaget, J. (1960) [1929]. *The Child's Conception of the World*. Savage, MD: Littlefield, Adams, Patterson.

Piaget, J. (2013) [1947]. *La Représentation du Monde chez l'enfant*. Paris: Presses Universitaires de France.

Pigliucci, M. (2002). *Denying Evolution: Creationism, Scientism, and the Nature of Science*. Sunderland, MA: Sinauer Associates.

Pigliucci, M. (2008). Is evolvability evolvable? *Nature Reviews Genetics* **9**, 75–82.

Pigliucci, M. (2009). Phenotypic plasticity. In G. Müller and M. Pigliucci (Eds.) *Evolution: The Extended Synthesis*. Cambridge, MA: MIT Press, 355–378.

Pigliucci, M. (2012). *Answers for Aristotle: How Science and Philosophy Can Lead Us to a More Meaningful Life*. New York: Basic Books.

Pigliucci, M. (2013). The nature of evolutionary biology: at the borderlands between historical and experimental science. In K. Kampourakis (Ed.) *The Philosophy of Biology: A Companion for Educators*. Dordrecht: Springer 87–100.

Pigliucci, M. and Boudry, M. (2011). Why machine-information metaphors are bad for science and science education. *Science & Education* **20**, 453–471.

Pigliucci, M. and Kaplan, J. (2006). *Making Sense of Evolution: The Conceptual Foundations of Evolutionary Biology*. Chicago, IL: University of Chicago Press.

Pigliucci, M. and Müller, G. (Eds.) (2010). *Evolution: The Extended Synthesis*. Cambridge, MA: MIT Press.

Poiani, A. (Ed.) (2012). *Pragmatic Evolution: Applications of Evolutionary Theory*. Cambridge: Cambridge University Press.

Posner, G. J., Strike, K. A., Hewson, P. W., and Gertzog, W. A. (1982). Accommodation of a scientific conception: toward a theory of conceptual change. *Science Education* **66**, 211–227.

Potochnik, A. (2013). Biological explanation. In K. Kampourakis (Ed.) *The Philosophy of Biology: A Companion for Educators*. Dordrecht: Springer, 49–65.

Prothero, D. R. (2007). *Evolution: What the Fossils Say and Why it Matters*. New York: Columbia University Press.

Psillos, S. (2007). *Philosophy of Science A–Z*. Edinburgh: Edinburgh University Press.

Ptacek, M. B. and Hankinson, S. J. (2009). The pattern and process of speciation. In M. Ruse and J. Travis (Eds.) *Evolution: The First Four Billion Years*. Cambridge, MA: Harvard University Press, 177–207.

Radick, G. (2009). Is the theory of natural selection independent of its history? In J. Hodge and G. Radick (Eds.) *The Cambridge Companion to Darwin* (2nd edn.). Cambridge: Cambridge University Press, 147–172.

Raff, R. A. (2000). Evo-devo: the evolution of a new discipline. *Nature Reviews Genetics* **1**, 74–79.

Raff, R. A. (2007). Written in stone: fossils, genes, and evo-devo. *Nature Reviews Genetics* **8**, 911–920.

Rambaut, A., Posada, D., Crandall, K. A., and Holmes, E. C. (2004). The causes and consequences of HIV evolution. *Nature Reviews Genetics* **5**, 52–61.

Ravetz, J. R. (1996). The Copernican revolution. In R. C. Olby, G. N. Cantor, J. R. R. Christie, and M. J. S. Hodge (Eds.) *Companion to the History of Modern Science*. London and New York: Routledge, 201–216.

Reeve, H. K. and Sherman, P. W. (1993). Adaptation and the goals of evolutionary research. *Quarterly Review of Biology* **68**(1), 1–32.

Reiss, J. O. (2009). *Not by Design: Retiring Darwin's Watchmaker*. Berkeley, CA: University of California Press.

Restif, O. (2009). Evolutionary epidemiology 20 years on: challenges and prospects. *Infection, Genetics and Evolution* **9**(1), 108–123.

Reuter, M. A., Pecora, N. D., Harding, C. V., Canaday, D. H., and McDonald, D. (2010). *Mycobacterium tuberculosis* promotes HIV *trans*-infection and suppresses MHC-II antigen processing by dendritic cells. *Journal of Virology* **84**(17), 8549–8560.

Richards, R. A. (2010). *The Species Problem: A Philosophical Analysis*. Cambridge: Cambridge University Press.

Richards, R. J. (1992). *The Meaning of Evolution: The Morphological Construction and Ideological Reconstruction of Darwin's Theory*. Chicago, IL: University of Chicago Press.

Roger, A. J., Sandblom, O., Doolittle, W. F., and Philippe, H. (1999). An evaluation of elongation – factor-1α as a phylogenetic marker for eukaryotes. *Molecular Biology Evolution* **16**(2), 218–233.

Rogers, A. R. (2011). *The Evidence for Evolution*. Chicago, IL: University of Chicago Press.

Rokas, A. and Carroll, S. B. (2006). Bushes in the Tree of Life. *PLoS Biology* **4**, e352.

Rosenberg, A. (2005). *Philosophy of Science: A Contemporary Introduction* (2nd edn.). New York: Routledge.

Rosenberg, A. and McShea, D. W. (2008). *Philosophy of Biology: A Contemporary Introduction*. New York: Routledge.

Rosengren, K. S., Gelman, S. A., Kalish, C. W., and McCormick, M. (1991). As time goes by: children's early understanding of growth in animals. *Child Development* **62**(6), 1302–1320.

Rosengren, K. S., Brem, S. K., Evans, E. M., and Sinatra, G. M. (Eds.) (2012). *Evolution Challenges: Integrating Research and Practice in Teaching and Learning About Evolution*. Oxford: Oxford University Press.

Rughinis, C. (2011). A lucky answer to a fair question: conceptual, methodological, and moral implications of including items on human evolution in scientific literacy surveys. *Science Communication* **33**(4), 499–530.

Ruse, M. (1975a). Charles Darwin and artificial selection. *Journal of the History of Ideas* **36**(2), 339–350.

Ruse, M. (1975b). Darwin's debt to philosophy: an examination of the influence of the philosophical ideas of John F.W. Herschel and William Whewell on the development of Charles Darwin's theory of evolution. *Studies in the History and Philosophy of Science* **6**, 159–181.

Ruse, M. (1999). *Mystery of Mysteries: Is Evolution a Social Construction?* Cambridge, MA: Harvard University Press.

Ruse, M. (2000). Darwin and the philosophers: epistemological factors in the development and reception of the Origin of Species. In R. Creath and J. Maienschein (Eds.) *Biology and Epistemology.* Cambridge: Cambridge University Press, 3–26.

Ruse, M. (2001). *Can a Darwinian be a Christian? The Relationship between Science and Religion.* Cambridge: Cambridge University Press.

Ruse, M. (2004). The argument from design: a brief history. In W.A. Dembski and M. Ruse (Eds.) *Debating Design: From Darwin to DNA.* Cambridge: Cambridge University Press, 13–31.

Ruse, M. (2005).*The Evolution/Creation Struggle.* Cambridge, MA: Harvard University Press.

Ruse, M. (2009). *Monad to Man: The Concept of Progress in Evolutionary Biology.* Cambridge, MA: Harvard University Press.

Ruse, M. (2010). *Science and Spirituality: Making Room for Faith in the Age of Science.* Cambridge: Cambridge University Press.

Ruse, M. (2012). Science and the humanities: Stephen Jay Gould's quest to join the high table. *Science & Education* **22**(9), 2317–2326.

Ruse, M. (Ed.) (2013). *The Cambridge Encyclopedia of Darwin and Evolutionary Thought.* Cambridge: Cambridge University Press.

Ruse, M. and Richards, R. J. (Eds.) (2009). *The Cambridge Companion to the "Origin of Species."* Cambridge: Cambridge University Press.

Ruse, M. and Travis, J. (Eds.) (2009). *Evolution: The First Four Billion Years.* Cambridge, MA: Belknap Press of Harvard University Press.

Salmon, W. C. (1984). *Scientific Explanation and the Causal Structure of the World.* Princeton, NJ: Princeton University Press.

Salmon, W. C. (1989). Four decades of scientific explanation. In P. Kitcher and W. C. Salmon (Eds.) *Minnesota Studies in the Philosophy of Science*, Vol. **13**. Minneapolis, MN: University of Minnesota Press, 3–219.

Sansom, R. S., Gabbott, S. E., and Purnell, M. A. (2010). Non-random decay of chordate characters causes bias in fossil interpretation. *Nature* **463**, 797–800.

Sarkar, S. (2007). *Doubting Darwin? Creationist Designs on Evolution.* Malden, MA: Blackwell.

Sato, A., Tichy, H., O'Huigin, C., Grant, P. R., Grant, B. R., and Klein, J. (2001). On the origin of Darwin's finches. *Molecular Biology and Evolution* **18**(3), 299–311.

Schweber, S. S. (1980). Darwin and the political economists: divergence of character. *Journal of the History of Biology* **13**(2), 195–289.

Scriven, M. (1959). Explanation and prediction in evolutionary theory. *Science* **130**, 477–482.

Scriven, M. (1969). Explanation in biological sciences. *Journal of the History of Biology* **2**(1), 187–198.

Sears, K. E., Behringer, R. R., Rasweiler, J. J., and Niswander, L. A. (2006). Development of bat flight: morphologic and molecular evolution of bat wing digits. *Proceedings of the National Academy of Sciences USA* **103**(17), 6581–6586.

Secord, J. A. (1985). Darwin and the breeders: a social history. In D. Kohn (Ed.) *The Darwinian Heritage*. Princeton, NJ: Princeton University Press, 519–542.

Secord, J. A. (2000). *Victorian Sensation: The Extraordinary Publication, Reception, and Secret Authorship of Vestiges on Natural History of Creation.* Chicago, IL: University of Chicago Press.

Sereno, P. C. (2005). The logical basis of phylogenetic taxonomy. *Systematic Biology* **54**(4), 595–619.

Shapin, S. (1996). *The Scientific Revolution*. Chicago, IL and London: University of Chicago Press.

Shtulman, A. (2006). Qualitative differences between naive and scientific theories of evolution. *Cognitive Psychology* **52**, 170–194.

Shtulman, A. and Schulz, L. (2008). The relationship between essentialist beliefs and evolutionary reasoning. *Cognitive Science* **32**, 1049–1062.

Shubin, N. (2008). *Your Inner Fish: The Amazing Discovery of our 375-Million-Year-Old Ancestor*. London: Penguin Books.

Shubin, N. H., Daeschler, E. B., and Jenkins, F. A., Jr. (2006). The pectoral fin of *Tiktaalik roseae* and the origin of the tetrapod limb. *Nature* **440**, 764–771.

Shubin, N., Tabin, C., and Carroll, S. (2009). Deep homology and the origins of evolutionary novelty. *Nature* **457**, 818–823.

Sinatra, G. M., Southerland, S. A., McConaughy, F., and Demastes, W. (2003). Intentions and beliefs in students' understanding and acceptance of biological evolution. *Journal of Research in Science Teaching* **40**(5), 510–528.

Skipper, R. A. (2006). Stochastic evolutionary dynamics: drift versus draft. *Philosophy of Science* **73**(5), 655–665.

Skoglund, P., Götherström, A., and Jakobsson, M. (2011). Estimation of population divergence times using non-overlapping genomic sequences: examples from dogs and wolves. *Molecular Biology and Evolution* **28**(4), 1505–1517.

Smit, J. and Hertogen, J. (1980). An extraterrestrial event at the Cretaceous–Tertiary boundary. *Nature* **285**, 198–200.

Smith, M. U. (2010). Current status of research in teaching and learning evolution: II. Pedagogical issues. *Science & Education* **19**(6), 539–571.

Sober, E. (1993) [1984]. *The Nature of Selection: Evolutionary Theory in Philosophical Focus*. Chicago, IL and London: University of Chicago Press.

Sober, E. (1997). Two outbreaks of lawlessness in recent philosophy of biology. *Philosophy of Science* **64**(4), 467.

Sober, E. (2003, November 30). Contingency or inevitability? What would happen if the evolutionary tape were replayed? A review of Simon Conway Morris's life's solution – inevitable humans in a lonely universe. *New York Times*.

Sober, E. (2007). What is wrong with Intelligent Design? *Quarterly Review of Biology* **82**(1), 3–8.

Sober, E. (2009). Did Darwin write the *Origin* backwards? *Proceedings of the National Academy of Sciences USA* **106**, 10048–10055.

Sober, E. (2011). *Did Darwin Write the Origin Backwards? Philosophical Essays on Darwin's Theory*. Amherst, NY: Prometheus Books.

Soshnikova, N., Dewaele, R., Janvier, P., Krumlauf, R., and Duboule, D. (2013). Duplications of *hox* gene clusters and the emergence of vertebrates. *Developmental Biology* **378**, 194–199.

Spencer, N. (2009). *Darwin and God*. London: SPCK.

Springer, K. (1999). How a naive theory of biology is acquired. In M. Siegal and C. Peterson (Eds.) *Children's Understanding of Biology and Health*. Cambridge: Cambridge University Press, 45–70.

Stamos, D. N. (2008). *Evolution and the Big Questions*. Oxford: Blackwell Publishing.

Stauffer, R. C. (Ed.) (1975). *Charles Darwin's Natural Selection: Being the Second Part of His Big Species Book Written From 1856 to 1858*. Cambridge: Cambridge University Press.

Sterelny, K. (2005). Another view of life (essay review). *Studies in the History and Philosophy of Biological and Biomedical Sciences* **36**, 585–593.

Sterelny, K. and Calcott, B. (Eds.) (2011). *The Major Transitions in Evolution Revisited.* Cambridge, MA: MIT Press.

Sterelny, K. and Griffiths, P. E. (1999). *Sex and Death: An Introduction to the Philosophy of Biology.* Chicago, IL and London: University of Chicago Press.

Stern, D. L. (2011). *Evolution, Development, and the Predictable Genome.* Greenwood Village, CO: Roberts and Company Publishers.

Strevens, M. (2009). *Depth: An Account of Scientific Explanation.* Cambridge, MA: Harvard University Press.

Szathmáry, E. (2005). Life's solution: inevitable humans in a lonely universe: Simon Conway Morris. *Biology & Philosophy* **20**, 849–857.

Tattersall, I. (1998). *Becoming Human: Evolution and Human Uniqueness.* Oxford: Oxford University Press.

Tattersall, I. (2009). Human origins: out of Africa. *Proceedings of the National Academy of Sciences USA* **106**(38), 16018–16021.

Taylor, R. S. and Ferrari, M. (Eds.) (2011). *Epistemology and Science Education: Understanding the Evolution vs. Intelligent Design Controversy.* New York: Routledge.

Thagard, P. (1978). The best explanation: criteria for theory choice. *Journal of Philosophy* **75**(2), 76–92.

Thagard, P. (1992). *Conceptual Revolutions.* Princeton, NJ: Princeton University Press.

Thagard, P. (2012). *The Cognitive Science of Science: Explanation, Discovery, and Conceptual Change.* Cambridge, MA: MIT Press.

Thagard, P. and Findlay, S. (2010). Getting to Darwin: obstacles to accepting evolution by natural selection. *Science & Education* **19**, 625–636.

Thewissen, J. G., Cohn, M. J., Stevens, L. S., Bajpai, S., Heyning, and, J., Horton, W. E., Jr. (2006). Developmental basis for hind-limb loss in dolphins and origin of the cetacean bodyplan. *Proceedings of the National Academy of Sciences USA* **103**(22), 8414–8418.

Trut, L. N. (1999). Early canid domestication: the farm-fox experiment. *American Scientist* **87**, 160–169.

Turner, J. S. (2007). *The Tinkerer's Accomplice: How Design Emerges from Life Itself.* Cambridge, MA and London: Harvard University Press.

UNAIDS (2012). *UNAIDS Report on the Global Aids Epidemic 2012.* Geneva: UNAIDS.

Van Dijk, E. and Reydon, T. (2010). A conceptual analysis of evolutionary theory for teacher education. *Science & Education* **19**, 655–677.

Van Wyhe, J. (2007). Mind the gap: did Darwin avoid publishing his theory for many years? *Notes and Records of the Royal Society* **61**, 177–205.

Van Wyhe, J. and Pallen, M. J. (2012). The "Annie Hypothesis": did the death of his daughter cause Darwin to "give up Christianity"? *Centaurus* **54**(2), 105–123.

Van Wyhe, J. and Rookmaaker, K. (2012). A new theory to explain the receipt of Wallace's Ternate Essay by Darwin in 1858. *Biological Journal of the Linnean Society* **105**, 249–252.

Vermeij, G. (2010). *The Evolutionary World: How Adaptation Explains Everything from Seashells to Civilization.* New York: Thomas Dunne Books, St. Martin's Press.

Vilà, C., Savolainen, P., Maldonado, J. E., *et al.* (1997). Multiple and ancient origins of the domestic dog. *Science* **276**, 1687–1689.

Virgin, H. W. and Walker, B. D. (2010). Immunology and the elusive AIDS vaccine. *Nature* **464**, 224–231.

Von Holdt, B. M., Pollinger, J. P., Lohmueller, K. E., *et al.* (2010). Genome-wide SNP and haplotype analyses reveal a rich history underlying dog domestication. *Nature* **464**, 898–902.

Vorzimmer, P. (1969). Darwin, Malthus and the theory of natural selection. *Journal of the History of Ideas* **30**(4), 527–542.

Vosniadou, S. (Ed.) (2008). *International Handbook of Research on Conceptual Change.* New York and London: Routledge.

Vosniadou, S. (2012). Reframing the classical approach to conceptual change: preconceptions, misconceptions and synthetic models. In B. J. Fraser, K. G. Tobin, and C. J. McRobbie (Eds.) *Second International Handbook of Science Education.* Dordrecht: Springer, 119–130.

Wagner, G. (2007). The developmental genetics of homology. *Nature Reviews Genetics* **8**, 473–479.

Wagner, G. P. and Draghi, J. (2009). Evolution of evolvability. In G. Müller and M. Pigliucci (Eds.) *Evolution: The Extended Synthesis.* Cambridge, MA: MIT Press, 379–399.

Wake, D. B. (1997). Incipient species formation in salamanders of the *Ensatina* complex. *Proceedings of the National Academy of Sciences USA* **94**, 7761–7767.

Wake, D. B., Wake, M. H., and Specht, C. D. (2011). Homoplasy: from detecting pattern to determining process and mechanism of evolution. *Science* **331**, 1032–1035.

Walsh, D. (2006). Evolutionary essentialism. *British Journal for the Philosophy of Science* **57**(2), 425–448.

Walsh, D. (2008). Teleology. In M. Ruse (Ed.) *The Oxford Handbook of Philosophy of Biology.* Oxford: Oxford University Press, 113–137.

Walsh, D. M. (2007). The pomp of superfluous causes: the interpretation of evolutionary theory. *Philosophy of Science* **74**, 281–303.

Walsh, D. M., Lewens, T., and Ariew, A. (2002). The trials of life: natural selection and random drift. *Philosophy of Science* **69**, 452–473.

Walters, S. M. and Stow, E. A. (2002). *Darwin's Mentor: John Stevens Henslow, 1796–1861.* Cambridge: Cambridge University Press.

Waters, C. K. (2009). The arguments in *The Origin of Species.* In J. Hodge and G. Radick (Eds.) *The Cambridge Companion to Darwin* (2nd edn.). Cambridge: Cambridge University Press, 120–143.

Wayne, R. K., Leonard, J. A., and Vilà, C. (2006). Genetic analysis of dog domestication. In M. A. Zeder, D. Decker-Walters, D. Bradley, and B. D. Smith (Eds.) *Documenting Domestication: New Genetic and Archaeological Paradigms.* Los Angeles, CA and London: University of California Press 279–293.

West-Eberhard, M. J. (1992). Adaptation: current usages. In E. F. Keller and E. A. Lloyd (Eds.) *Keywords in Evolutionary Biology.* Cambridge, MA: Harvard University Press, 13–18.

West-Eberhard, M. J. (2005). Developmental plasticity and the origin of species differences. *Proceedings of the National Academy of Sciences USA* **102**(1), 6543–6549.

White, T. D., Asfaw, B., Beyene, Y., *et al.* (2009). *Ardipithecus ramidus* and the paleobiology of early hominids. *Science* **64**, 75–86.

Whiteside, A. (2008). *HIV/AIDS: A Very Short Introduction.* Oxford: Oxford University Press.

WHO (2012). *Global Tuberculosis Report.* Geneva: WHO.

Wilberforce, S. (1860). [Review of] On the Origin of Species, by means of Natural Selection; or the Preservation of Favoured Races in the Struggle for Life. By Charles Darwin, M. A., F.R.S. London, 1860. *Quarterly Review* **108**, 225–264.

Wilkins, J. S. (2009). *Species: A History of the Idea.* Berkeley, CA: University of California Press.

Wilkins, J. S. (2011). Are Creationists rational? *Synthese* **178**(2), 207–218.

Wilkins, J. S. (2013). Essentialism in biology. In K. Kampourakis (Ed.) *The Philosophy of Biology: A Companion for Educators*. Dordrecht: Springer, 395–420.

Williams, G. C. (1996) [1966]. *Adaptation and Natural Selection: A Critique of Some Current Evolutionary Thought*. Princeton, NJ: Princeton University Press.

Williams, G. C. (2001) [1996]. *Plan and Purpose in Nature: The Limits of Darwinian Evolution*. London: Phoenix.

Wilmut, I., Schnieke, A. E., McWhir, J., Kind, A. J., and Campbell, K. H. S. (1997). Viable offspring derived from fetal and adult mammalian cells. *Nature* **385**, 810–813.

Wilson, D. S. (2002). *Darwin's Cathedral: Evolution, Religion and the Nature of Society*. Chicago, IL: University of Chicago Press.

Wilson, E. O. (1975). *Sociobiology: The New Synthesis*. Cambridge, MA: Harvard University Press.

Winslow, M. W., Staver, J. R., and Scharmann, L. C. (2011). Evolution and personal religious belief: Christian university biology-related majors' search for reconciliation. *Journal of Research in Science Teaching* **48**(9), 1026–1049.

Woese, C. (1998). The universal ancestor. *Proceedings of the National Academy of Sciences USA* **95**, 6854–6859.

Woese, C. R., Kandler, O., and Wheelis, M. L. (1990). Towards a natural system of organisms: proposal for the domains Archaea, Bacteria, and Eucarya. *Proceedings of the National Academy of Sciences USA* **87**, 4576–4579.

Wood, B. (2005). *Human Evolution: A Very Short Introduction*. Oxford: Oxford University Press.

Wood, B. (2010). Reconstructing human evolution: achievements, challenges, and opportunities. *Proceedings of the National Academy of Sciences USA* **107**(2), 8902–8909.

Woodward, J. (2003). *Making Things Happen: A Theory of Causal Explanation*. Oxford: Oxford University Press.

Woodward, J. (2008). Explanation. In S. Psillos and M. Curd (Eds.) *The Routledge Companion to Philosophy of Science*. New York: Routledge, 171–181.

Woodward, J. (2011). Scientific explanation. In E. N. Zalta (Ed.) *The Stanford Encyclopedia of Philosophy*. http://plato.stanford.edu/archives/win2011/entries/scientific-explanation.

Index

acquired immune deficiency syndrome (AIDS), 23
adaptation, 172
  ahistorical definition of, 174, 176–7
  historical definition of, 173, 176–7
adaptations, 19, 44
agnosticism, 50
allele, 9
allopatric speciation, 193
allopatry, 194
anagenesis, 196, 198
apomorphies, 144, 146
argument from design, 33
arguments in the *Origin*, 118
artifact thinking, 91
  artifact-thinking argument, 41
artifacts, 14, 73
artificial selection, 103–5
atheism, 47

being justified in believing, 53
blind watchmaker, 42

character, definition of, 2
clades, 134
cladogenesis, 196, 198
common descent, 118, 120, 127
conceptions, 63
concepts, 63
conceptual change, 63
  Darwin, 100, 108, 114
conceptual change in evolution, 93
conceptual conflict, 65, 92, 95, 100
conceptual obstacles, 64
contingency, 189
  causal dependence version, 190
  unpredictability version, 190
convergence, 151
Creationism, ix, 31

Darwin's analogies
  division of labor, 106
  overview, 107
  selection, 104
  struggle for existence, 102

Darwin's conceptual change, 108
Dawkins' argument, 41
deep homology, 156
Deism, 50
descent with modification, 1, 120, 158
design teleology, 72, 90–1
developmental repatterning, 161
digit formation in tetrapods, 140
distinguishing between knowing and believing,
    45, 51–2
DNA sequences, 2
domains, 136
domestication, 19
domestication of dogs, 20
domestication of silver foxes, 21
dominance, 9
drift, 186

end-Cretaceous extinction, 202
essence
  of artifacts, 81, 83
  definition of, 80
  of organisms, 83
evolution, definition of, 1
evolution of dogs, 20
evolution of drug resistance, 28
evolution of HIV, 26
evolutionary developmental biology, 157
evolutionary explanations, 14, 200
evolutionary network of life, 129, 134
evolutionary novelty, 165
evolutionary theory, 1
evolutionary trees, 134–5
evolvability, 166
exaptation, 178
explanandum, 66
explanans, 66
explanatory aims, 19
extensive drug-resistance (XDR) to tuberculosis, 27
extinction, 196

Galápagos finches, 11
Galápagos Islands, 109
genes, 2

genetic accommodation, 165
genetic draft, 188
genetic drift, 185
genotype, 10
geocentric model, 67, 69, 71
geographic isolation, 192
God of the gaps argument, 32
gradualism, 200

heliocentric model, 67, 69, 71
heterochrony, 161
heterometry, 161
heterotopy, 161
heterotypy, 161
heterozygote, 9
historicity, 200
homology, 19, 135, 139–41, 150
homoplasy, 19, 135, 148, 150–1
homozygote, 9
horizontal DNA transfer, 185
human evolution, 4, 6
human immunodeficiency virus (HIV), 23
hypotheses, 16

ignorance, 212
independent lines of evidence, 203
inference to the best explanation, 18, 55, 200–1
Intelligent Design, ix, 31
intuitions, 62–3
irreducibly complex systems, 31

justified belief, 53

K-Pg boundary, 202
K-Pg extinction, 201

limited perfection, 110
low probability arguments, 31

macroevolution, 164, 198
macroevolutionary processes, 11
major transitions in evolution, 171
Malthus' *Essay*, 101
mass extinctions, 196
metaphysical naturalism, 59
methodological naturalism, 57, 59
microevolution, 164, 198
microevolutionary processes, 11
migration, 10
militant modern atheism, 33
misallodoxy, 215
misconceptions, 63
modules, 144
monophyletic group, 138
multicellularity, 156
multi-drug resistance (MDR) to
    tuberculosis, 27

natural selection, 10, 107, 118–20, 172
    definition of, 1
natural theology, 33, 108

Old-Earth Creationists, 31
orthologous sequences, 137
overdetermination of causes by effects, 203

Paley's argument, 39, 41
Paley's argument vs. Dawkins' argument, 42
paralogous sequences, 137
parapatric speciation, 194
parapatry, 194
perfect adaptation, 110–11
phenological isolation, 195
phenotype, 10
phenotypic accommodation, 165
plasticity, 83, 157, 164
plesiomorphies, 144, 146
polyphenism, 195
preconceptions, 63
predictions, 6
principle of divergence, 106–7
proximate explanations, 13
psychological essentialism, 80, 85, 91
punctualism, 200
punctuated equilibrium, 200

radiometric dating, 169
randomness in evolution, 188
regulatory sequences, 15
relative adaptation, 110–11
reproductive isolation, 193
research questions, 17
reviews of the *Origin of Species*, 121
robustness, 83, 157, 165

sampling
    discriminate, 185
    indiscriminate, 186
    indiscriminate gamete sampling, 187
    indiscriminate parent sampling, 186
scientific explanation, 66
scientific ignorance, 57
scientific knowledge, 56
scientific method, 16
scientific perspectivism, 56
selection against, 181–2
selection for, 179, 181–2
selection of, 179, 181
social context of Darwin's theory, 100
speciation, 192
species, definition of, 1, 192
species selection, 198
stasis, 196, 200
stochastic events, 184
stochastic processes, 184

struggle for existence, 102
symbiogenesis, 136
sympatric speciation, 194
sympatry
    mosaic, 194
    pure, 194
symplesiomorphies, 144
synapomorphies, 144

taming, 19
taxon, 134
teleological explanations, 73–4
teleology, 35, 72
    artifacts, 76
    organisms, 76
temporal scale unobservability, 11
testability, 57
Theism, 50
theodicy, 45
theory, 1
Tiktaalik, 7
transmutation, 98, 101, 104, 109, 118
true causes, 105
tuberculosis, 27

underdetermination of effects by causes,
    203
uniformitarianism, 104
unnatural death, 45

verae causae, 105
virtues of evolutionary theory
    consilience, 210
    durability, 211
    empirical fit, 209
    explanatory power, 212
    external consistency, 210
    fertility, 211
    internal coherence, 209
    internal consistency, 209
    optimality, 211
    simplicity, 210

xenologous sequences, 137

Young-Earth Creationists, 31

zero-force evolutionary law (ZFEL),
    189

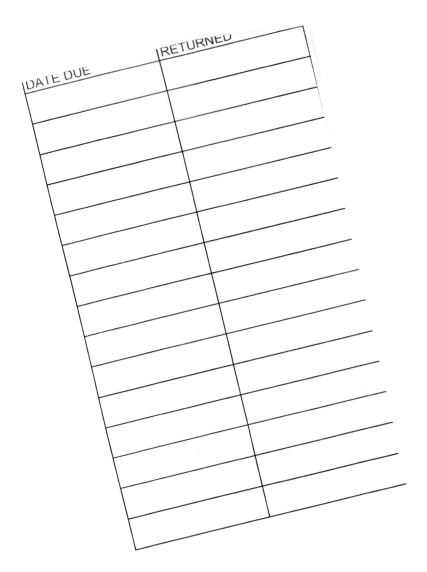

DATE DUE

RETURNED